Green Energy and Technology

Climate change, environmental impact and the limited natural resources urge scientific research and novel technical solutions. The monograph series Green Energy and Technology serves as a publishing platform for scientific and technological approaches to “green”—i.e. environmentally friendly and sustainable—technologies. While a focus lies on energy and power supply, it also covers “green” solutions in industrial engineering and engineering design. Green Energy and Technology addresses researchers, advanced students, technical consultants as well as decision makers in industries and politics. Hence, the level of presentation spans from instructional to highly technical.

Indexed in Scopus.

More information about this series at http://www.springer.com/series/8059

Mohammad Jawaid · Mohammad Asim ·
Paridah Md. Tahir · Mohammed Nasir
Editors

Pineapple Leaf Fibers

Processing, Properties and Applications

Springer

Editors
Mohammad Jawaid
Laboratory of Biocomposite Technology, INTROP
Universiti Putra Malaysia
Serdang, Selangor, Malaysia

Mohammad Asim
Laboratory of Biocomposite Technology, INTROP
Universiti Putra Malaysia
Serdang, Selangor, Malaysia

Paridah Md. Tahir
Laboratory of Biocomposite Technology, INTROP
Universiti Putra Malaysia
Serdang, Selangor, Malaysia

Mohammed Nasir
College of Forestry
Banda University of Agriculture and Technology
Banda, Uttar Pradesh, India

ISSN 1865-3529 ISSN 1865-3537 (electronic)
Green Energy and Technology
ISBN 978-981-15-1415-9 ISBN 978-981-15-1416-6 (eBook)
https://doi.org/10.1007/978-981-15-1416-6

This Springer imprint is published by the registered company Springer Nature Singapore Pte Ltd.
The registered company address is: 152 Beach Road, #21-01/04 Gateway East, Singapore 189721, Singapore

The editors, Dr. Mohammed Nasir and Dr. Mohammad Asim dedicated this book to their beloved grandparents late Abdul Qayyum khan and late Quddusiya bano.

They inspired us for higher studies, and we attribute them whatever we have achieved.

Preface

Natural fibres are under intensive study due to their ecofriendly nature, peculiar properties, and some other advantages such as availability, easy and safe handling, and biodegradability.

Natural fibres have admirable physical and mechanical properties, though it varies with the plant source, species, geography, and climatic conditions. Pineapple leaf fibre (PALF) is one of the abundantly available waste materials of South East Asia, India, and South America until now not explored full potential of it. From the socioeconomic prospective, PALF can be a new source of raw material to the industries and can be a potential replacement of synthetic fibre. This book will study the anatomical structure, source, and variety of PALF which will further elaborate physical, mechanical, and fibre/matrix interfacial bonding and composites. This sustainable material penetrates in the market segment and has significant potential in automotive, marine, aerospace, construction and building, wind energy and consumer goods, etc. The book contains extensive examples and real-world products that will be suitable as per the need of markets.

This book covered versatile topics such as cultivation of anatomical structure of pineapple as future material for versatile applications, extraction process of pineapple leaf fibres, improvement of pineapple leaf fibres by various treatments, comparative study of natural fibres, design and fabrication of green biocomposites, conceptual design of biocomposites, green biocomposites for automotive components, structural purposes and aircraft application.

We are highly thankful to all authors who have contributed chapters and provided their valuable ideas and knowledge in this edited book. We attempt to gather all the scattered information of authors from diverse fields around the world (Malaysia, Brazil, and India) in the areas of green composites and biocomposites and finally complete this venture in a fruitful way. We greatly appreciate contributors' commitment for their support to compile our ideas in reality.

We are highly thankful to Springer Singapore team for their generous cooperation at every stage of the book production.

Serdang, Malaysia Mohammad Jawaid
Serdang, Malaysia Mohammad Asim
Serdang, Malaysia Paridah Md. Tahir
Banda, India Mohammed Nasir

Contents

About the Editors

Dr. Mohammad Jawaid is currently working as High Flyer Fellow (Professor), at Biocomposite Technology Laboratory, Institute of Tropical Forestry and Forest Products (INTROP), Universiti Putra Malaysia, Serdang, Selangor, Malaysia and also Visiting Professor at Department of Chemical Engineering, College of Engineering, King Saud University, Riyadh, Saudi Arabia since June 2013. He is also Visiting Scientist to TEMAG Laboratory, Faculty of Textile Technologies and Design at Istanbul Technical University, Turkey. Previously he worked as Visiting Lecturer, Faculty of Chemical Engineering, Universiti Teknologi Malaysia (UTM) and also worked as Expatriate Lecturer under UNDP project with Ministry of Education of Ethiopia at Adama University, Ethiopia. He received his Ph.D. from Universiti Sains Malaysia, Malaysia. He has more than 15 years of experience in teaching, research, and industries. His area of research interests includes Hybrid Reinforced/Filled Polymer Composites, Advance Materials: Graphene/Nanoclay/ Fire Retardant, Lignocellulosic Reinforced/Filled Polymer Composites, Modification and Treatment of Lignocellulosic Fibres and Solid Wood, Nano Composites and Nanocellulose fibres, Polymer blends. So far he has published 32 books, 60 book chapters, and more than 300 International Journal Papers and five published review papers under Top 25 hot articles in science direct during 2015–2019. He is also the Guest Editor Special issue-International Journal of Polymer Science, Current Organic Synthesis, Current analytical chemistry, SN Applied Sciences and Editorial board member and Journal of Polymers and The Environment. Beside that he is also reviewer of several high impact ISI journals of Elsevier, Springer, Wiley, Saga, etc. Presently he is supervising 15 Ph.D. students and five master students in the field of hybrid composites, Green composites, Nanocomposites, Natural fibre reinforced composites, etc. 17 Ph.D. and 8 Master students graduated under his Supervision in 2013–2019. Recently he received International Grant- Newton-Ungku Omar as Malaysian Head supported by MIGHT-Innovative-UK (RM 5.3 Million). He has several Research grant at University and National level on polymer composites of around RM 700,000 (USD 175,000). He also delivered Plenary and Invited Talk in

International Conference related to composites in India, Turkey, Dubai, China, Thailand, Saudi Arabia, UK, France, and Malaysia. Beside that he is also member of Technical committee of Several National and international conference on Composites and Material Science.

Dr. Mohammad Asim is currently a post-doctoral fellow in the Laboratory of Biocomposite Technology at the Institute of Tropical Forestry and Forest Products (INTROP), Universiti Putra Malaysia. He was born on the 2nd August 1988 in Mau, Uttar Pradesh, India. He completed his bachelor's degree in forestry from C. S. Azad University of Agriculture and Technology, Kanpur, India in 2011. Afterward, he continued his study and obtained his MSc in wood science and technology from Forest Research institute, Dehradun, India in 2013 and finally Ph.D. degree in the field of biocomposite technology from Universiti Putra Malaysia in 2017. His main research areas are: Treatment and modification of natural fibres, Hybrid Reinforced/ Filled Polymer Composites, Advance Materials: Nanoclay/Fire Retardant, Ligno-cellulosic Reinforced/Filled Polymer Composites, Nano Composites and Nanocellulose fibres, thermosets and thermoplastics. Dr. Asim has published more than 25 international journal papers, four review papers, five book chapters and six conference proceedings. Dr. Asim is the regular reviewers of different international journals published by Elsevier, Wiley, Springer, etc.

Prof. (Dr.) Paridah Md. Tahir is a Professor at the Faculty of Forestry, Universiti Putra Malaysia. Malaysia. She has nearly 30 years of experience as a lecturer specializing in Wood/ Fibre bonding, Surface coating and Biocomposites. She served as the Director of Institute of Tropical Forestry and Forest Products (INTROP) since April 2009–March 2018 and was instrumental in making INTROP as one of the national Higher Institution Centres of Excellence (HICoE) in 2016. She is well known for her involvement in developing standards for timber and timber products, oil palm trunk, bamboo and kenaf. She has led the Malaysian delegations in various international standards plenary meetings and is a convener for WG 4—Test methods in ISO TC 218. She has helped developing and reviewing more than 100 standards for timber, wood-based panels, structural timbers and wood finishing. In recognition to her work, the Department of Standards Malaysia has awarded her with a STaR Award in 2015, a prestigious award given by the government to individuals in recognition of their significant and excellent contribution to standards development in Malaysia. To date, Paridah have published more than 200 articles in numerous journals, co-authored six books and more than 100 chapters in book, proceedings, technical and consultation reports. Paridah has been involved extensively in RDC&I, securing more than RM 10 million worth of research funds and consultancy projects from the public, industry and international sources. From these projects, she had filed 10 patents, one trade secret and two copyrights. Her study on oil palm plywood has been applied in several plywood mills in the country. She is currently engaged as a consultant by an international paper mill to develop tannin from eucalyptus bark as phenolic bioresin. In addition, she and her team are assisting the Malaysian Timber Industry Board in determining the maximum limit for formaldehyde emission from wood-based panels and developing OPT-based products industry. Her expertise has been recognized by the Science Academy of Malaysia (ASM) through two significant awards she received, Top Research Scientists Malaysia (TRSM) 2014 for her outstanding research work, and Fellow of ASM 2018 for her teaching, research, networking and community involvements throughout her career.

Dr. Mohammed Nasir serving as assistant professor in forest products division in college of forestry, at Banda University of Agriculture and technology, Banda (UP) India. He was born on 17 June 1982, in Mau Uttar Pradesh, India. He obtained his BSc. Degree in Forestry Hons, from CSAUA&T Kanpur, M.Sc. in Wood Science and Technology from FRI Dehradun and moved to Malaysia to complete his Ph.D. in Chemical Engineering (wood composites) from Universiti Malaysia Pahang, Malaysia. After Ph.D., He worked as a postdoc fellow at School of Industrial Technology, Universiti Sains Malaysia for two years. During postdoc his focus research was to develop an ecofriendly method of nano-cellulose synthesis from oil palm trunk waste. Afterwards he joined as a National postdoc fellow at Forest Research Institute Dehradun and worked on binderless fiberboards fabrication. His main research interests are: (1) Lignin based Bio adhesive preparation, (2) Binderless Fiberboard fabrication, (3) Nano-cellulose synthesis through enzyme hydrolysis methods and (4) Hybrid composite. He has published many research paper and book chapters in various international journals. Furthermore, he has applied for two Malaysia patents for adopting a new method and preparing a composite from lignin based bio-adhesive.

Pineapple Leaf Fibre: Cultivation and Production

Pintu Pandit, Ritu Pandey, Kunal Singha, Sanjay Shrivastava, Vandana Gupta and Seiko Jose

Abstract A pineapple leaf fibre (PALF) is classified according to the sources in plants, where they occur and from which they are extracted. PALF is considered to be superior in texture than any other vegetable fibre. It helps in climate restoration and soil quality by preventing soil erosion. This chapter includes pineapple cultivation practices, plant anatomy, varieties, diseases, nutritional needs, usefulness and its production at a global level. Plant distribution, varieties, fruit and fibre yield potential are also envisioned in this chapter. Post-harvest operations, decorticating practices, fibre retting, finishing, chemical composition and physico-chemical properties are reported. It also explains plant benefits to farmers, consumers and the environment.

Keywords Pineapple · Leaf fibre · Environment · Cultivation · Production · Application

1 Introduction

Pineapple (*Ananas comosus*), a perennial plant belongs to the family Bromeliaceae. Its height and circumference range between three and six ft. The plant consists of scaly fruit and radiating leaves arranged spirally around the single butt. Pineapple cultivation is done outdoors in fields and also indoor in pots, containers and tissue culture. The plant flourishes well in the tropical or subtropical region under humid climate. Pineapple cultivation originated in Central and South America and later

P. Pandit (✉) · K. Singha · S. Shrivastava
Department of Textile Design, Ministry of Textilesm, Govt. of India, National Institute of Fashion Technology, NIFT Campus, Mithapur Farms, Patna 800001, India
e-mail: pintupanditict@gmail.com; pintu.pandit@nift.ac.in

R. Pandey
Chandra Shekhar Azad University of Agriculture and Technology, Kanpur 208002, India

V. Gupta
Department of Fashion and Design, Chandigarh University, Punjab, India

S. Jose
Central Sheep and Wool Research Institute, Indian Council of Agricultural Research, Govt. of India, Avikanagar, Rajasthan 304501, India

M. Jawaid et al. (eds.), *Pineapple Leaf Fibers*, Green Energy and Technology,
https://doi.org/10.1007/978-981-15-1416-6_1

spread throughout the world perhaps by European travellers. Cultivation is primarily for fruit, which symbolizes sociability and friendliness among all communities, particularly during auspicious occasions. Earliest references of cultivating pineapple (*A. comosus*) date back in 1399 in Brazil and Paraguay, however, PALF was first mentioned by Filipinos in 1591 [1–3]. Pineapple is one of the most important commercially grown fibre crops since it yields fruit and textile fibre both. Other fibre crops competing with pineapple in terms of yielding nutrition as well as textile grade fibres are banana [4, 5], corn, bamboo [6, 7], flax [8], jute [6, 9, 11] and lotus [12]. However, pineapple surpasses all other fibre crops in terms of global yield/ha [13]. The fineness of PALF is comparable to most of the leaf and bast fibres. PALF is characterized with gleaming whiteness, length and strength making it qualitatively and aesthetically second to none of the common leaf and bast fibres. Interestingly, the plant is usually referred to as pineapple in English literature but is popularly known by its genus name *Ananas* in more than 30 languages especially in European countries and in different Indian vernacular languages.

Looking at the top ten countries producing pineapple, Costa Rica takes the first position showcasing production capabilities of 68.15 ton/ha of fruits and 300 ton of pineapple leaf stubbles/ha [14]. During post-fruit harvesting, the pineapple leaf bunches are largely discarded as agro-wastes, which are then majorly utilized for textiles, paper and composite materials by extracting the fibre from discarded/abandoned leaves. The leaves of the pineapple yield strong, white silky fibres which can be spun into fine textile grade yarn on jute as well as cotton spinning system [15]. Pineapple leaf fibre (PALF) fabric is marketed as pina fabric in the Philippines. Awareness about PALF advantages over synthetic fibres will improve its prospects in the textile market prompting the farm owners to utilize the pineapple leaves. The worldwide area under pineapple harvest in the year 2017 was 1,098,705 ha which can supply approximately 1318 thousand ton of PALF considering 40 ton/h usable fresh leaves and 3% PALF yield [16]. Freshly harvested green leaf bundles are used for fibre extraction. Fibre length is dependent on leaf length of the cultivar. Long leaves of Perolera cultivar of the Caribbean region are considered the most suitable for fibre extraction. Studies are being carried out to establish the stage of leaf harvesting for obtaining quality fibres [17]. The fibres are extracted either manually or mechanically using decorticator. Decorticated fibres contain waxy matter and fleshy leaf parts. To separate fibres from leafy components and gummy substances, fibres are water retted in a tank in material to liquor ratio of 1:10 at 28 °C temperature. Optimum retting time for maximum fibre yield is seven days yielding 2.8% clean PALF fibres [2].

2 Botanical Description, Varieties and Distribution of Pineapple

Pineapple is a terrestrial plant with a height of 0.75–1.5 m and 0.9–1.2 m leaf spread. Central stem stick contains flower buds forming fruit at the tip. Fruit is scaly outside and contains phyllotaxies leaves on top. The plant consists of a bunch of 0.5–1.8 m long concave and 0.52–0.055 m wide pointed tip leaves emerging from the soil and single central stalk. The colour of the leaf ranges from green to red, blue and purple depending upon the cultivar [1, 2].

Four major categories of pineapple varieties are Smooth Cayenne, Queen, Spanish and Abacaxi. Common phenotypic characteristics of all the pineapple varieties are thick skin and juicy pulp of the fruit. Different cultivars of the plant differ in fruit sweetness, flavours as well as phenotypic characteristics. Plant growth characteristics also display variation with varietal differences. Smooth Cayenne variety is resistant to mealy bugs, fruit collapse and heart rot diseases. The leaves are characterized by smooth spineless leaves. Queen varieties are cultivated for fresh consumption due to its sweetness and flavour. Spanish group cultivated in coastal areas of Central and South America is characterized by spiny purplish leaves. Small fruit weighing 1–2 kg are sweet, aromatic and largely used for canning. Pineapple in Brazil is called Abacaxi. Phenotypic attributes of Abacaxi group varieties are tall oblong-shaped fruit and narrow spiny leaves (refer Table 1). The plant largely cultivated in the Caribbean region grows up to 1.5 m with a spread of 1.2 m [1, 18, 19].

3 Classification of Vegetable Fibres

Vegetable fibres are classified according to the sources in plants, where they occur and from which they are extracted. These classifications are as follows:

(a) *Seed hair fibres*: These fibres are obtained from the seeds or seed pods, e.g. cotton, kapok.
(b) *Bast fibres*: These fibres occur in the bast tissue or bark of dicotyledonous plants, e.g. jute, flax, hemp, ramie, sunn, kenaf, etc.
(c) *Leaf fibres*: These fibres are produced using the leaves of selected monocotyledonous plants, e.g. pineapple, banana, Manila hemp, etc.
(d) *Fruit or husk fibres*: These fibres are obtained from the husk of fruit, e.g. coconut (coir), beetle nut, etc.

4 Pineapple Cultivation: Preparation and Propagation

4.1 Climate and Soil

Pineapple flourishes well in sandy loam in a temperature range of 18–45 °C at an altitude below 800 m above sea level. Extreme temperatures and higher altitudes result in low yield, quality and size of pineapple fruit. Rainfall is required throughout the period of plant growth for better performance of the crop. In general, a moderate

Table 1 Pineapple varieties and distribution [16, 20, 21]

Pineapple varieties	Producing countries/states	Fruit weight and salient features
Smooth Cayenne		
Giant Kew, Charlotte Rothschild	India	2.75–4.5 kg, oval-shaped, juicy, mildly acidic, aromatic, used for canning
Hilo	Hawaii	1 to 1 1/2 kg, small crown
Cayenne Lisse	Martinique, Ivory Coast	1 to 1 1/2 kg
Sarawak, Samarahan, Nanas Durian, Nanas Paun	Malaysia	2.5 kg, juicy, acidic
Perolera, Bumanguesa, Santa Maria	Venezuela, Colombia	3–4 kg, long leaves are ideal for fibre extraction, cylindrical, yellow skin
St. Michael	Azores	2.25–2.75, small crown and core, sweet but flavourless
Kona Sugarloaf	Caribbean, Florida, Hawaii, Central and South America	2.3–2.7 kg, large, white to golden flesh, extra sweet
Baron Rothschild	Guinea	0.8–2 kg
Criolla	Peru	
Monte Lirio	Mexico, Costa Rica	Spineless leaves, rounded fruit, white flesh, flavourful with the aroma
Esmeralda	Mexico, Florida	
Fu Mu, Cherimoya, Perfume, Sugarcane, Sugar honey, Golden diamond, Milk	Taiwan	1.3–1.6 kg, cylindrical, yellow flesh, fragrant
Cayenne Guadeloupe	Guadeloupe	Sweet, disease resistant
Queen cultivar		
Mauritius (Queen Malacca)	Malaysia, India, Sri Lanka	1.36–2.25 kg, small plants, sold fresh for juice
Del Monte Gold (MD 2), Hawaiian Gold	Hawaii, Costa Rica, Ghana, Cuba, France	1.5–2 kg, super sweet, 30 days shelf life
Nanas Moris, Sarikei	Malaysia	Short plant with spiny and purplish-green leaves, cold and disease resistant
McGregor	South Africa, Queensland	Medium size
James Queen	South Queen	Large size, square shoulder
Ripley Queen	Florida, Queensland	1.4–2.7 kg, takes unusual 22 weeks from flowering stage to mature fruit, very sweet and non-fibrous

(continued)

Table 1 (continued)

Pineapple varieties	Producing countries/states	Fruit weight and salient features
Kallara local	India	1.3–1.6 kg, Pleasant aroma, used for table purpose
Natal Queen	South Africa, El Salvador	0.75–0.9 kg, Sweet and flavorous
Perolera, Pernambuco, Mordilona, Perola	Venezuela, Brazil, Ecuador, Colombia, Peru, Venezuela,	1.5–3 kg, Smooth spineless leaves, yellow cylindrical fruit, white flesh, tender and juicy
Spanish		
Hybrid 36	Malaysia	Cross between Gandul (Spanish) and cayenne
Josephine	Malaysia	Cross between Johor (Spanish) and Sarawak (cayenne), strong aroma and sweetness
Maspine	Malaysia	1.8 kg, high yielding variety (56 ton fruits/h), excellent canning quality
Cabezona, Cumanesa, Castilla	Venezuela, West Indies, Mexico, Puerto Rico, El Salvador	1.36–2.7 kg, large plant, fragrant, round-shaped, yellow–orange skin, spiny green leaves, resistant to fruit rot, valued for canning
Valera, Morada,	Colombia, Venezuela	1.5–2.5 kg, Conical fruit, Purple–green foliage, purple skin fruit with white juicy flesh
Pineapple Panare	Venezuela	0.45–0.7 kg, bottle-shaped, fragrant, small core and non-fibrous
Abacaxi		
Sugarloaf, Black Jamaica, Montufar	Central and South America, Puerto Rico, Cuba, the Philippines	0.68–1.36 kg, ultra-sweet, blue–green foliage, disease resistant, canning and fresh fruit

rain of 700 mm per year, less water retentive soil (pH 5.5–6) mixed with farmyard manure (FYM) is most suitable for plant growth and fruit yield.

Well-drained, evenly moist soil is required for pineapple plantation. Land should be well ploughed, levelled and also free from weed, stones and plant stubs of previously harvested crops. Curing of planting materials such as suckers and slips is required for 8–10 days in the shade to avoid decaying of fresh green plants sown in the moist soil. Lower leaves of suckers, slips and crown are removed before plantation to facilitate the formation of root system into the soil. The cut end of the plant may attract contaminants and soil pathogens; therefore, it is also recommended to treat the plantlets with monocrotophos (0.15%) or any other systemic insecticide solution for 20 min. Biopriming with bio-fungicides (0.2%) for 5 h or treatment with carbendazim solution (0.1%) for 15 min protects against fungi and improves the resistant power of the plantlets [2, 22].

4.2 Propagation by Suckers, Slips and Fruit Crown

The propagation can be carried out with either by means of sucker, slips or fruit crown [14]. Pineapple is a sucralose plant that produces several suckers or plant sprouts at the time of inflorescence. When the mature plant starts declining after fruit harvest, suckers are separated from the mother plant and planted for growing individual plants. Plantation of suckers to the ripening of the fruits takes about 22 months. A mature plant at fruit development stage yields several slips which are removed from the mother plant for sowing. Planted slips get mature and produce fruit within two years. Pineapple fruit crown with detached lower leaves is also used as a planting material on well-drained evenly moist soil. Within two to three weeks, the root system develops, and the plant starts nutrition from the soil to continue its growth and develops to a mature plant within two and half years to produce its own flowers and fruit.

4.3 Sowing

Pineapple plantlets are sown in flatbeds, furrows, contours and trenches. Its normal time of sowing differs from region to region and is largely dependent on climatic conditions of the area. In the coastal and tropical humid region, it is generally sown between April and July; whereas, in subtropical plains and low elevated areas, the crop is sown in August–November. Pineapple growers use mechanical planters to sow rows of disinfected plant materials in large fields which are faster and capable of planting 50,000 plants in a day. Plantlets cannot stand much rain during sowing as it results in bud rot [1, 2, 18, 23].

4.4 Spacing

The spacing between plants is associated with plant density. The optimum spacing for commercial viability is 0.3 m between plants and 0.4 m between lines. Advantages of high density include high fruit yield (75–105 tonne/h) and propagules/unit area. High density also protects the crop from weed infestation and sunburn. Plant spacing and density vary place to place as per agro-geo-climatic conditions. Plant density for a tropical region, hot climate and at higher elevations is 63,400, 53,300 and 31,000 plants/h, respectively.

4.5 *Hoeing and Weeding*

Three to four weeding in a year is sufficient for optimum yields. Spraying of weedicides (diuron and atrazine) of 2.5–3.5 kg/ha twice in a year is advantageous to remove weeds and also eliminate strenuous hand hoeing operations.

4.6 *Nutrient Management*

Application of N, K_2O rate of 498 and 384 kg/ha has been found optimum for fruit enlargement and maximum fruit yield. Higher doses of *N* result in an increase in fruit yield and size by 3–50% but deteriorate the fruit quality by reducing soluble solids and titrable acidity. In Queensland, urea spraying (151 kg/ha) improved the crop and offshoot yield by 8%. Experiments indicated no significant response of P_2O_5 application on fruit size and yield. Half of the potassium fertilizer should be applied at the time of planting. The remaining *K* could be applied six months after planting.

4.7 *Irrigation*

The pineapple is primarily a rainfed crop and is grown in areas which receive an average rainfall of 700 mm. However, additional irrigation just before flowering improves fruit size and grading. Irrigation once in 25 days is ideal for good production if grown offseason under low rainfall or in hot weather conditions.

5 Fruit of Pineapple Plant

The time between pineapples planting to flowering may take 16–28 months due to varietal and altitude differences and also a method of propagation. Flowering stage is followed by fruit formation when individual flowers, bracts and sepals fuse to form an oval-shaped fleshy seedless syncarp. Flowering stage to the ripening of pineapple fruit takes six months causing its outer scales to turn yellowish. Fruits of pineapple from the plant are generally used as fresh fruit, canning and juice concentrate with distinctive necessities of size, form, colour, smell and taste [24]. Pineapples are processed into a number of value-added products like jam, jelly, cheese, chutneys and canned halves. Besides, fruit pineapples are an amalgamation of flowers fused together around a central core. Each fruitlet can be recognized by an "eye", the irregular spiny pattern on the outward. The fibrous fleshy part of pineapple looks yellowish in colour. The base part of the fruit area is richer in sugar, and therefore, has a sweetened taste. Fruits require about six months from growing to harvest. Total production time is approximately 15–18 months from transplanting or around 12 months for a ratoon crop [2, 25].

Pineapples are naturally drought-tolerant since they are crassulacean acid metabolism (CAM) plants. Pineapple plants need large amounts of nitrogen that can be supplied by urea and sulphate of ammonia. Nitrogen is also important to the weight of the fruit and should be applied before flowering as nitrogen applied after may result in a reduction in fruit juice acid [25].

On the basis of five-year (2013–2018) average pineapple fruit population, Costa Rica ranks first sharing 11.4% in the world; whereas, the area under pineapple cultivation is highest in Nigeria followed by India. The distribution and production of pineapple in main growing countries are given in Table 1 [25, 26].

6 Diseases of the Pineapple Plant

Diseases of pineapple are related to microorganisms like fungi, bacteria, virus and pest. They damage and spoil different parts of a plant and affect its growth during pre- and post-stages of harvesting. If proper care is not given to the plant, then the ripe fruit may also get infected with these unwanted members such as fungi; associated with diseases like Phytophthora heart (top) rot (The oomycetes *Phytophthora cinnamomi* and *Phytophthora nicotine* causes the Phytophthora heart (top) rot in pineapple plant); Phytophthora root rot (The Phytophthora root rot is caused by a pathogen "*P. cinnamomi*"); Base (butt) rot (The fungus Chalaraparadoxa); Fruitlet core rot (green eye); Fusariosis; Green fruit rot; Inter fruitlet corking; Leathery pocket; Water blister; White leaf spot (Chalaraparadoxa is common in pineapple plantations. The fungus will only invade wounds and is most active in warm, wet weather); Fruit rot by yeast and Candida species (The disease may occur before or after harvest); Nematodes associated diseases (refer Table 2) [18, 23, 27].

Bacteria and phytoplasmas associated diseases are Marbling (The bacteria enter through the open flower and natural growth cracks on the fruit surface. Infected fruits are usually low in both acid and sugars); Pink disease (The bacteria are thought to be carried by nectar-feeding insects and mites to open flowers from infected, decaying fruit near flowering fields); Virus associated diseases such as Mealybug wilt disease; Yellow spot (the disease is rarely seen) [19].

Table 2 Diseases of the pineapple plant [18, 19, 23, 27, 28]

Causal organisms of diseases	Disease symptoms	Remedial measures
Thielaviopsis paradoxa (plant pathogen) stem rot	Rotting of plant material for planting and of fruits post-harvest	Prevention is done by immersion of the stem in benzoic acid or bio-fungicides in bio-fungicide solution before planting, and proper drainage is a must
Penicillium funiculosum, mites, fruit flies, moth, scale insects	Fruit infection, fruitlet core rot, inter fruitlet corking, corky tissues on fruit scales, fruitlet turn brown and sunken as the fruit ripens, malformed fruit	Mulching is recommended to avoid fruit contact with soil-borne pathogens and insects. Spraying with fungicides and insecticides
Dysmicoccus brevipes, *Pseudococcus brevipes*, *P. neobrevipes* (mealy bugs)	Wilted plant, reddening of leaves, root rot	Spraying with insecticides
Cottony woodlouse	Plant fading, yellow spot on plant tip	Spraying with insecticide parathion
Chalara paradoxa (plant pathogen; fungus)	Brown spot on leaves	Spraying with bio-fungicides
Nematodes	Root swelling and infection, hinder plant growth	Before planting, remove vegetable matter from the soil which possibly hosts nematodes, crop rotation and fumigation of soil prior to planting. Spray nematocides
Myriapods	Root rot, plant decay	Fumigation, insecticide spray
Phytophthora cinnamomic, *P. parasitica*, (fungus)	Heart rot, root rot, rotten leaves at the base	Mulching on the raised bed is recommended, insecticide spray; Captafol (2%) on 3,500 L/h, applied after planting, one month later and one week after the treatment for flowering induction
Butt Rot	Plant decay at ground level	Spraying with bio-fungicides
Thecla Basilides (fruit borer)	Fruit damage	Spraying with fungicides, smooth handling of fruits before and after harvest to avoid cracks and injury
Dickeya dadantii.	Wilting, soft rot, stunted plant growth, plant cell degradation	Prevention through sanitation maintenance and exclusion of infected plant materials

7 Production of Pineapple Leaf Fibre (PALF)

World pineapple production was approximately 51 million ton in the year 2016. Pineapple agriculture cultivation produced large amounts of leaves beside fruits. Pineapple and similar plants such as *Ananas erectifolius* and *Ananas lucidus* leaves might be used for their high quality of fibre or as feedstuff. The pineapple plants are herbaceous monocots about 2–4 ft tall, 3–4 ft wide with short stems and unnoticeable rosette of long leaves. The strap-like leaves have spines at tips and margins are spirally arranged on the stem and have axillary lateral buds at their base known as suckers used as planting stock in propagation for the next crop production. One sucker will be left to grow in place of the original plant yield is known as the second crop using the same agricultural ground [1]. Pineapple can be a useful species of agroforestry approximately produced in ranging from 30 ton/ha to 60–80 ton/ha. Optimum production yield is 60–80 t/ha for the first harvest. Optimum first crop produced approximately 10% less compared with the second ratoon crop which is approximately 30% less. The Philippines and Taiwan are the principal producers of the PALF followed by Brazil, Hawaii, Indonesia, West Indies and India. Only in India, the yield of fibres could be around six lakhs tonne in one year if the proper method for extraction process is adopted. Pina clothing made of PALF was popular before the nineteenth century not only in the Philippines but also in Europe, North America and Africa. Subsequently, the unexpected rise of cotton fibre caused the pina fabric to disappear from the world market and was limited to few places in Asia where PALF is still processed for eco-conscious consumers [13, 25] as shown in Table 3.

Table 3 Pineapple production, area and average yield [13, 25]

Pineapple producing countries (Top 10)	Fruit production (thousand ton)	Harvested area		Average fruit yield		Approximate PALF production potential (thousand ton)
		Hectare	Rank	Metric ton/hectare	Rank	
Costa Rica	3056.445	44,500	8	68.15	2	53
the Philippines	2671.711	66,088	5	39.22	14	79
Brazil	2253.897	62,116	6	40.05	13	74
China	2129.936	80,115	4	18	41	96
Thailand	2123.177	86,454	3	24.29	28	103
India	1861.000	111,000	2	8.12	66	133
Indonesia	1795.986	15,500	15	27.48	24	18
Nigeria	1642.376	200,010	1	132.14	1	240
Colombia	1091.042	26,140	11	45	9	31
Mexico	945.210	20,006	13	41.11	12	24

However, at present, the pineapple leaves which go mostly as agriculture waste might be utilized for fibres extraction from fresh leaves for making of ropes, twines, composites and clothing [25]. The pineapple plant is largely used as a source of fibre in the Philippines, Taiwan and India. With the current interest in natural fibres, improved extraction technologies and scarcity of cotton, the revival of pina clothing is certain in coming decades. In 2008, global production of pineapple was about 19 million tonne which rose from 16.6 million tonne in 2004 to approximately 21 million tonne in 2007. However, in 2008, the production of pineapple declined [16].

8 Pineapple Fibre

- Pineapple fibre is considered as more delicate in texture among all vegetable fibres.
- Approximately 60 cm long, white and lustrous-like silk can easily take and retain different classes of dyes.
- The fibre is ten times coarser than cotton.
- It is multicellular lingocellulosic fibre acquired from the leaf of the plant *A. comosus*.
- The yield of fibre is 2.5–3.3% of the weight of green leaves.
- Its main constituents are α-cellulose, hemicelluloses and lignin.
- Application of pineapple fibre as end fabric is lightweight, easy to care, elegant and looks like linen.

8.1 Methods of Pineapple Fibre Extraction

Hand Stripping/scrapping: *(*https://core.ac.uk/download/pdf/82556402.pdf*)*

- This process utilizes a broken porcelain plate and manually scraping the fibres of the Spanish or native variety pineapple.
- Two types of fibres are produced in this method, namely liniwan and bastos.
- After scrapping the fibres, they are washed thoroughly with tap water and air-dried.

The pectic substances in the soft cells are dissolved by means of microorganisms, which free the fibre bundles and make it possible to separate them from the woody core material.

Pineapple fibre extracted from the fresh leaf of the pineapple plant alike sisal plant. The length of pineapple fresh leaves varies in the range of 55–75 mm; width of the leaves in the range of 3–6 mm with an average weight of each leaf is 15–50 g. The fibre yield of pineapple is in the range of 1.55–2.5%. In most cases, the fresh pineapple leaves are a by-product of fruit production, which provide an added revenue source to the producers. As pineapple fibres are environmentally friendly being a natural source plant; their use is expected to advance in many fields.

8.2 Extracting Fibre Using Decorticating Machines

The leaves are fed manually in the decorticating machine for scraping by the revolving blades. Decorticated fibres after washing with water are sun-dried. Fibres produced are a mixture of liniwan and bastos [29].

Extraction methods for pineapple leaf fibre nowadays fall under the waste products of pineapple cultivation. A special purpose machine having metal knife scrapper roller and the serrated roller is used to scrap out the waxy layer, and at the same time with retting process, the pineapple leaf fibre is also extracted. Study conducted by Yusof et al. [30] revealed that PALF produced by decorticating machine was more soft, bright and had creamy white colour instead of brown when compared to the conventional method. Chemical constituents of various pineapple fibres include α-cellulose, pentosans, lignin, fat and wax, ash content, nitrogenous matter and pectin.

After extraction, splitting up of fibrous strands, which are coarser due to generic reasons, is carried out by retting and degumming in one of the following two ways:

a. Biological natural retting, in which bacteria or fungi (dew retting) are the active ingredients.
b. Chemical retting or degumming, in which dilute acids, bases or enzymes are used as active ingredients.

8.2.1 Extraction of Fibre by Retting

Retting is defined as the separation of the fibre bundles from the cortex or wood, effecting digestion of the cementing material between the fibres in the bundles. This loosening of the fibres is due to the removal of various cementing tissue components presumably of pectic nature. Retting is a microorganism process. This is a two-stage process: (1) physical stage (swelling and extraction of some soluble substances) and (2) the growth of microorganisms like fungus or bacteria. The scratched leaves are tied and immersed in a water tank. Urea or diammonium phosphate is added for quick retting. At the end of retting, leaves are taken out and washed mechanically by pond water. Using ceramic plate over the pineapple leaf with pressure and fast movement of it will give the fibre beneath the leaf. This is the easy way to do the extraction of the fibre from long leaf. Pre- and post-harvesting metabolism: As a part of the grading standard, not only pineapple fruit but also the crown quality is a vital characteristic of economic concern. As such, leaf damage, occurring as brown spots on the crown leaves, causes appreciable economic losses. The pineapple crown is being a continuation of the vegetative stem, and the spirally arranged leaflets have similar morphology. The photosynthesis activity of crown leaves remained unexplored. Pineapple fibre is white, creamy and lustrous as silk fibre and is coarser as cotton and fibre can easily retain dyes.

For the purpose of the rope and twine manufacturer, the pineapple fibre is usually processed on the jute processing system. In blended yarns, the quality of the yarns and their spinnability during processing is found to improve with the increase of the pineapple fibre content.

8.2.2 Extraction of Fibre by Chemicals

The chemical degumming of pineapple can be carried out according to the following procedures: preparation (immersion in acid, H_2SO_4) → washing → boiling in NaOH solution → washing → bleaching → water extraction → oiling → drying. It should be noted that the degumming process must avoid the complete removal of the gums because the single fibres, if separated from each other (without the gum), cannot be spun due to their short length (refer Fig. 1).

Chemical degumming of pineapple fibres is accompanied by subjecting the fibres to the solution of acids, alkalies or enzymes at varying levels of temperature and duration of treatment in the absence of air. The hemicellulose which is made up of largely mixed polysaccharides is converted to their soluble simple products of sugars;

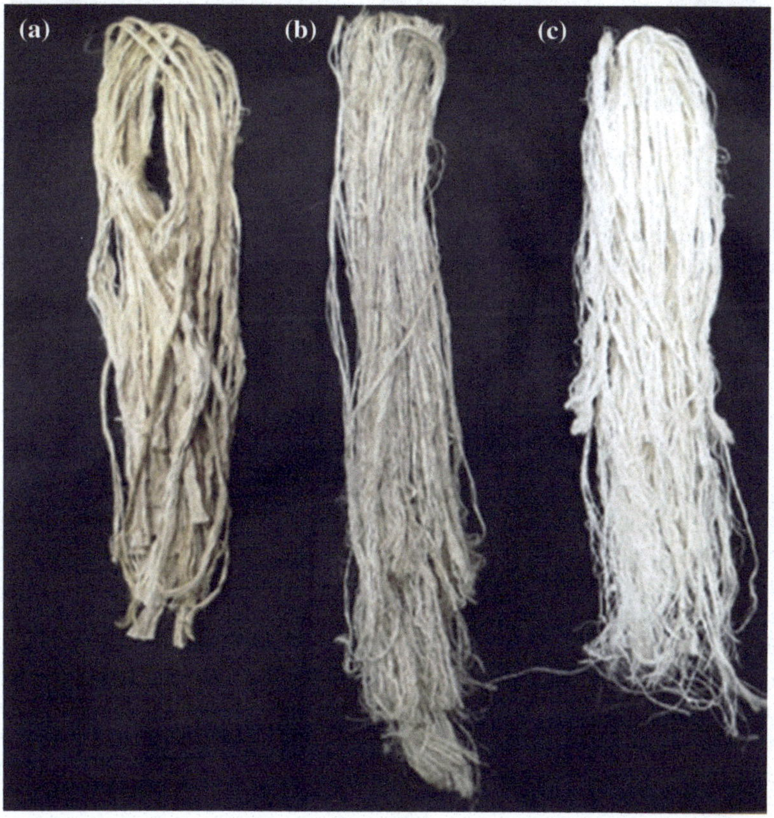

Fig. 1 **a** Retted, **b** degummed, **c** bleached PALF [31]

Fig. 2 **a** Degummed PALF yarn, **b** bleached PALF yarn [31]

saponifiable gums and waxes are converted to soluble soaps, and unsaponifiable oils are emulsified by these soaps and wetting agents. Retting with 5% sodium hydroxide for 12 h at boil can produce the finest fibres (refer Fig. 2).

8.3 Physical and Chemical Properties

8.3.1 Physical Properties

- ***Physical structure***: Under the microscope, pineapple fibre is found to be scaly, a cellular structure with its vegetable matter intact. They have a high degree of crystallinity with a spiral angle of 15°
- ***Effect of Moisture***: Pineapple fibres lose strength and elongation in wet condition. The loss in strength may be due to the penetration of water molecule into the multicellular lignin cellulosic fibres and subsequently swells it up to some extent; thus, it loosens the binding of the ultimate cells which results in cell slippage when the load is applied, and wetting extension is also reduced by 7 and 12% in untreated and degummed fibres, respectively.

Table 4 Physical characteristics of pineapple fibre [32]

Single cell	
Length (mm)	3–8
Diameter (μm)	7–18
Fineness (tex)	2.5–4
Fibre bundle	
Length (mm)	10–90
Fineness (tex)	2.5–5.5
Tenacity (cN/tex)	30–40
Elongation (%)	2.4–3.4
Initial modulus (cN/tex)	570–700
Density (g/cm^3)	1.543

- ***Soft fibre***: Pineapple fibre is a very soft fibre. One can feel the softness of pineapple fibre.
- ***White with good lustre***: Pineapple fibre is white in colour, and on extracting it is slightly dull yellowish in colour and has good lustre properties.
- ***Flexural and torsional rigidity***: Pineapple fibre is having high flexural and torsional rigidity than that of cotton fibre.
- ***Crystallinity***: Pineapple fibre is more ordered, i.e. it is more crystalline. The strength and elongation are comparable with cotton fibre.

For other physical properties of pineapple fibres, refer Table 4 [32].

8.3.2 Chemical Properties

- Treatment with 18% sodium hydroxide imparts crimp and enhances the breaking elongation of fibre. Shrinkage is more in length way direction.
- Peroxide bleaching improves the fineness by 5–6% but reduces the tensile strength by 40–45%. During bleaching, the fibre loses its original shape and feel and becomes hard.
- The reduction in strength may be due to the loss of hemicelluloses and lignin as they are directly related to the alteration in physical properties of fibres such as changes in the angle of orientation, decrease in crystallinity and change from cellulose-I to cellulose-II.
- The degree of brightness or bleached pineapple fibres is about 78%, whereas that for raw pineapple fibres is 70%. Hypochlorite bleaching is ineffective due to fibre degradation.
- It was found that peroxide bleaching reduces lignin, hemicelluloses and pectin present in raw pineapple fibres by 27.3%, 52.8% and 100%, respectively.
- Pineapple fibres dissolve in 60% sulphuric acid in 5 min. 2 h soda boiling increases absorbency with a marginal loss in tensile strength and weight loss.

Table 5 Chemical content of pineapple fibre bundle [32, 33]

Cellulose	55–68%
Hemicellulose	15–20%
Pectin	2–4%
Lignin	8–12%
Water-soluble material	1–3%
Fat and wax	4–7%
Ash	2–3%

- Pineapple fibres can be successfully dyed with direct, reactive, vat and azo dyes with better fastness properties as compared to that of cotton. Also, dye absorption tendency of the fibre is more than that of cotton. This may be due to the relatively high moisture content of the fibre and low reflectance value of fibre due to natural greenish yellow colour present.
- Presence of –OH and –COOH group in the molecular chain enhances the fixation of reactive dyes.
- It was also studied and suggested that pineapple fibre can be dyed conveniently with basic dyes at room temperature. This is due to the presence of lignin and hemicellulose, which are more than 15% amorphous in character with acidic nature.

Almost all vegetable fibres contain one or more of the following components (refer Table 5) [32, 33].

- *Cellulose*: Principal constituent.
- *Hemicelluloses*: Amorphous short-chained isotropic polysaccharides of polyuronides.
- *Pectin*: Water insoluble, calcium, magnesium and iron salt of pectic acid.
- *Fats and waxes*: These are found on the surface and can be extracted with benzene.
- *Lignin*: Short-chained isotropic and non-crystalline polymer.
- *Colouring matters*: In cortical cells.

9 Applications of Pineapple Fibres

Pineapple fibre is used for making cloth and also at times combined with silk or polyester to manufacture textile fabrics. Pineapple fibre is also used for table linens, bags, mats and other clothing items. It makes different uses across the various parts of the world. The huge potential for pineapple fabric makes it for diverse uses and eco-friendly properties. Weaving, sewing and other activities lead to commercial products manufacturing. The scope of a huge market in Assam and outside of the north-east region of India is there. The natural fibres with different crops like jute, coir, ramie, flax and hemp in comparison with PALF already established themselves in the market worldwide. Products can be made from PALF are handbags, coasters and many more products for interior design. PALF copolymer and composites are used in automobiles and railway coaches.

Due to very high initial modulus, it can be used in industrial textiles. It can be used in the manufacturing of conveyor belt cord, V-belt cord, lightweight duck cloth, etc. It is also used for other table linens, ropes, bags, mats and other clothing items, or anytime that a lightweight, but the stiff and sheer fabric is needed. Pineapple fibres are also used in paper industries as pulping material.

10 Economic Importance for Farmers

Pineapple provides the raw material for food, textile and pharmaceutical industry, and thus, enables employment to millions world over. Innumerable health benefits of pineapple make it a regular feature at friendly get-together and restaurants in the form of fruit chunks, cake, juices, smoothness and barbecue. Besides fresh pineapple consumption, the fruit is widely preserved in the form of fruit slices, jam, jellies, marmalades, ice cream and fermented vinegar for salad dressing, wine and desserts and is available in food stores across the globe. Pineapple crop is uprooted for taking out the fruit. After harvesting the fruits, leaf bunches are cut manually using the sharp sickle. Freshly harvested green leaf bundles are used for fibre extraction. Pineapple leaf fibre (PALF)-based products including apparel furnishing, yarn and footwear have penetrated the market and made their mark as the consumers are looking for alternative eco-form involving cleaner production techniques. Biopulping of PALF yields fine quality papers with considerable whiteness. Plant butt is used to extract bromelain used in food processing, medicine and paint industry. Pineapple plant parts: fruit, leaves, butt and propagules are providing livelihood to farmers and many small–medium industries, thus sustaining the economy. In Costa Rica alone, pineapple production employs 23 thousand people directly and 92 thousand indirectly. Thus, post-harvest handling and marketing of fruit and leaves are empowering farm women and men by improving their livelihood. Discarded fruits, as well as waste materials, could be utilized for other industrial purposes, viz. fermentation, extraction of bioactive components, extraction of functional ingredients, etc. They can also be utilized for extraction of bromelain enzyme and secondarily as low-cost raw material for the production of ethanol, phenolic antioxidants, organic acids, biogas and fibre production [34].

11 Environmental Aspects of Pineapple Cultivation

Pineapple supplements soil nutrients and helps in improving the nutritive value of intercropped plants. Pineapple crop also helps in preventing soil erosion and run-off during flood [35] which depends on the farming technique used [36] and suggested benefits of contour cultivation of pineapple as increased plantation of suckers (30% more suckers per hectare), reduction in soil loss provided better soil nutrient status, less use of fertilizers and healthier and better fruit yield [36]. Numbers of environmental issues are related to the production and cultivation of pineapple plants. These can be listed as follows:

- Deforestation
- Agro-toxics
- Air pollution
- Biodiversity loss (wildlife, agro-diversity),
- Food insecurity (crop damage)
- Health issues

The major reason for such issues is the fact that pineapple production depends on monoculture; which is single crop farming leading to negligible habitat for other species in the fields. This results in ecological imbalance and makes the pineapple vulnerable to pests. Pests and diseaseses are naturally regulated only with changing types of fruits, mixed cultures or with the use of pesticides [1]. Most of the chemicals used are very toxic and have a harmful effect on the environment especially aquatic ecosystem, groundwater and creating health problems of the surrounding community [20, 37]. In Costa Rica where monoculture cultivation of pineapple is prevalent [38], the villagers of surrounding areas reported health issues related to headache, body aches, nausea and leukaemia. Deforestation is common in order to develop huge pineapple plantation which affects the biodiversity by reducing the area for their habitat (http://www.bananalink.org.uk/the-problem-with-pineapples). Dole Food Company has replaced rainforests in the Philippines with vast pineapple plantations. Such widespread deforestation has put innumerable Asian species on the edge of extinction [1].

Apart from the negative aspects of pineapple, it has a positive role in sustaining the mother earth. The waste produced by the jam and juice industries in the form of fruit peel and leaves is rich in lignin and cellulose, and thus, forms a very good raw material for allied fibres. Recycling of the pineapple waste as a substitute for cotton/artificial fibres production will minimize the use of forest resources and blending with other allied fibres to improve their quality and application. Cellulosic natural fibre from pineapple leaves is considered as a green alternative to the conventional polyethene (PE) soil cover in agro-industry. The use of pineapple leaf fibres soil cover can result in disposal problem of the conventional plastic covers which take hundreds of years to degrade [39, 40].

12 Conclusions and Future Aspects

PALF is considered to be more delicate in texture than any other vegetable fibre. Pineapple cultivation is now spread throughout the world since the plant propagation and cultivation practices are easy to adopt by farm owners and also due to its economic importance. The plant does not require high fertilization and also some disease-resistant varieties of the plant are developed by scientists. Different pineapple varieties characterized with specific phenotypic traits are spread throughout the world. Spineless pineapple leaf bunches are preferable for fibre production due to ease of handling during post-harvest processing, as compared to the plant with spiny leaves. PALF production potential is highest in Nigeria due to the highest cultivation area followed by India, Thailand and China. Pineapple leaves are retted using water, microbes and chemical. PALF was characterized with qualities such as length, lustre, strength, softness, whiteness and spinnability. Elongation per cent of the PALF was found to be improved with alkaline treatment. Fibre is mainly composed of cellulose (55–68%), and it is also resistant to alkali and environmental factors. At present, PALF is used to prepare various utility articles as well as high-end fashion garments. With its lustrous silk-like quality, this soil born white gold is gaining popularity as a substitute for more expensive cotton that is now in short supply. It hopes to create greater economic benefits for Indian farmers and also provides more employment opportunities in textile industries. Innumerable advantages of eco-friendly PALF make it an undisputed choice for domestics and technical textiles benefiting not only farmers but also consumers and the environment.

References

1. Collins JL (1949) History, taxonomy and culture of the pineapple. Econ Bot 3(4):335–359
2. Collins JL (1961) The pineapple: botany, cultivation and utilization. Pineapple Bot Cultivation Util
3. Mukherjee PS, Satyanarayana KG (1986) Structure and properties of some vegetable fibres. J Mater Sci 21(1):51–56
4. Johnson MO (1935) The pineapple. Pineapple
5. Mukhopadhyay S, Fangueiro R, Arpac Y, Şentürk Ü (2008) Banana fibers–variability and fracture behaviour. J Eng Fiber Fabr 3(2):155892500800300200
6. Jadhav AC, Pandit P, Gayatri TN, Chavan PP, Jadhav NC (2019) Production of green composites from various sustainable raw materials. In: Green composites. Springer, pp 1–24
7. Nongdam P, Tikendra L (2014) The nutritional facts of bamboo shoots and their usage as important traditional foods of northeast India. Int Sch Res Not
8. Pandey R (2016) Fiber extraction from dual-purpose flax. J Nat Fibers 13(5):565–577
9. Das S (2017) Mechanical properties of waste paper/jute fabric reinforced polyester resin matrix hybrid composites. Carbohydr Polym 172:60–67
10. Das S, Basak S, Pandit P, Singha AK (2018) Hybrid bast fibre strengthened thermoset composites. Therm Compos Prep Prop Appl 38:112–127
11. Teli MD, Pandit P, Basak S (2017) Coconut shell extract imparting multifunction properties to ligno-cellulosic material. J Ind Text 1528083716686937
12. Pandey R, Sinha MK, Dubey A (2018) Cellulosic fibers from Lotus (*Nelumbo nucifera*) peduncle. J Nat Fibers 1–12
13. FAOSTAT FAO (2018) Available online: http://www.fao.org/faostat/en/#data.QC. Accessed 2017 Jan
14. Mathew S (2017) Economic analysis of production and disposal of pineapple (*Ananas Comosus* L.). In: Dodamarg Tahsil of Sindhudurg District (Ms). Dbskkv., Dapoli
15. Asaolu VO, Binuomote RT, Oyelami OS (2016) Assessment of feeding value of vegetable-carried pineapple fruit wastes to Red Sokoto goats in Ogbomoso, Oyo State of Nigeria. Afr J Biotechnol 15(31):1648–1660

16. Jose S, Salim R, Ammayappan L (2016) An overview on production, properties, and value addition of pineapple leaf fibers (PALF). J Nat Fibers 13(3):362–373
17. Luo S, Netravali AN (1999) Mechanical and thermal properties of environment-friendly "green" composites made from pineapple leaf fibers and poly (hydroxybutyrate-co-valerate) resin. Polym Compos 20(3):367–378
18. Sanewski GM, Bartholomew DP, Paull RE (2018) The pineapple: botany, production and uses. CABI
19. Martínez R, Torres P, Meneses MA, Figueroa JG, Pérez-Álvarez JA, Viuda-Martos M (2012) Chemical, technological and in vitro antioxidant properties of mango, guava, pineapple and passion fruit dietary fibre concentrate. Food Chem 135(3):1520–1526
20. Pavithran C, Mukherjee PS, Brahmakumar M, Damodaran AD (1987) Impact properties of natural fibre composites. J Mater Sci Lett 6(8):882–884
21. Medina JD, García HS (2005) Pineapple: post-harvest operations. Instituto Tecnologico de Veracruz
22. Mwaikambo L (2006) Review of the history, properties and application of plant fibres. Afr J Sci Technol 7(2):121
23. Lobo MG, Paull RE (2017) Handbook of pineapple technology: production, postharvest science, processing and nutrition. Wiley
24. Joy PP, Sindhu G (2012) Diseases of pineapple (*Ananas comosus*): pathogen, symptoms, infection, spread and management. Consultado Agosto
25. Asim M, Abdan K, Jawaid M, Nasir M, Dashtizadeh Z, Ishak MR et al (2015) A review on pineapple leaves fibre and its composites. Int J Polym Sci
26. Spironello A, Quaggio JA, Teixeira LAJ, Furlani PR, Sigrist JMM (2004) Pineapple yield and fruit quality effected by NPK fertilization in a tropical soil. Rev Bras Frutic 26(1):155–159
27. Prasetyo J, Aeny TN (2014) Pineapple fruit collapse: newly emerging disease of pineapple fruit in Lampung, Indonesia. J Hama dan Penyakit Tumbuh Trop 14(1):96–99
28. Rot PHT (2012) Diseases of pineapple (*Ananas comosus*)
29. Uddin MG (2014) Effects of different mordants on silk fabric dyed with onion outer skin extracts. J Text
30. Yusof Y, Yahya SA, Adam A (2015) Novel technology for sustainable pineapple leaf fibers productions. Procedia CIRP 26:756–760
31. Hazarika D, Gogoi N, Jose S, Das R, Basu G (2017) Exploration of future prospects of Indian pineapple leaf, an agro waste for textile application. J Clean Prod 141:580–586
32. Franck RR (ed) (2005) Bast and other plant fibres, vol 39. CRC Press
33. Doraisswamy I, Chellamani P (1993) Textile progress, vol 24, no 1. Textile Institute, Manchester, UK
34. Bhat NV, Upadhyay DJ, Deshmukh RR, Gupta SK (2003) Investigation of plasma-induced photochemical reaction on a polypropylene surface. J Phys Chem B 107(19):4550–4559
35. Abbasi MA, Jamal T (1999) Soil loss and runoff measurement from banana-pineapple intercropping system. Pak J Biol Sci 2(3):689–692
36. Bhuiyan AA (2006) Benefits of contour cultivation of pineapple. Policy Brief 9. MACH (Management of aquatic ecosystems through community husbandry), House# 2, Road# 23/A, Gulshan 1, Dhaka 1212, Bangladesh. Available at htpp. 2006
37. Mohamed AR, Sapuan SM, Shahjahan M, Khalina A (2009) Characterization of pineapple leaf fibers from selected Malaysian cultivars. J Food Agric Environ 7(1):235–240
38. Fagbemigun TK, Fagbemi OD, Buhari F, Mgbachiuzo E, Igwe CC (2016) Fibre characteristics and strength properties of Nigerian pineapple leaf (*Ananas cosmosus*), banana peduncle and banana leaf (*Musa sapientum*)–potential green resources for pulp and paper production. J Sci Res Rep 12(2):1–13
39. Sarah S, Rahman W, Majid RA, Yahya WJ, Adrus N, Hasannuddin AK et al (2018) Optimization of pineapple leaf fibre extraction methods and their biodegradabilities for soil cover application. J Polym Environ 26(1):319–329
40. Thirumal Y (2015) Advantages of pineapple fiber. fiber2fabric

Anatomical Structure of Pineapple Leaf Fiber

Kunal Singha, Pintu Pandit and Sanjay Shrivastava

Abstract The use of natural fibers, such as pineapple, sisal, banana, coir, sun hemp, mesta, or jute, in polymer composite materials has expanded fundamentally in recent years. Today, pineapple fiber is enormously popular among the composite research community due to its various advantages including its smoothed and scaled morphology, low thickness, firmness, reduced weight, and superior mechanical properties. In addition, pineapple fiber is completely/partially biodegradable and recyclable, cheap to produce, and easy to make. Its various mechanical testing characterization values, including tensile strength, spilt tensile strength, flexural strength, impact strength, peeling test, and compressive strength, represent benchmarks compared with other, currently available natural fibers. In this chapter we will extensively discuss the various anatomical structures of pineapple leaf fiber and the effects these have on thermal and mechanical characteristics—observed via scanning electron microscope imaging of surface morphology and the mechanical fracture patterns identified via Fourier-transform infrared spectroscopy and XRD. Consideration is given to external loading and molecular characterization and crystallography of pineapple fiber to better understand its mechanical and thermal behavior.

Keywords Pineapple fiber · Composite · Surface morphology · Mechanical properties · FTIR

1 Introduction

Pineapple is a regular herbaceous tree that grows 1–2 m high and has a spread of similar dimensions. Pineapple is in the family Bromeliaceae. It is principally cultivated in tropical and coastal provinces, primarily for fruit. In India, it is grown continually on sites that are approximately 9–10 km^2 in size. Pineapple grows in grassland and forms on an initial deep-green colored stem. The sprout of the leaf is initially decorative and goes on to grow to 3 ft in length, with 2–3 in. wide blades

K. Singha (✉) · P. Pandit · S. Shrivastava
National Institute of Fashion Technology, Mithapur Farms, Patna 800001, India
e-mail: kunal.singha@nift.ac.in

M. Jawaid et al. (eds.), *Pineapple Leaf Fibers*, Green Energy and Technology,
https://doi.org/10.1007/978-981-15-1416-6_2

and countless spirally organized fibrous leaf edges that are curved near their cross areas to sustain leaf rigidity.

Each pineapple leaf fiber (PALF) has an identical number of hexagonal areas on its exterior layer that are not dependent on size or even contour. Currently Malaysia and Hawaii are the largest producers of PALF. Such production creates a large volume of waste material—approximately 3846.73 lakh kg in the 2008 season [1]. Production of PALF is required for many manufacturing functions. In Malaysia pineapple is widely available and is known as Nanas—people eat it and associate it with wealth and prosperity. Commercially, pineapple is available on the market in various colors and varieties: green pineapple, red pineapple, Sarawak pineapple, as well as Morris pineapple. The waste produced during production is non-toxic. It represents a resource of bioactive elements, especially proteolytic enzymes. Pineapple has a quite rich supply of bromelain along with additional cysteine proteases that exist in its various parts. Bromelain is one of the main sources of protein used for supplements [2, 3] and it continues to be utilized in the food industry, for the production of makeup, and in nutritional supplements since materials like bromelain are viscous in nature and therefore have the power of gelation with other food ingredients, skin, or even plasma [4, 5]. Over the time, pineapple has become indigenous in America and was initially noticed by Columbus in 1493 on an island off the West Indies. This new area was found to facilitate ample pineapple production due its humid environment. The name pineapple originated from the word 'pina,' meaning cone-shaped object, and 'ananas,' meaning fresh fruit. The pineapple is a classic symbol of wealth and can be observed in many embossed embellishments. In the seventeenth century, Americans shipped pineapples from the Caribbean due to their seemingly unusual functions as well as rareness—they soon started to be viewed as an icon for rich individuals in America. The Portuguese played a critical role in providing fresh fruit throughout many exotic areas as well as important communities located on the east and south coast of Africa, the Philippines, Java, China, southern India, Madagascar, and Malaysia [6]. Nowadays, varieties of pineapple plants can be found that are utilized in numerous ways, such as in the production of non-toxic therapeutics and in industrial processes. Examples of its diverse use include combining pineapple juice with sand as an effective cleaner for boat decks and using dehydrated surplus waste material from pineapple as bran feed for livestock, pigs, chicken, etc. [7, 8].

Each year a significant mass of pineapple fibers is generated, some parts of which are utilized for feed and power production. In recent times, the manufacturing waste from pineapple leaves is used in sustainable materials like biocomposites. Pineapple is thereby associated with a non-food-based farming sector [9–11].

2 Characteristics of PALF

- Colored white with a sleek appearance.
- Shiny as silk.
- Moderate in length.

- High tensile strength.
- Softer surface compared with some other organic fibers.

Pineapple leaf fiber has high strength and stiffness and is hydrophilic in nature because of its high cellulose content [9]. Extraction processes are performed using a few easy physical procedures, one of which being the retting technique, as exhibited in Fig. 1. New leaves yield approximately 2–3% of the volume of total fiber in a tree. The fibrous cells of PALF are made up of the vascular bundle in the type of mixtures that are acquired after the physical elimination of the whole top level of a plant after harvesting. Pineapple leaf fiber is made up of many chemical substances. It contains multicellular lignocellulosic fiber with polysaccharides, lignin in huge quantities, and several minor chemical substances, such as excess fat, inorganic substances, color pigments, pentosan, anhydride, uronic acid, pectin, and wax [10]. Pineapple fiber is comprised of tiny cell-like multicellular fibers that are firmly linked together with the aid of pectin. Fibers are comprised of cellulose (70.82%), and their orientation is similar to cellulosic fiber cotton (82.7% cellulose). Hence, most of its physical, mechanical, or even chemical properties are closely aligned to cotton fiber [12–14].

Pineapple leaf fiber has a cell wall system that, under transmission electron microscopy, shows various unique layers, such as major (P), secondary, and tertiary (S1, S2, and S3) layers. The chemical structure of PALF is depicted in Tables 1 and 2. It has several chemical-based constituents, such as cellulose, pentosans, lignin, wax and fat, pectin, nitrogenous material, ash, all of which affect the level of polymerization, the crystallinity of cellulose, and its antioxidant ability [15–17]. PALF has a large amount of cellulose (81.27%), minimal quantities of hemicellulose (12.31%), as well as lignin (3.46%) [18, 19]. Pineapple leaf fiber has a greater cellulosic content

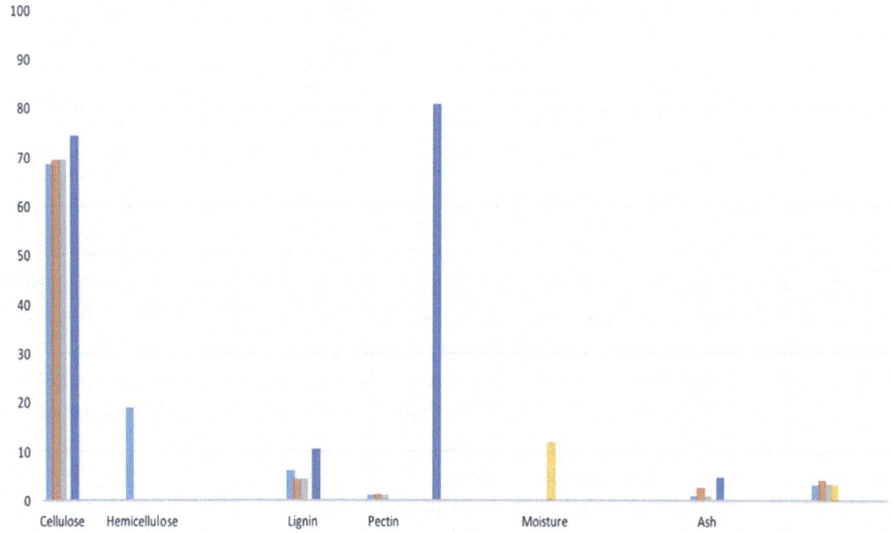

Fig. 1 Percentage based composition in percentage in pineapple fiber [3]

Table 1 Annual production of natural fiber and its sources [1]

Fiber source	World production (Tons)	Origin
Abaca	70,000	Stalk
Bamboo	10,000,000	Stalk
Banana	200,000	Fruit
Broom	Plentiful	Stalk
Coir	100,000	Stalk
Cotton lint	18,500	Stalk
Elephant grass	Plentiful	Stalk
Flax	810,000	Stalk
Hemp	215,000	Stalk
Jute	2,500,000	Stalk
Kenaf	770,000	Stalk
Linseed	Plentiful	Fruit
Pineapple	Plentiful	Foliage/leaf
Caroa	–	Foliage/leaf
Nettles	Plentiful	Stalk
Oil palm fruit	Plentiful	Fruit
Palm rah	Plentiful	Stalk
Ramie	100	Stalk
Roselle	250	Stalk
Rice husk	Plentiful	Fruit/grain
Rice straw	Plentiful	Stalk
Sisal	380,000	Stalk
Sun hemp	70,000	Stalk
Wheat straw	Plentiful	Stem
Wood	1,750,000	Stem
Sugarcane bagasse	75,000	Stem
Cantala	–	Leaf
China jute	–	Stem

compared with many other organic fibers, such as petroleum palm frond, coir, and banana-based fibers. The heavy percentage of cellulose found PALF helps to carry out the bigger weight of the berry or pineapple leaf.

Considered altogether, PALF is very similar to most natural fibers in terms of chemical composition. However, PALF has superior physical strength compared to jute and is therefore good for constructing yarn [20, 21]. The cellulosic molecular cells of PALF form a 3D framework parallel to the crystalline area of the fiber. In addition to the crystalline area there is also an amorphous area. Pineapple leaf fiber is an essential organic fiber with substantial toughness and rigidity, including torsional and flexural rigidity—comparable to jute fibers. Due to these extraordinary qualities,

Table 2 Percentage chemical composition of pineapple fiber [2]

C	O	N	Ca	P	Fe	K	Mg	Cu	O/C ratio
73.13	24.17	2.70	0.00	–	–	–	–	0.00	0.33%
–	–	6.4–10	2.5–10	0.1–0.18	0.06–0.11	2.89	0.33	0.002–0.02	–

Table 3 Chemical composition of PALF [4]

Fiber	Density (kg/m^3)	Elongation (%)	Tensile strength (MPa)	Moisture absorption (%)	Young's modulus (GPa)
Cotton	1500–1600	3.0–10.0	287–597	8–25	5.5–12.6
Jute	1300–1460	1.5–1.8	393–800	12	30-Oct
Flax	1400–1500	1.2–3.2	345–1500	7	27.6–80
Hemp	1480	1.6	550–900	8	70
Ramie	1500	2.0–3.8	220–938	17-Dec	44–128
Sisal	1330–1500	2.0–14	400–700	11	9.0–38
Coir	1200	15.0–30.0	175–220	10	4.0–6.0
Softwood kraft	1500	–	1000	<0.01	40
E-glass	2500	2.5–3.0	2000–3500	0	70
S-glass	2500	2.8	4520	0	86
Aramide (normal)	1400	3.3–3.7	3000–3150	<0.1	63.0–67.0
Carbon (standard)	1400	1.4–1.8	4000	<0.1	230.0–400.0

industries are able to utilize PALF as an excellent option for reinforcing composite matrixes [22] (Table 3).

2.1 *Anatomy of Pineapple Fiber*

Pineapple fruit is very famous and considered glamorous due to its self-sterile nature and its self-fertilization power of germination. Fruit is rich in ascorbic acid and carotene, has lower acidity, and is highly translucent due to their clonal selection. Pineapple fiber is fully enriched with herbicide (bialaphos) and Polyphenol oxidases (PPO) gene expression which can regulate the 1-aminocyclopropane-1-carboxylate (ACC) synthase gene inside the human body with the help of ACACS2 enzymes. Pineapple fiber is a nematode-type root-knot material containing significant pathogen-killing capability and antioxidants. Pineapple fibers are a typical example of a monocot root from a family of plants with strong root forces (Table 4).

Pineapple fiber originates from a herbaceous plant that is approximately 1–2 m tall and 25–50 cm long, 2–5 cm wide at the base, and 5–8 cm wide at the top—containing nodes and internodes. Fully grown pineapple fibers can number 68–82 in each plant, being dispersed over a dense compact rosette. The spine is in the middle of the pineapple fruit and the upper and lower surfaces of the leaves are covered with hairs up to two thirds the circumference of the stem. The crown, suckers, and slips are used for reproduction (Fig. 2) [11].

Table 4 List of nutrients and minerals present in pineapple fibers and the syndromes and outcomes resulting from their deficiency [23, 24]

Nutrient	Amount (mg)
Manganese	2.56
Vitamin C	23.87
Vitamin B1(thiamine)	0.14
Copper	0.17
Dietary fibre	1860.0
Vitamin B6 (pyridoxine)	0.13
Calcium	6.2–37.2
Nitrogen	38.0–98.0
Phosphorous	6.6–11.9
Iron	0.27–1.05
Ascorbic acid	27.0–165.2
Carotene	0.003–0.055
Thiamine	0.048–0.138
Riboflavin	0.011–0.04
Niacin	0.013–0.267

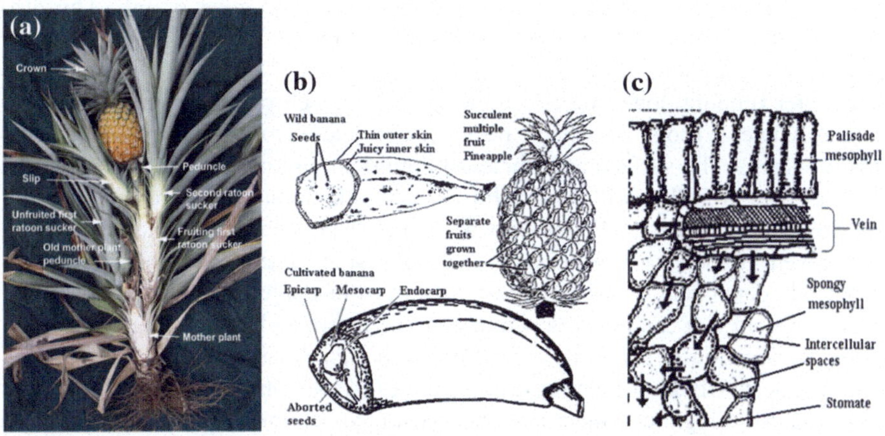

Fig. 2 **a** Morphology of a pineapple plant. **b** Anatomical morphology of a pineapple fruit. **c** Anatomical morphology of a pineapple fiber [24]

Pineapple plants have the following important parts:

- Slips—the leafy branches below the pineapple fruit on the peduncle, producing the basal eye of the fruit.
- Collars—structures produced within 14–16 months after planting.
- Side shoots or suckers/stem shoots—parts generated above the ground after 18–20 months of planting.

- Crown—the short stem and leaves that bud from the apex of the fruit after 24 months of planting, typically producing the pineapple fruit and leaf [24, 25].

Pineapple fibers have very high proportions of metal ions and nutrients, such as phosphorous (P), potassium (P), calcium (Ca), copper (Cu), sulfur (S), manganese (Mn), magnesium (Mg), chlorine (Cl), molybdenum (Mo), zinc (Zn), boron (B), and iron (Fe). Controlling both temperature stress and water stress are crucial to producing a thick cutinized epidermis, a multicellular hypodermis, and a small portion of stomatal pores. Half of the pineapple leaf contains water held in storage parenchymatic cells. This type of special anatomical structure of pineapple fiber ensures minimal evapotranspiration.

2.2 Root System

Pineapple roots have special characteristics in terms of the soil and their auxiliary (aerial) root system. Bromeliads have weakly developed, delicate root systems. The main system identified for pineapple is generally shallow rooting that does not regenerate if damaged by disease or pests. Using urea can give a polished look to the pina leaves and also can act as a pesticide repelling rodents and mites or ants. This also helps to reduce copper deficiency in the plant's nutrition. The auxiliary roots, along with the adrenal gland of the leaves, play a significant role in absorbing nutrients from water running down the leaves. Roots that branch off the primary roots are known as lateral roots. Auxiliary, primary, and lateral roots have root hairs that greatly increase the surface area increasing nutrient and moisture uptake. Pineapple tree roots are filled with hard white tissue with stomata that are multicellular and mushroom-shaped, and with scaly trichomes (leaf hairs). These have evolved to help prevent of the loss of water from the root system. The creation of root hairs is completely dependent on air supply to the roots. Pineapple trees have a root development system that can adapt depending on the native soil and underground water supply at a location. Crowns create far more roots than slides or suckers. Root growth commences shortly after planting and continues to blossoming. Root growth slows but proceeds throughout winter, associated with temperature and soil moisture. The longest principal roots disperse laterally from the bottom of the plant not far beneath the soil surface. Roots can penetrate to a depth of 1.5 m, but most root improvement is inside the bed where the soil is not compacted. Smooth Cayenne plants may create a total of approximately 70 m of primary roots within 12 months. Each plant may have approximately 450 chief roots (Fig. 3).

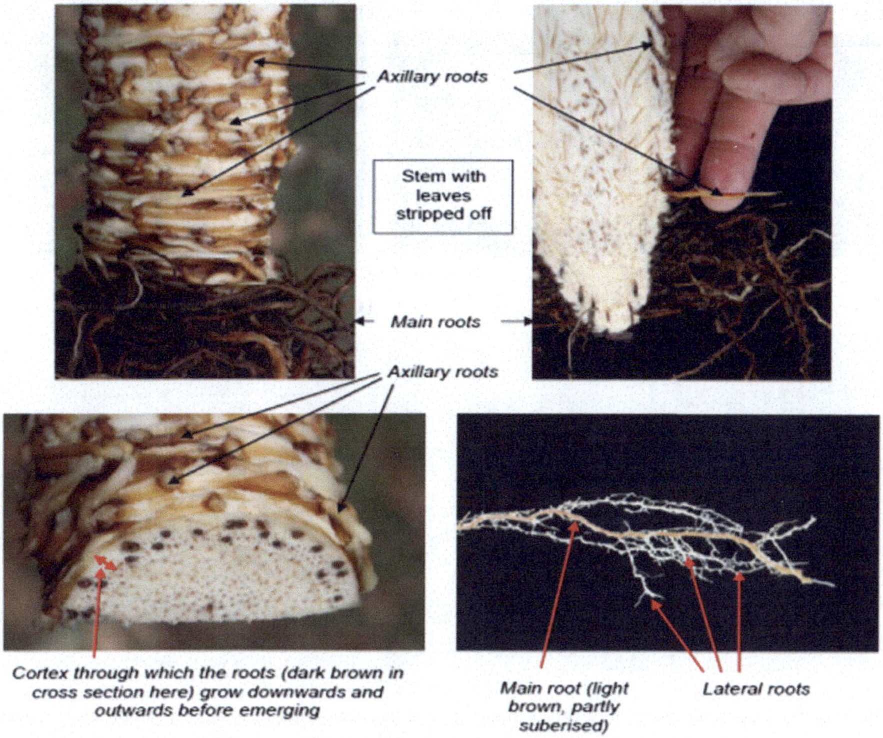

Fig. 3 The root system of pineapple plants [24]

2.3 *Stem (Butt) Anatomy*

The stem is the source of stored water and approximately 11% of a plant's starch—generally used as an excess form of energy during photosynthesis during the daytime. Fruiting first, ratoon suckers are where this extra food and energy source are reserved. This sucker part is only visible after stripping off all leaf follicles. One important feature of pineapple leaves is that they vary in sizes and diameter:

- Broad leaves are associated with large stem diameters and large fruit diameters.
- Narrow leaves are associated with small stem diameters and small fruit diameters.
- Fruit with a larger diameter are heavier and produce higher yields (Fig. 4).

2.4 *Pineapple Leaf Anatomy*

The pineapple is in Bromeliaceae family, which is comprised of 50 genera and 2500 known species of plant from Central and South America. Ananassoides (a really

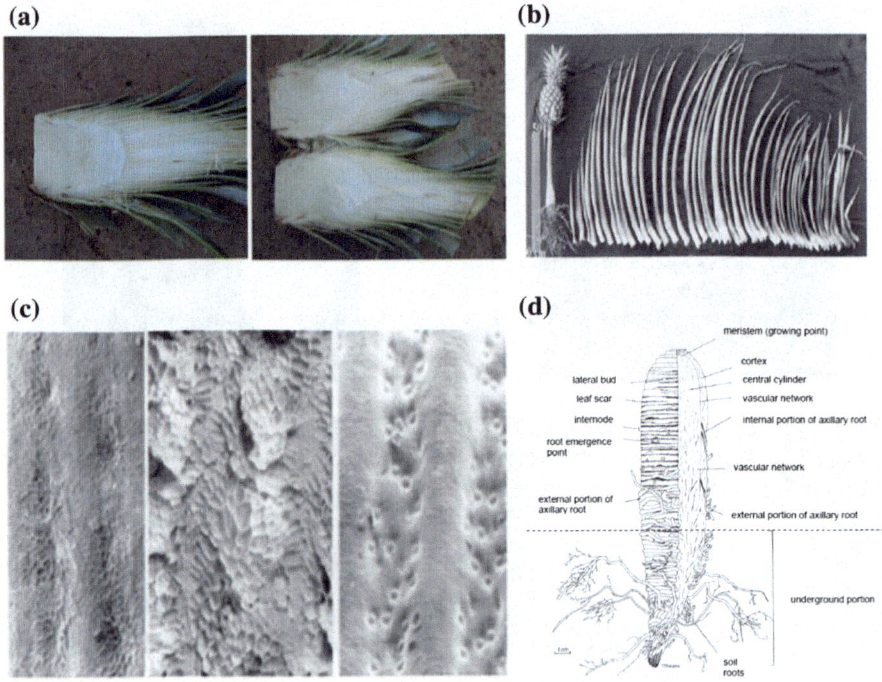

Fig. 4 **a** The flowering portion of pineapple plants; **b** the root stomata system; **c** scanning electron microscope photographs of a root stomata system; **d** anatomical tissue view of a root system [24, 26]

small, seeded berry with fruit leaves) is regarded as the pineapple's wild ancestor—growing best in a warm to hot climate (16–33 °C) with rainfall of about 50 mm/month. It cannot withstand frosts that can cause bloating to its delicate root system and special xylem and phloem vascular cambium. Pineapple is a tropical plant having the following botanical and physiological adaptations:

- Cannot tolerate frost.
- At high temperatures (in excess of 40 °C) bloating damage to fruit and plants can be acute.
- It has a delicate root system which needs well-drained conditions.
- It has a lot of special characteristics that let it survive and flourish under low-rainfall conditions.
- Pineapple leaves can absorb sunlight at maximum level during the daytime and can also hold on to moisture with maximum efficiency. This ability has evolved over time. The leaf can as a reservoir for nutrient water and can assist in absorbing nutrients via auxiliary roots at the base of the leaf.
- Inside pineapple leaves water tissue can comprise around half the foliage depth and can be used during periods of low rainfall to help maintain growth.

- A specialized metabolic system for capturing carbon dioxide at night, to be used throughout the day, significantly reduces water loss.
- Its adaptation to arid conditions originates not just from growing in a humid climate but also from its own epiphytic ancestry (epiphytes develop other plants) [24, 26].

2.5 Leaf Shape and Arrangement

Pineapple has strong, rough, horizontal leaves that are tapered from base to tip. They are spirally arranged for maximum sunlight absorption. Leaves are arranged in a Fibonacci series around the stem of the fruit, thus ensuring maximum exposure of leaves under sunlight. Pineapple leaves spirally overlap to produce a final and matured pineapple leaf of approximately 2.2 m^2. Plant weights are often around 3.6 kg.

2.6 Water Storage Tissue

The cross section of a leaf has an almost 4-mm-thick water storage area ('smooth cayenne') and if the health of the plant is good with adequate external moisture then this stored water reservoir can act as a catalyst to plant growth, fruit nutrients, and even plant development. During extended dry periods, this juicy portion decreases to leave an almost dry or empty water space—a sign of drought (Fig. 5).

Fig. 5 Leaf water reservoir of a pineapple plant [24]

2.7 CAM Photosynthesis

Pineapple fiber has a special photosynthesis pathway know as CAM (Crassulacean Acid Metabolism) which is specially designed by nature to preserve moisture and absorb maximum carbon dioxide through open stomata during photosynthesis under sunlight. Stomata are also designed in a special manner, filled with abundant malic acid, which allows them to be open during cooler and humid periods in the night. In the day, the secretion of excessive malic acid helps to start the plant photosynthesizing (Table 5).

3 The Pineapple Fiber Extraction Process

Pineapple organic fiber shows outstanding manual strength but because of a lack of understanding its potential has not been fully utilized until now. It can be used for numerous purposes, such as for producing man-made fibers, as a good absorber, and as an insulator. The PALF extraction technique has different stages:

1. *Scraping method of extraction*. The scraping of pineapple leaf fiber is done with the help of a scraping machine. The device is made of three feed rollers, a leaf-scraping roller, and a serrated roller. The feed rollers are operated to feed leaves into the machine. Pine leaves are added at the next roller—known as the scratching roller. The top level of the leaf is subsequently scratched, thereby eliminating the waxy cuticle. Next, leaves arrive at the enclosed mower blade that forms the serrated roller. This crushes leaves and also tends to break leaf surfaces permitting the entry of retting microbes [22, 27].
2. *Retting of pineapple leaves*. For the retting procedure, little bunches of scratched pineapple foliage are submerged into a water container of liquor with a 1:20 ratio and a urea content of 0.5% (diammonium phosphate (DAP) is used for quick retting results). Use of warm water is really helpful to ensure that fibers are loosened allowing them to acquire several chemical-based constituents, such as pentosans, lignin, wax and fats, nitrogenous matter, ash content, and pectin. Subsequently, fibers are segregated mechanically through cleaning with pond water. The extracted fibers are dehydrated in a hanging site by air and ball mill. A disc mill can also be utilized to acquire PALF from cut new pineapple leaves [28]. These techniques are not only simple but additionally give greater fiber yields plus small-sized fibers compared with traditional methods. Of all the physical grinding approaches studied identified that damp heel milling is significantly slower but offers PALF with a better variety of elementary fibers [29].

Table 5 Elements found in pineapple and the consequences of deficiency [23, 24]

Essential element	Level required	Effect of element deficiency
N	High requirement Optimal soil level is 120 parts per million (ppm) Deficiency occurs at 50 ppm or below	Reduced leaf size, leaf number, and crown mass—crowns sometimes absent Leaves turn from green to greenish yellow and yellow—fruit quality reduced
K	High requirement Optimum soil level is 150 ppm Deficiency occurs at 60 ppm or below	Reduces plant growth and fruit mass Affects slip production Fruits with reduced fruit acidity and aroma and are prone to sunburn • Short leaves, leaf tip death and necrosis
P	Low requirement Optimal soil level is 20 ppm Deficiency occurs at 5 ppm or below	All plant parts depressed in growth Erect, long narrow leaves; older leaf tips show die back and chlorosis
Ca	High requirement Optimal soil level is 100 ppm Deficiency occurs at 25 ppm or below	Abnormal leaves (short) with grey–green coloration—severe deficiency affects growth Fruit aroma and translucence reduced Death of growth tips
Mg	Low requirement Optimal soil level is 50 ppm Deficiency occurs at 10 ppm or below	Reduces chlorophyll concentration, photosynthesis, and growth Leaves turn bright yellow Short stems Weak root system -Fruits low in acid and sugar content and lack aroma
S	Low requirement Rare deficiency in pineapple plants	Leaf yellowing and narrowing—plants stunted Fruits reduced in size
Fe	Low requirement Optimal soil level is 27–78 ppm Deficiency occurs at 3 ppm or below	Interveinal chlorosis, yellowing and mottling of leaves Small, hard, and red-colored fruit Crown light creamy or white color
Zn	Low requirement Optimal soil level of 4 ppm Deficiency occurs at 3 ppm or below	Centre cluster of leaves curved in young plants—yellow–brown blister-like spots on leaves of old plants

4 Mechanical and Physical Properties

Reinforced all-natural fiber composites constitute an enormous share of biocomposites and scientific components. Pineapple leaf fibers have been shown to be an excellent substitute for artificial fibers due to their renewable nature and economics.

Table 6 A comparison of the mechanical properties of PALF and glass fiber [23, 31–34]

Property	Natural fiber	Glass fiber
Density	Low	Double
Cost	Low	High
Renewability	Yes	No
Recyclability	Yes	No
Energy consumption	Low	High
Distribution	Wide	Wide
Disposal	Biodegradable	Nonbiodegradable

They support the improvement of the mechanical and physical power of polymer matrices, without needing any extra processing. The advantage of PALF's physical qualities is connected to its high content of alpha-cellulose material as well as its minimal microfibrillar structure. The physicomechanical qualities of any organic fibers rely on fiber matrix adhesion, the volume portion of fiber, and its orientation, aspect ratio, and load transfer effectiveness at an interface as well as its other remarkable characteristics that may be employed for reinforcing composite matrix materials [17].

The outcomes of primarily PALF-based polymer composites show high-quality strength and stiffness compared with other primarily cellulose-based composite materials. Abnormal qualities of PALF are noticed when damp—with bundles exhibiting 50% reduced mechanical modulus power. However, yarn has shown as much as a 13% increment in tensile strength. Table 3 shows the mechanical and physical strength of PALF. The PALF exhibits a modulus selection from 34.5 to 82.51 GN/m^2, with tensile strength ranging from 413 to 1627 MN/m^2, and elongation at breaking ranging from 0.8 to 1.6%. PALF can be used as an abrasive due to its high mechanical strength [16]. Past studies posit a lot of diverse behavior and properties in terms of the morphology of area and tensile behavior, alongside dielectric property. Pineapple fiber exhibits the excellent flexible properties of cellulose type I structures in contrast to many other organic fibers PALF—providing it with a high level of strength. Top anisotropy is shown by its electric properties (Tables 6 and 7) [30].

4.1 FTIR Spectra

Fourier-transform infrared spectroscopy (FTIR) is utilized to observe purposeful groups in organic fibers, such as hydroxyl, carbonyl, vinyl, and ketone classes. It helps to determine the changes in the chemical combination of healthy fibers prior and subsequent to chemical treatments. Table 8 shows the common FTIR spectra of different untreated organic fibers such as hemp, sisal, jute, kapok, kenaf, and oil palm fiber along with PALF. The distinctive O–H class is typical in the intensity range 3338–3450 cm^{-1}. Other peaks identify C–H stretching and C–O stretching at 2924.2 cm^{-1}

Table 7 Physical and mechanical strength of PALF variants [6, 7, 35]

Density (kg/m^3)	Tensile strength (MPa)	Young's modulus (GPa)	Specific strength ($GPa/g/cm^3$)	Specific modulus ($GPa/g/cm^3$)	Elongation at break (%)	Dia. (m)	Microfibril Angle (°)
1520	413–1627	34.5–82.51	0.3–1.1	22.7–54.3	1.6–3	20–80	–
1526	170	62.10	1.1	40.70	3	–	–
1440	413–1627	34.5–82.51	–	–	1.6	–	–
1526	413	62.10	–	–	1.6	50	–
1070	126.60	4.405	–	–	2.2	–	–
	413–1627	34.5–82.5	–	–	1.6	20–80	14
1526	170	6.260	–	–	3	–	–
1520	170	6.21	–	–	3	–	–
1070	126.60	4.405	–	–	2.2	–	–
1526	413	6.5	–	–	1.6	30–60	–
1526	170	62.10	–	–	3	–	–
1440	413–1627	34.5–82.51	–	–	–	20–80	8–14
1440	413–1627	34.5–82.51	–	–	1.6	20–80	–
1440	170	6.26	–	–	1.6	5–30	12
1526	413	4.2	–	–	3.0–4.0	50	14
1440	–	–	–	–	–	1.56–4.5	8–15
–	293.08	18.934	–	–	1.41	150–300	–

Table 8 Infrared transmittance peaks (cm^{-1}) of untreated natural fibers [23, 33, 37]

Bond/stretching	PALF (cm^{-1})	Hemp (cm^{-1})	Sisal (cm^{-1})	Jute (cm^{-1})	Kapok (cm^{-1})	Kenaf (cm^{-1})	Oil palm fiber (cm^{-1})
– OH	3349.9	3448	3447.2	3447.9	3419.7	3338	3450
C–H	2903.8	2920.5	2924.2	2918.8	2918.1	2899	2850
C=O	1737.4	–	1736.5	1737.2	1741.1	1736	1735
C=C	1608.3	1654	1653.9	1653.8	1596.1	–	1606
C–H	1374.2	1384.1	1384.1	1384.1	1383.6	–	–
C–H	–	–	1259.9	1255.6	1245.5	–	–

and 1741.1 cm^{-1} respectively [17] in the case of untreated fibers. According to most recent literature for kenaf, the wide peak in the spectra at 3338 cm^{-1} is due to the O–H frequency, while the peaks at 2899 cm^{-1} mostly occur from C–H stretching. Sreekala [30] observed that untreated oil palm fiber demonstrates peaks corresponding to C–O stretching at 770 cm^{-1} as well as extending C–H at 2850 cm^{-1}; petroleum palm fiber shows yet another at 3450 cm^{-1} because of O–H stretching (Table 8) [36].

The FTIR spectra of hemicellulose and cellulose samples, totally free after the extraction of PALF, are provided in Fig. 2. The wavelength peak found in FTIR experiment at 3343 cm^{-1} signifies O–H groups providing of cellulose test and hydroxyl stretching frequency shown at 3296 cm^{-1} and 3327 cm^{-1} respectively for free-extractive samples and holocellulose. In the case of free extractive and holocellulose, the peak frequency at 1728 cm^{-1}, along with that at 1733 cm^{-1}, corresponds to carbonyl peak wavelengths. The clear band found at 1733 cm^{-1} is due to the absorption of the carbonyl stretching of ester as well as carboxyl groups and abundant pineapple leaf hemicelluloses [37, 38].

5 Applications of PALF

Pineapple leaf fiber has been, and is currently being, used to generate yarns for textile materials. The diversified uses of PALF include textiles, mats, cabinets, automobiles, baggage, sports items, etc. Surface-modified PALF can be used for making machinery components, such as conveyor belt cords, belt cables, transmission cloths, airbag tying cords, and several specialist types of clothing. Its is great for carpet manufacturing due to its easy chemical processing and behavior during the dyeing process. The utilization of PALF may be viewed as fairly interesting in the newspaper production business in Malaysia. In addition, it could be ideal for some other uses, such as beauty products, medication, and biopolymer coverings for chemical substances [11]. Pineapple leaf fiber is one of the organic fibers with high cellulosic content, at about 80%. It represents a great option to all the fiber scientists due to its lower and skin-friendly density that has a good tensile modulus strength compared with many other organic fibers. These attributes are ideal for new applications and creations as well as for use in furniture production, automotive component production, and construction materials. Currently, PALF has only been analyzed when reinforced with polypropylene and unsaturated polyester. Therefore, it is necessary to study its behavior with various resins too, in relation to fabricated biocomposites as well as hybrid composites. Pineapple leaf fiber is commonly recognized in the textile sectors and further analysis will improve its range of applications in some other exciting fields [25].

Thermoplastic matrix substances are the most crucial components of a composite. Polypropylene is an amorphous thermoplastic polymer and it is popular as an engineering thermoplastic due to its transparency, dimensional stability, and flame resistance. Such materials have excessive heat distortion temperatures and substantial impact strengths. Polypropylene is additionally extremely ideal for reinforcing, filling, and blending. Polypropylene with fibrous, all-natural polymers, originating from biomass, is one of the most promising routes to producing natural synthetic polymer composites [31].

Nevertheless, the main limitation of healthy fiber–polymer hybrid combinations is the incompatibility between the hydrophobic thermoplastic matrices and hydrophilic organic fibers. Thus, it is extremely crucial to alter the purely natural fiber surface area to enhance the fiber–matrix interaction. This can be done by chemical treatments or graft copolymerization, etc. The utilization of coupling elements on natural fibers has been investigated by a selection of researchers with a view to enhancing the

Table 9 Pineapple leaf fiber characterization in thermosets and thermoplastics [31, 32, 35]

Natural fiber	Resin
PALF	Vinyl ester
PALF	Epoxy
PALF	Polycarbonate
PALF	Polypropylene
PALF	Polyester
PALF	Low-density polyethylene
PALF	Polyethylene
PALF + banana	Epoxy
PALF + disposable Chopstick hybrid fiber	Polylactic acid and polybutylene succinate
PALF + kenaf	High-density polyethylene
PALF + glass fiber	Polyester

fiber–matrix interaction. Nevertheless, mercerization of organic fibers using alkali is currently regarded as the preferred method to enhance fiber–matrix interactions by decreasing the hydrophilicity of the organic fiber. Today, sodium hydroxide is utilized for the surface area adjustment of PALF while making natural fiber–based hybrid composites.

To date, PALF has been investigated by a selection of researchers where composites reinforced with fibers have received several alkali therapies with the purpose of enhancing the physical qualities of composites. The target of such analysis is the manufacture of PALF–polypropylene composites with better physical strengths. The consequences of fiber loading on the physical qualities of PALF–polypropylene composites are being examined. Finally, the consequences of this focus on alkali remedies for the physical qualities of composites are also being investigated (Table 9) [32, 34].

6 Conclusions

Pineapple fiber offers a good choice for making various reinforcement composites or even home-based decorative items. Mechanical strength–based statistical analysis of PALFs have discovered an inverse correlation between tensile strength and fiber diameter. This correlation suggests that a feasible hyperbolic mathematical equation might predict PALF anatomy, based on the correlation between pineapple fiber diameter and mechanical stiffness. Scanning electron microscopy observations provide sufficient evidence that a fuller PALF with increased microfibrils, along with porosity and fewer defects compared with a finer ones, encounter ruptures under conditions of comparatively reduced stress. Statistically, the bigger division of microfibrils physical a resistance of the fuller fiber additionally offers a greater likelihood that a weaker fibril to rupture shortly compared to the fewer fibrils of the finer fiber. The diverse uses of PALF include textiles, mats, cabinets, automobiles, baggage, sports items, etc. Surface-modified PALF can be used for making machinery components, such as conveyor belt cords, belt cables, transmission cloths, airbag tying cords, and several specialist types of clothing.

References

1. Allen RG et al (1998) Guidelines for computing crop water requirements-FAO Irrigation and drainage paper 56. FAO-Food and Agriculture Organisation of the United Nations, Rome (http://www.fao.org/docrep); ARPAV (2000) La caratterizzazione climatica della Regione Veneto, Quaderni per. Geophysics 156:178
2. Agarwal BD, Broutman LJ, Chandrashekhara K (2017) Analysis and performance of fiber composites. Wiley
3. Ketnawa S, Rawdkuen S, Chaiwut P (2010) Two phase partitioning and collagen hydrolysis of bromelain from pineapple peel Nang Lae cultivar. Biochem Eng J 52(2–3):205–211
4. Bernfeld P (1951) Enzymes of starch degradation and synthesis. In: Advances in enzymology and related areas of molecular biology, vol 12. Wiley Online Library, pp 379–428
5. Uhlig H (1998) Industrial enzymes and their applications. Wiley
6. Heinicke RM, Gortner WA (1957) Stem bromelain—a new protease preparation from pineapple plants. Econ Bot 11(3):225–234
7. Cathcart DJ (1995) The importance of maintaining bromeliad imports. In: Florida entomologist. JSTOR, pp 16–21
8. Pandit P et al (2018) Applications of textile materials using emerging sources and technology: a new perspective. Green Sustain Adv Mater Appl 2:49–83
9. Bengtsson M, Gatenholm P, Oksman K (2005) The effect of crosslinking on the properties of polyethylene/wood flour composites. Compos Sci Technol 65(10):1468–1479
10. Cherian BM et al (2011) Cellulose nanocomposites with nanofibres isolated from pineapple leaf fibers for medical applications. Carbohydr Polym 86(4):1790–1798
11. Py C, Lacoeuilhe JJ, Teisson C (1987) The pineapple. Cultivation and uses. G.-P. Maisonneuve et Larose
12. Elango G, Govindasamy R (2018) Analysis and utilization of temple waste flowers in Coimbatore District. Environ Sci Pollut Res
13. Ghosh SK et al (1982) Processing of pineapple leaf fibre (PALF) in cotton machinery. Text Trends 22(10):1–5
14. Omojasola PF, Jilani OP, Ibiyemi SA (2008) Cellulase production by some fungi cultured on pineapple waste. Nat Sci 6(2):64–79
15. Imandi SB et al (2008) Application of statistical experimental designs for the optimization of medium constituents for the production of citric acid from pineapple waste. Bioresour Technol 99(10):4445–4450
16. Kumar D et al (2003) Utilisation of fruits waste for citric acid production by solid state fermentation. Process Biochem 38(12):1725–1729
17. Nadirah WOW et al (2012) Cell wall morphology, chemical and thermal analysis of cultivated pineapple leaf fibres for industrial applications. J Polym Environ 20(2):404–411
18. Dey SK, Bhattacharyya GK, Bhattacharyya SK (2005) Magic yarns from ramie and pineapple—a new dimension in 21st century. In: Proceedings of the 20th Indian engineering congress, p 69
19. Jadhav AC et al (2019) Production of green composites from various sustainable raw materials. In: Green composites. Springer, Berlin, pp 1–24
20. Das S et al (2018) Hybrid bast fibre strengthened thermoset composites. In: Thermoset composites: preparation, properties and applications, vol 38. Materials Research Forum LLC, p 112

21. Satyanarayana KG, Guimarães JL, Wypych F (2007) Studies on lignocellulosic fibers of Brazil. Part I: Source, production, morphology, properties and applications. Compos Part A Appl Sci Manuf 38(7):1694–1709
22. Kannojiya R et al (2013) Extraction of pineapple fibres for making commercial products. J Environ Res Dev G. SEED (Glob Earth Soc Environ Energy Dev) 7(4):1385
23. Arib RMN et al (2006) Mechanical properties of pineapple leaf fibre reinforced polypropylene composites. Mater Des 27(5):391–396
24. Johnson MO (1935) The Pineapple. In: The pineapple. Paradise of the Pacific Press, Honolulu
25. Morton JF, Dowling CF (1987) Fruits of warm climates. JF Morton Miami, FL
26. Hossain MF (2016) World pineapple production: an overview. Afr J Food Agric Nutr Dev. Afr Sch Sci Commun Trust (ASSCAT) 16(4):11443–11456
27. Banik S et al (2003) Ribbon retting of jute—a prospective and eco-friendly method for improvement of fibre quality. Ind Crops Prod 17(3):183–190
28. Kengkhetkit N, Amornsakchai T (2012) Utilisation of pineapple leaf waste for plastic reinforcement: 1. A novel extraction method for short pineapple leaf fiber. Ind Crops Prod 40:55–61
29. Sreekala MS, Thomas S (2003) Effect of fibre surface modification on water-sorption characteristics of oil palm fibres. Compos Sci Technol 63(6):861–869
30. Uma Devi L et al (2004) Ageing studies of pineapple leaf fiber–reinforced polyester composites. J Appl Polym Sci 94(2):503–510
31. Joffe R, Andersons J, Wallström L (2003) Strength and adhesion characteristics of elementary flax fibres with different surface treatments. Compos Part A Appl Sci Manuf 34(7):603–612
32. John MJ, Anandjiwala RD (2008) Recent developments in chemical modification and characterization of natural fiber-reinforced composites. Polym Compos 29(2):187–207
33. Mangal R et al (2003) Thermal properties of pineapple leaf fiber reinforced composites. Mater Sci Eng A 339(1–2):281–285
34. Mohamed AR, Sapuan SM, Khalina A (2010) Selected properties of hand-laid and compression molded vinyl ester and pineapple leaf fiber (PALF)-reinforced vinyl ester composites. Int J Mech Mater Eng 5(1):68–73
35. Van de Weyenberg I et al (2006) Improving the properties of UD flax fibre reinforced composites by applying an alkaline fibre treatment. Compos Part A Appl Sci Manuf 37(9):1368–1376
36. Jones FR (1994) Moisture absorption-anomalous effects. In: Handbook of polymer fibre composites. Longman Scientific and Technical Group, Ltd. UK, pp 371–375
37. Lopattananon N, Payae Y, Seadan M (2008) 'Influence of fiber modification on interfacial adhesion and mechanical properties of pineapple leaf fiber-epoxy composites. J Appl Polym Sci 110(1):433–443
38. Pandit P, Nadathu GT (2018) Characterization of green and sustainable advanced materials. Green Sustain Adv Mater Process Charact 1:35–66

Effect of Extraction on the Mechanical, Physical and Biological Properties of Pineapple Leaf Fibres

A. Rafiqah, K. Abdan, M. Nasir and M. Asim

Abstract Natural fibres have been acknowledged as potential material in many countries and widely used in vast application due to its specific properties and positive environmental impact. Selection of natural fibres for research or applications is categorized as per availability in particular region. Pineapple leaf fibres (PALF) are well-known fibre in South-East Asia. Pineapple leaf contains only 2.5–3.5% fibre, covered by a hydrophobic waxy layer. Suitable extraction method is the main challenged to obtain good quality PALF for future applications. The methods for PALF extraction were classified into three main categories, manual, mechanical and retting method. Physical and mechanical properties of PALF may differ from the other PALF due to the different extraction method. Extraction of thousands of tons of PALF can be done only after harvesting the fruit. The extraction method was chosen based on different criteria that involve cost of manufacturing, PALF stiffness, physical appearances and time consumption. This topic aims to indicate different extraction methods to obtain PALF and discussed on its physical and mechanical properties.

Keywords Pineapple leaf · Pineapple leaf fibres · Extraction method · Retting · Mechanical and physical properties

1 Introduction

The environmental threat and future sustainability have forced to use natural redeemable materials [5]. Natural resources are suitable substitute for synthetic materials. Natural fibres-based composites are known for specific characteristics such as better flammability, renewability, cost effectiveness, lightweight, high specific strength and modulus, possess tool wearing rates, thermal insulation properties,

A. Rafiqah · K. Abdan · M. Asim (✉)
Institute of Tropical Forestry and Forest Products (INTROP), Universiti Putra Malaysia, 43400 Serdang, Selangor, Malaysia
e-mail: khanfatehvi@gmail.com

M. Nasir
Forest Products Utilization College of Forestry, Banda University of Agriculture and Technology (BUAT), Banda, U.P. 210001, India

M. Jawaid et al. (eds.), *Pineapple Leaf Fibers*, Green Energy and Technology,
https://doi.org/10.1007/978-981-15-1416-6_3

acoustic properties, sufficient energy requirements and are safer towards health [6]. These natural fibres are natural resources, and it imparts in biodegradable qualities to polymer composites which provide advantages over synthetic fibre-based composites especially from the environmental sustainability point of view [7]. Some manufacturing industries are selecting natural fibres as filler or reinforcement materials in composites such as Hemp, jute, sisal, pineapple leaf fibre (PALF), kenaf and ramie [4].

Recently, industry has been widely used fibre-reinforced materials in structural plastics for various applications [12]. It is remarkable how natural fibres reinforcements have gained renewed interest, especially as a glass fibre substitute in automotive industries [30]. Synthetic fibre has been replaced with natural fibre for high-performance materials such as in automotive and aerospace industry. In Germany, automotive industry aims to produce biodegradable and recyclable components which give opportunity in the use of natural fibre [13].

Since 2000, natural fibres have been used in automotive industry for European cars. From 2000 to 2005, the demand of natural fibre for automotive industries increased from 10,000 to 19,000 ton. In automotive, they use natural fibre composite to produce lightweight parts such as door panels and noise absorber [15]. Meanwhile, in US, natural fibre was applied variously in decking application for building [28]. Natural fibre got attention due to its great properties such as renewable character, lightweight, high specific strength and modulus, availability in a variety of forms, reactive surface and the possibility to generate energy, [1]. It also has good thermal and acoustic insulating properties [9]. Natural fibre also acts as filler or reinforcement in plastic to reduce cost, increase productivity and improve mechanical properties of product. The major drawback of natural fibre is hydrophilic in nature which easily absorbs moisture. Natural fibres also had limited maximum processing temperature which limits for high temperature processing. Besides that, its quality varies depending on several factors such as weather. However, chemicals treatments of natural fibres were able to modify the fibres surface properties to improve adhesion between fibre and matrix [9, 10, 21]. Generally, higher mechanical strength of natural fibres had achieved due to higher cellulose content and cellulose microfibrils aligned more in the fibre direction. The components of the fibre cell wall normally act as main binder [33]. Other than that, chemical composition and structure, which relate to fibre type as well as growing conditions, harvesting time, extraction method, treatment and storage procedures also influence the fibre properties. It is observed that strength of natural fibre has been reduced by 15% over 5 days after optimum harvest time. However, manually extracted fibres have been found to have strength 20% higher than those extracted mechanically [29].

PALF is also known as one of the highest textile grade commercial fibres. It is generally extracted by two ways: by manual extraction and mechanical extraction through decorticator machine. The process of extracting long vegetable fibres is of great importance since the quality as well as the quantity of extracted fibres is strongly influenced by the method of extraction employed. The manual extraction process of pineapple leaf fibre is time consuming and labourious. On the other hand, the automatic decorticator machine is costly, and it needs skilled operator [2, 19, 20].

The aim of this chapter is to review the extraction method, process of PALF and also the effect of extraction process on physical and mechanical properties of PALF.

2 Pineapple Leaf Fibre

Among several natural fibres, pineapple leaf fibre can be used as a viable and abundant substitute for the expensive synthetic fibres. Pineapple leaf fibre (PALF) is one of the natural fibres that exhibits high specific strength and stiffness due to high cellulose content (70–80%) and relatively low microfibrillar angle which has been used widely in composite industry [22]. Meanwhile, PALF tensile modulus and tensile strength are also as high as glass fibre [20]. High mechanical strength of PALF is due to its hydrophilic nature and high cellulose content [3]. From other researcher study, found out the chemical composition of PALF constitutes holocellulose (70–82%), lignin (5–12%) and ash (1.1%) [3, 29]. Pineapple leaf fibres are silky and fine which has fibre content approximately 2.5–3.5% of total leaf biomass. Thus, the PALF extraction is not economically viable [8]. The size of PALF leaf usually 1–1.60 m long, and its leaves can produce silky fibre which yields approximately 2.5–3.5% of the weight of fresh leaves [11]. Thailand is currently the world's leading pineapple producer with about 240,000 acres of cultivation area. After harvesting, bulky amount of pineapple leaf waste remains causing many problems for farmers to deal with. Although methods for fibre extraction are known for a long time and have been using to the present day, the major problem with PALF is low fibre content. Fibre extraction processes (scrapping, retting or decorticating) also need many labour or problematic to the environment. In addition, these methods are only suitable for extracting long fibre of a limited volume for niche textile applications but not suitable for mass produced plastic industry [20]. The aim of this topic is to highlight the method of extraction PALF and the effect on mechanical and physical properties of different extraction method.

3 Extraction Method of Pineapple Leaf Fibres

PALF fibre was extracted in few ways such as scrapping, retting and decorticating. These methods are suitable for extracting long fibre of a limited volume for niche applications, but not suitable for mass produced plastic industry in which long fibre may not be necessary. PALF has a ribbon-like structure, consists of vascular buddle system and is present in the form of bunches of fibrous cells. PALF can be extracted by removing epidermal tissue of the leaves [18].

3.1 Manual Extraction

Scrapping process is a conventional method that was used to extract PALF. This process was conducted on a long bench. Scrapping tool called 'ketam' as illustrated in Fig. 1 is well-known tool to extract PALF. This method also known as hand

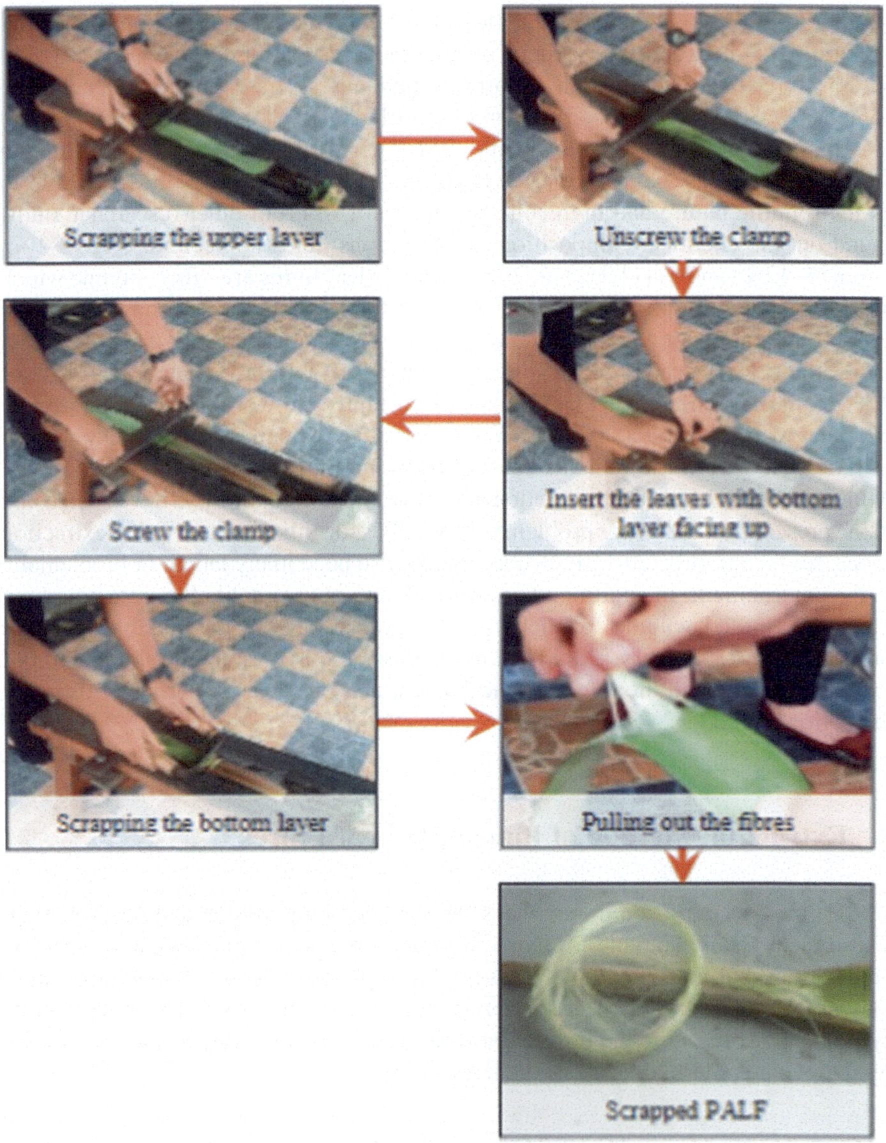

Fig. 1 Hand scrapping method using 'ketam' [36] (with permission)

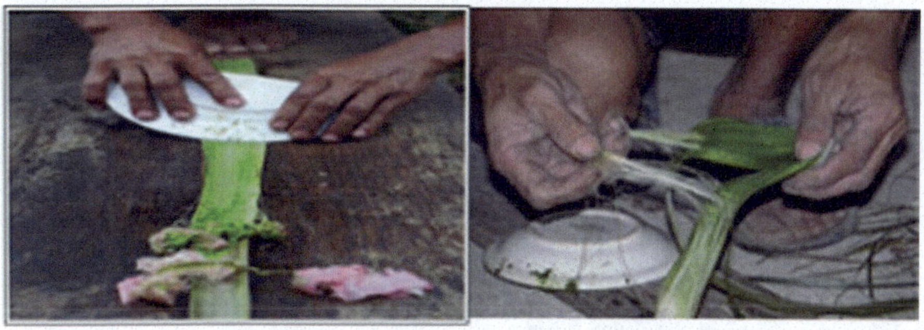

Fig. 2 Extraction of PALF using ceramic [2]. (Adopted)

scrapping method used to process manually. The process started with scrapping pineapple leaf then the extracted fibre, then washed under running tap and dried directly under the sun [36]. By using ceramic plate over the pineapple leaf with pressure and fast movement give the fibre beneath the leaf. However, this process only applicable for only long leaves due to easy processing compared with short leaf [24]. Another method was used in manual extraction of PALF is immersion process. In this process, micro-organisms play an important role to remove sticky substances which surround the pineapple leaves. The fibres were decomposed and easily separated from each other throughout this process. This process is done by soaking the leaves of the pineapple into the water for certain period of time. Then, pineapple leaf spread on flat surface to remove outer skin of leaf using ceramic as shown in Fig. 2. Finally, extracted PALF was washed with water and dried under the sun or using the oven [2]. The manual extraction allows two types of fibres to be obtained from the leaves which are 75 wt% of large vascular bundles present in the top lamina and 25 wt% of fine fibre strands in the bottom lamina [31]. The process of extracting long fibres is great importance since the quality as well as the quantity of extracted fibres is strongly influenced by the method of extraction employed. Manual extraction process of pineapple leaf fibre is time consuming and labourious. This process requires manpower of 30 people per ton of leaves. Manual method produces very low yield and also difficult to scale up. However, manual extraction produced good quality fibres compared with mechanical extraction [9, 22].

3.2 Mechanical Extraction

In mechanical extraction, there is few methods had been conducted. The conventional machine as shown in Fig. 3, is known as pineapple decorticator machine. From recent studies, new machine had been developed to use for scrapping process of PALF. This machine was installed three different rollers which are feed roller, leaf scratching roller and serrated roller as shown in Fig. 4 [8]. The process started with feeding the leaves through feed roller and then scratched by another roller to remove the

Fig. 3 Conventional pineapple decorticator machine [8]. (Adopted)

Fig. 4 Pineapple scratching machine [8]. (Adopted)

waxy layer followed by serrated roller creating space. Then, leaves pass through the dense attached blade serrated roller, which crushes leaves for the entry passage for the retting microbes [19]. This process able to collect at least 15 kg green fibre per day [11]. The mechanical processes like breaking, scotching and hackling may cause damage to fibres. To overcome this, special design machines were required for the extraction of PALF that can reduce the fibres damage [17]. Raspador machine is one of mechanical extraction machines that was used to crush green pineapple leaves. The fibres used to collect by washing of the soft green parts of leaf then, collected fibres brushed with a comb and obtained fine threads after removal of the spongy parts. However, by using this process, the extracted fibres are coarser than manually extracted fibres even though this process takes shorter time than manually PALF extracted [18].

From another research, PALF extraction is illustrated in Fig. 5 by using novel technology. This machine used crush technology to extract PALF. Waxy layer on the PALF surface was removed by blades designed in the machine. The number of blades used, sizes and certain angle of the two blades needed to ensure that the leaf will not snap during the process plays crucial part of the extraction process and also this technology shows more efficiency but time consuming [36].

Fig. 5 Pineapple leaf machine (novel technology) [36]. (Adopted)

3.3 *Retting Process*

Generally, during retting process, scratched pineapple leaves were immersed in a water tank which contains substrate which full of liquor in 1:20 ratio, diammonium phosphate (DAP) or urea 0.5%, to improve retting process. Towards the end of process, leaves were removed and washed mechanically by pond water retting [3, 19]. The retting process took around 15–18 days for immersed PALF in water tank until the unwanted materials sticking on the PALF surface were successfully detached, and in a result, this process obtained clean fibres [18]. The aim for retting process conducted is to separate the fibre bundles from reinforcing materials. This process also resulted in finer fibre bundles. Retting process can be performed in various ways such as in ponds, ditches, tanks or river, and this process needs large amount of water. Additionally, this process needs longer time and non-eco-friendly process [18]. There have several retting methods namely dew, water, enzymatic, mechanical and chemical retting. However, there is no single method that can give optimum results in terms of cost, fibre strength and environmental pollution [27]. Besides water retting, bacteria also were used as alternative for PALF extraction. By using bacterial retting process helps to produced cleaner fibres with higher tensile strength and eliminate non-fibrous components. Extraction of PALF using bacterial action was conducted within 4–5 days for fibres easily come out from the fleshy mass of leaves. The existence of bacteria in this process also may help to reduce retting time compared with conventional water retting process. During retting process, the bacteria were grown and multiplied to produce extracellular pectinases by dissolving pectin in the PALF. However, this technology is not yet commercialized [18, 20, 27].

4 Effect of Extraction Process on Physical Properties

4.1 *Manual Extraction*

Manual extraction method consume time and also productivity is lesser than expected. It was reported, there was about 85–88% of pineapple leaves have been wasted since year 2008 until 2010. However, PALF waste has been started to apply into several purposes included vermicomposting and animal pellets. Two types of fibres can be obtained by manual extraction process which are 75 wt% of large vascular bundles present in the top lamina and 25 wt% of fine fibre strands in the bottom lamina [31]. PALF only has 2.5–3.5% fibre covered by a hydrophobic waxy layer [19]. Physical appearance of PALF extracted from manual extraction method shows brownish in colour as shown in Fig. 6 [36]. In manual extraction, it was reported the fibre yield was up to 3–4%. In scrapping method, average fibre diameter is approximately 57.36 μm with wide range between 5 and 129 μm [18]. Another researcher extracted PALF by using hand scrapping method and reported average diameter of PALF (90.7 μm) [36]. Meanwhile, another study showed that fibre yield

Fig. 6 Physical appearance manual PALF extraction [36]. (Adopted)

was near about 2%, when hand scrapping technique was applied. Another comparative study conducted on wastage materials while using various extraction techniques, it is found out that manual scrapping method produce 50% more waste materials than mechanical extraction process [34].

4.2 Mechanical Extraction

For mechanical PALF extraction, the fibres appearance looks more attractive in creamy white as illustrated in Fig. 7. By using different types of extraction machines provide different type of fibres fineness structure as tabulated in Table 1. The fineness of PALF were improved by using mechanical extraction process as compared with manual extraction method [26]. Other research showed fibre yield of mechanical extraction significantly increased as compared with manual extraction. Table 2 shows a comparison of fibre yield with different mechanical extraction method. Other study showed that pineapple leaves waste was reduced due to mechanical extraction process. Mechanical extraction gives positive impact and also reduces PALF processing time. Figure 8 shows the PALF waste reduced 2–8% started in year 2008–2013 [34]. The previous study shows the percentage of fibre yield for mechanical extraction also influenced by fibre ages. The result showed matured pineapple leaves give higher fibre yield and easier to extract [23].

Fig. 7 Physical appearance mechanical PALF extraction [36]. (Adopted)

Table 1 Comparative fineness of PALF by different mechanical extraction techniques [18]. (Adopted)

Machine type	Fineness (tex)
Machine scrapping (one side)	5.00
Machine scrapping (both side)	4.30
Machine scrapping (both side) plus serration	5.70
Machine scrapping (one side) plus serration	3.40

Table 2 % Fibre yield from different mechanical extraction method [34]. (Adopted)

Machine	Fibre yield (%)
Ball milling	2.90
Milling	2.80
PALF M1	3.75

4.3 Retting Extraction

PALF having good quality, can be extracted through retting process from retting tank with 0.5% urea under 35 °C [8]. In this retting process, the fibre yield reached approximately 1.8% of leaf. Meanwhile, the average fibre diameter is 58.98 μm and the range in between 5 and 166 μm [18]. A researcher studied the PALF diameter after retting process, was 60 μm and the fibre fineness was 4.3 tex. The percentage of water holding capacity during retting process was 203% and water absorption 24%

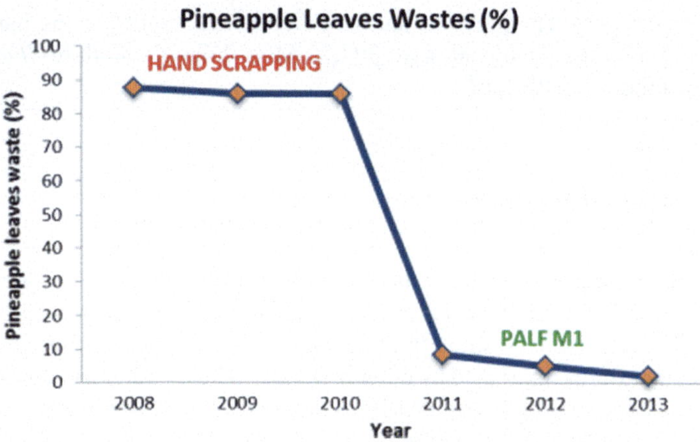

Fig. 8 Percentage of pineapple waste from 2008 to 2013 [34] (with permission)

[14]. Other than that, the bundle strength of PALF decreases by 50% when in a wet condition. However, the yarn strength increases by about 13% shows interesting characteristics of PALF that has been found out from previous researcher [25]. In retting process, the PALF produced is smooth and clean surface due to other non-cellulosic substances are removed. This can help improve appearance and the processibility of PALF in further processing [14]. In other research was found that chemical retting method such as 5% sodium hydroxide and sodium benzoate affected the colour of fibres and the fibre colour become darker. The chemical retting process is very efficient to eliminate lignin content from the fibre surface and helps to improve the stiffness of fibre [27].

5 Effect of Extraction Process on Mechanical Properties

5.1 Manual Extraction

The method of extraction is one of factors that influenced the mechanical properties of PALF. For manual extraction, the mechanical properties show lower result compared with retting and mechanical method. This is due to some forces are being applied continuously at same point in order to scrap the fibre. The force applied is not evenly distributed along the fibre's length which can lead to fibre damage and broken at certain point. As a result, fibre's tensile properties were affected [36]. High cellulose content and low microfibrillar angle are main factor that contribute to good mechanical properties of PALF. Besides that, the fibre strength also influenced by the extraction process [3]. The previous study on fibre extraction by using hand scrapping method found the tensile strength 393.70 MPa, Young's modulus 7254.2 GPa and strain of failure 3.24% [36].

5.2 Mechanical Extraction

The previous researches showed some mechanical processes may induce damage to natural fibres. These mechancial process include breaking, scotching and hacking actions that effect tensile strength of the elementary fibres to be only marginally higher than other fibres (extracted by other methods). The overall mechanical properties were reduced by 40–45% [26]. PALF has higher mechanical properties compared with other natural fibres due to the presence of cellulose content and microfibrillar angle. In addition, PALF specific modulus and strength are close to or even higher than that of glass fibre. From previous research, they found out the method of extraction does not give significant result towards mechanical properties. The amount of reinforcement and size of fibrillated fibres were main concern to improve mechanical properties [20]. Some literature also acknowledge that PALF extracted by mechanical process provide better mechanical properties than hand scraped PALF. It is found that tensile strength of extracted PALF by mechanical extraction technique was much higher 613.75 MPa than hand scrapping technique 393.70 MPa. However some results of mechanically processed fibres were also recorded such Young's modulus 1,379,500 GPa, and strain to failure 6.67% [36]. Results of Kappa test showed lower lignin residue (3.39%) in PALF, extracted by using mechanical method which is lower than retting process. The lowest lignin content and other impurities give higher tensile strength of fibre due to closer cellulose chain packing order [32]. Other research was found that PALF extracted using decorticator machine showed tensile strength near about 288 MPa. Meanwhile, the modulus of elasticity was 20 GPa which due to hemicellulose content that act as cementing material that limits the slippage of the cellulose microfibrils, and keeping them in their original position [16].

5.3 Retting Extraction

Fibre extracted using NaOH chemical retting process also shows good mechanical properties. During retting process, NaOH solution gives strength to fibre and helps to reduce impurities such as wax and lignin on the fibre surface, thus improving cellulose chain structure in the fibre. For retting process, its specific ultimate strength was 28.54 MNm/kg. Meanwhile, specific modulus was approximately 30 MNm/kg [32]. From previous research it is found that enzymatic retting was the most suitable method to reduce the lignin content in fibres [35].

6 Conclusion

PALF can be extracted in various ways such as manual, mechanical and retting process. Manual extraction process of pineapple leaf fibre is time consuming and labourious. It also produced very low yield and also difficult to scale up. However, manual extraction produced good quality fibres compared with mechanical extraction. For mechanical extraction, the process is most efficient in terms of PALF production and time consuming. PALF physical appearance looks more attractive in creamy white. The fibre structure also looks finest and smooth. Mechanical properties of PALF, extracted by using mechanical machine, showed higher tensile strength compared to the extracted fibre through hand scrapping technique. Meanwhile, for retting method, it involves many chemicals and time consuming. PALF needs to be soaked in retting tank for about 10 days to remove lignin on the PALF surface before further to next processing. However, retting method could provide mechanical strength and stiffness to the fibres.

References

1. Abdelmouleh M, Boufi S, Belgacem MN et al (2007) Short natural-fibre reinforced polyethylene and natural rubber composites: effect of silane coupling agents and fibres loading. Compos Sci Technol 67:1627–1639
2. Adam A, Yusof Y, Yahya A (2016) Extraction of pineapple leaf fiber: josapine and moris. J Eng Appl Sci 11:161–165
3. Asim M, Abdan K, Jawaid M, et al (2015) A review on pineapple leaves fibre and its composites. Int J Polym Sci
4. Asim M, Jawaid M, Abdan K et al (2017) Effect of pineapple leaf fibre and kenaf fibre treatment on mechanical performance of phenolic hybrid composites. Fibers Polym 18:940–947
5. Asim M, Jawaid M, Abdan K et al (2018) Effect of hybridization on the mechanical properties of pineapple leaf fiber/kenaf phenolic hybrid composites. J Renew Mater 6:38–46
6. Asim M, Jawaid M, Nasir M et al (2018) Effect of fiber loadings and treatment on dynamic mechanical, thermal and flammability properties of pineapple leaf fiber and kenaf phenolic composites. J Renew Mater 6:383–393
7. Asim M, Jawaid M, Paridah MT, et al (2019) Dynamic and thermo-mechanical properties of hybridized kenaf/PALF reinforced phenolic composites. Polym Compos
8. Banik S, Nag D, Debnath S (2011) Utilization of pineapple leaf agro-waste for extraction of fibre and the residual biomass for vermicomposting
9. Biagiotti J, Puglia D, Kenny JM (2004) A review on natural fibre-based composites-part I: structure, processing and properties of vegetable fibres. J Nat Fibers 1:37–68
10. Bongarde U, Shinde V (2014) Review on natural fiber reinforcement polymer composites. Int J Eng Sci Innov Technol 3:431–436
11. Das P, Nag D, Debnath S, et al (2010) Machinery for extraction and traditional spinning of plant fibres
12. Devi LU, Bhagawan S, Thomas S (1997) Mechanical properties of pineapple leaf fiber-reinforced polyester composites. J Appl Polym Sci 64:1739–1748
13. George J, Sreekala M, Thomas S (2001) A review on interface modification and characterization of natural fiber reinforced plastic composites. Polym Eng Sci 41:1471–1485
14. Hazarika D, Gogoi N, Jose S et al (2017) Exploration of future prospects of Indian pineapple leaf, an agro waste for textile application. J Clean Prod 141:580–586

15. Heng JY, Pearse DF, Thielmann F et al (2007) Methods to determine surface energies of natural fibres: a review. Compos Interfaces 14:581–604
16. Jaramillo-Quiceno N, Ch EMC, Restrepo-Osorio A et al (2018) Improvement of mechanical properties of pineapple leaf fibers by mercerization process. Fibers Polym 19:2604–2611
17. Joffe R, Andersons J, Wallström L (2003) Strength and adhesion characteristics of elementary flax fibres with different surface treatments. Compos A Appl Sci Manuf 34:603–612
18. Jose S, Salim R, Ammayappan L (2016) An overview on production, properties, and value addition of pineapple leaf fibers (PALF). J Nat Fibers 13:362–373
19. Kannojiya R, Gaurav K, Ranjan R et al (2013) Extraction of pineapple fibres for making commercial products. J Environ Res Dev 7:1385
20. Kengkhetkit N, Amornsakchai T (2012) Utilisation of pineapple leaf waste for plastic reinforcement: a novel extraction method for short pineapple leaf fiber. Ind Crops Prod 40:55–61
21. Ku H, Wang H, Pattarachaiyakoop N et al (2011) A review on the tensile properties of natural fiber reinforced polymer composites. Compos B Eng 42:856–873
22. Lopattananon N, Panawarangkul K, Sahakaro K et al (2006) Performance of pineapple leaf fiber–natural rubber composites: the effect of fiber surface treatments. J Appl Polym Sci 102:1974–1984
23. Mazalan MF, Yusof Y (2017) Natural pineapple leaf fibre extraction on josapine and morris. MATEC Web of Conferences. EDP Sciences, 00043
24. Nasir Uddin SM, Abdul JM, Mazharul Islam M, Siddika A (2017) A review on extraction, characterization and application of pineapple leaf fiber (palf) in textiles and other fields. Int J Adv Res 5:112–116
25. Mishra S, Mohanty AK, Drzal LT et al (2004) A review on pineapple leaf fibers, sisal fibers and their biocomposites. Macromol Mater Eng 289:955–974
26. Mohamed A, Sapuan S, Shahjahan M et al (2010) Effects of simple abrasive combing and pretreatments on the properties of pineapple leaf fibers (palf) and palf-vinyl ester composite adhesion. Polym-Plast Technol Eng 49:972–978
27. Paridah MT, Basher AB, SaifulAzry S et al (2011) Retting process of some bast plant fibres and its effect on fibre quality: a review. BioResources 6:5260–5281
28. Pickering K (2008) Properties and performance of natural-fibre composites, Elsevier
29. Pickering KL, Efendy MA, Le TM (2016) A review of recent developments in natural fibre composites and their mechanical performance. Compos A Appl Sci Manuf 83:98–112
30. Puglia D, Biagiotti J, Kenny J (2005) A review on natural fibre-based composites—Part II: application of natural reinforcements in composite materials for automotive industry. J Nat Fibers 1:23–65
31. Sapuan S, Mohamed A, Siregar J et al (2011) Pineapple leaf fibers and PALF-reinforced polymer composites. Bio-and Nano-Polymer Composites. Springer, Cellulose Fibers, pp 325–343
32. Sarah S, Rahman W, Majid R et al (2018) Optimization of pineapple leaf fibre extraction methods and their biodegradabilities for soil cover application. J Polym Environ 26:319–329
33. Summerscales J, Dissanayake NP, Virk AS et al (2010) A review of bast fibres and their composites. Part 1–Fibres as reinforcements. Compos A Appl Sci Manuf 41:1329–1335
34. Yahya SA (2016) Characterizations of Malaysia PALF properties between hand scrapping and PALF M1. Universiti Tun Hussein Onn Malaysia
35. Yu H, Yu C (2010) Influence of various retting methods on properties of kenaf fiber. J Text Inst 101:452–456
36. Yusof Y, Yahya SA, Adam A (2015) Novel technology for sustainable pineapple leaf fibers productions. Procedia CIRP 26:756–760

Improving the Properties of Pineapple Leaf Fibres by Chemical Treatments

R. Siakeng, M. Jawaid, Paridah Md. Tahir, S. Siengchin and M. Asim

Abstract Natural fibres such as pineapple leaf fibre (PALF) have the advantages of low density, lightweight, low cost, biodegradability and renewability. Recently, many research works have been conducted all over the world on utilization of natural fibres as a reinforcing material for composites for variety of applications. Nonetheless, the inherent lack of good compatibility of natural fibre towards the polymer matrices limits the utilization of natural fibre for reinforcing agent in composites less attractive. For that reason, chemical treatments are considered in modifying the fibre properties which aimed at improving the compatibility of fibre and the polymer matrix. Chemical treatments of PALF can help clean the fibre surface, modify the surface and increase the surface roughness. Ultimately, moisture absorption is reduced in the treated fibre reinforced composites which results in enhancements of mechanical properties. In this article, effect of chemical treatment on the properties of PALF and PALF-reinforced polymer composites is discussed and different chemical treatment with their functions is listed. Recent studies dealing with chemical-treated PALF-reinforced composites have also been cited.

Keywords Pineapple leaf fibre · Treatments · Interfacial bonding · Mechanical properties · Physical properties · Thermal properties

1 Introduction

The focus towards natural fibres increased immensely in the past few decades due to their sustainability, economical and environmental advantages [26]. The concern of sustainable development is the driving force towards increasing efficient utilization

R. Siakeng · S. Siengchin
Department of Mechanical and Process Engineering, The Sirindhorn International Thai German, Graduate School of Engineering (TGGS), King Mongkut's University of Technology North Bangkok, Bangkok, Thailand

M. Jawaid · P. Md. Tahir · M. Asim (✉)
Laboratory of Biocomposite Technology, Institute of Tropical Forestry and Forest Products (INTROP), Universiti Putra Malaysia, 43400 Serdang, Selangor, Malaysia
e-mail: khanfatehvi@gmail.com

M. Jawaid et al. (eds.), *Pineapple Leaf Fibers*, Green Energy and Technology,
https://doi.org/10.1007/978-981-15-1416-6_4

of raw materials and reduction of waste products. Many of the commercial natural fibres are renewable resources in many producing countries [50]. Natural fibres, especially plant fibres are considered as worthy candidates to replace conventional glass/synthetic fibres due to their low cost, eco-friendliness, biodegradability and renewability [43]. Among the plant fibres, pineapple leaffibres (PALF) are nowadays extensively used in number of applications. PALF is a multipurpose plant fibre extracted from the leaves of pineapple plant (*Ananas comosus*). Pineapple is one of the most common fruit widely cultivated in tropical countries around the world especially for its fruits [1]. Pineapple leaves are the major part of the plant that is currently unused which needs large-scale attention for its commercial utilization. After pineapple fruits are harvested, the leaves usually go to waste which later on are disposed by burning or natural decomposition. Utilization of these leaves will lead to significant waste reduction therefore reduce environmental impact [50]. By this framework, utilization of agricultural by-products such as pineapple leaf fibre (PALF) as an alternatives raw material for research and engineering is recommended.

But these plant fibres enclose strongly polarized hydroxyl groups which are hydrophilic in nature. Therefore, they are innately incompatible with hydrophobic polymer matrices [44]. Moreover, water absorption of fibre is very high due to the presence of pendant hydroxyl and polar groups in various constituents of fibre, which leads to poor interfacial bonding with the reinforcing matrix [19]. For that reason, it is necessary to reduce the hydrophilicity of fibres by suitable chemical treatments [29]. Fibre properties can be modified and customized to different degrees by surface or chemical treatments [26]. There are several reports available on different aspects of pineapple leaf fibres (PALF). A number of researchers report on the properties of PALF such as chemical, physical and mechanical characterization. Whilst few researches have been done on the chemical treatments of PALF's mainly to increase interfacial adhesion with matrix and tensile properties. All these research findings showed that chemical treatments will improve PALF's mechanical properties. Almost all the studies done as regards to PALF were focussed on its properties, compositions and its utilizations in various applications. Studies show that surface treatments of PALF with silane and alkali make the fibres more hydrophobic [4, 48]. Alkaline and silane treatments also help improve interfacial bonding between PALF and polymer matrices [33, 41].

2 Pineapple Leaf Fibre

Pineapple leaves are produced in tonnes every year, though a very small portion utilized in feedstock and in energy production but big portion are used as land filling purposes [5]. Biocomposite industries are amplifying opportunity to reduce the wastage of biodegradable materials and improve the agriculture-based non-food industries [8]. Pineapple leaf fibres (PALF) have revealed unexpected excellent properties such as single fibre strength, flexural strength and impact (damping) properties because of its extraordinary chemical ratios such as cellulose (70–82%), lignin (5–12%) and ash (1.1%) [3]. PALF usually appears whitish in colour, regular smooth and glossy as silk, medium-length fibre (depend on leaf size). The outer surface of PALF is softer than other natural fibres and it has absorption capacity to maintain a good colour [1]. Though PALF has good specific strength and stiffness, it also has high hydrophilicity characteristic due to high cellulose content [9, 12]. Extraction of pineapple leaf fibres usually carried out by three procedures; mechanical, retting method and scrapping process. Mechanical process gives better yield than retting and scrapping process due to existence of vascular bundle system in pineapple leaf in form of bunches which need to remove the whole outer surface after harvesting. The appearance of PALF is thread like structure which tightly bound with pectin in bundle form and called multicellular lignocellulosic fibre [35], and PALF contains several chemicals constituents and impurities, the major constituents are cellulose, hemicellulose polysaccharides, lignin and some impurities or chemical in minorities are such as fat, wax, pectin, uronic acid, anhydride, pentosan, colour pigment, inorganic substance and so forth [1]. PALF and cotton fibres have same cellular arrangement as well as cellulose content (70–82%) and (82.7%), respectively [27, 32]. Among all natural fibres collections, PALF is more compatible lignocellulosic fibre resource due to its better chemical composition. PALF showed better mechanical properties than jute in form of making fine yarn [10]. The model of cellulosic molecules of PALF revealed its anatomical structure which is three-dimensional structure and parallel to crystalline region of the fibre, rest of the share of molecular structure are supposed to associate within amorphous regions. These desirable characteristics of PALF can be considered as a potential substitute for the industries as raw materials to produce natural reinforced polymer [24].

3 Chemical Treatments

A number of research studies were carried out on natural fibres and their chemical treatments for improving fibre properties and to achieve desired qualities [11, 17]. Hydrophilic property is reported to be the most stern drawback of these natural fibres which causes incompatibility in interfacial bonding between fibre and matrix in polymer composites [21]. Surface impurities and the hydroxyl groups present on the fibre surface are the fate of natural fibres as filler/reinforcement materials [46]. According to Belgacem and Gandini [7], the most promising approach of fibre's chemical modification are the ones that gave rise to continuous covalent bonds between fibre surface and the polymer matrix. The main hold up of natural fibre composites are the hydrophilic nature of natural fibre and hydrophobic nature of the polymer matrix [4].

Chemical treatments of natural fibres help in reduction of fibre hydrophilicity and ultimately improved its compatibility with the polymer matrix [13, 17, 23]. In this paper, we will focus on the fibres properties affected by chemical treatments which have been used on natural fibres. A few popular treatments of natural fibres are explained in Table 1. With the aim of improving fibre matrix compatibility and

Table 1 Chemical treatments of natural fibres [44]

Treatment	Effects/Results
Alkali treatment (mercerization)	It reduces fibre diameter by breaking the fibre bundle, thereby increases the surface area of fibre which results in good adhesion with the matrix and improves mechanical and thermal behaviours of the composite
Silane treatment	It is one of the most effective coupling agents for natural fibres surface modification. It is a multifunctional molecule which deposit on the fibre surface which makes better linkage with the matrix through a siloxane bridge. It improves the fibre matrix adhesion and stabilizes the composite properties
Acetylation treatment	This treatment is known as esterification process for plasticizing natural fibres. The reaction of acetyl group (CH_3CO) with the hydroxyl groups (–OH) reduce the hydrophilicity of natural fibre and improves dimensional stability of the composites
Benzoylation treatment	Benzoyl chloride is used as an agent to decrease hydrophilic nature of the natural fibres and improve its compatibility with the matrix, which, therefore, enhance the thermal stability and strength of the composite
Peroxide treatment	It improves the interfacial adhesion, thermal stability and reduces the moisture absorption of fibre and matrix
Maleated coupling agents	It gives proficient interaction with the functional surface of the fibre and matrix which reduces the melting temperature and lowers stiffness of the fibres
Sodium chlorite ($NaClO_2$) treatment	It is used for bleaching of fibres in acid solution. $NaClO_2$ is acidified and releases choleric acid ($HClO_2$), which undergoes an oxidation reaction and forms chlorine dioxide (ClO_2). ClO_2 reacts with lignin constituents and remove it from the fibre thus improving the adhesive properties of the fibre
Acrylation and acrylonitrile grafting	Acrylic acid reacts with the hydroxyl groups of the fibre and provides more free radical of reactive cellulose macro-radicals which helps to create good interfacial bonding. It reduces hydroxyl groups from the fibre and improves moisture resistance

(continued)

Table 1 (continued)

Treatment	Effects/Results
Calcium hydroxide $Ca(OH)_2$ treatment	It helps in degradation of the amorphous materials present in the fibre structure. It is connected with increasing the crystallinity index of cellulose and improved thermal stability
Isocyanate treatment	Act as a coupling agent in fibre surface modification and helps in better moisture resistance mechanism of the fibre and provides better bonding with the matrix to improve the composite properties
Triazine treatment	Triazine reacts with the hydroxyl groups of cellulose and lignin in the natural fibres that improves its moisture resistance properties
Stearic acid treatment	Stearic acid in ethyl alcohol solution is used for treating natural fibre surfaces to facilitate better interfacial bonding of fibre and matrix
Permanganate treatment	Potassium permanganate ($KMnO_4$) in acetone solution is used for treating fibre surface for enhancing interfacial adhesion between natural fibres and matrix by reduction of hydrophilicity of the fibre and increase thermal stability of the fibre
Fatty acid derivate (oleoyl chloride) treatment	Fatty acid derivative act as a coupling agent to alter natural fibre surface to improve its adhesion with the matrix and wettability properties of natural fibre in composites

composite properties, various chemical treatments were conducted on PALF by many researchers [4, 24, 33, 41, 48]. Most of these treatments helped in improving physical, mechanical and in few cases, thermal properties of the composites by strengthening fibre matrix bond.

4 FTIR of Treated Natural Fibres

Effects of chemical treatments on natural fibres can be explained by using FTIR spectra. Asim et al. [4] studied different chemicals and its composition on PALF to analyse the effect of chemicals on fibre's constituents. Untreated PALF and treated PALF with various chemicals NaOH and silane, and combination of NaOH and silane were investigated the FTIR spectra showed in Fig. 1. In untreated PALF, two vibration peaks showed presence of lignin and hemicellulose at 1200 cm^{-1} and 1500 cm^{-1}, respectively [36]. The peak at 1600 cm^{-1} showed presence of hydroxyl group [39], while another peak at 2900 cm^{-1} confirms presence of C–H group due to the presence

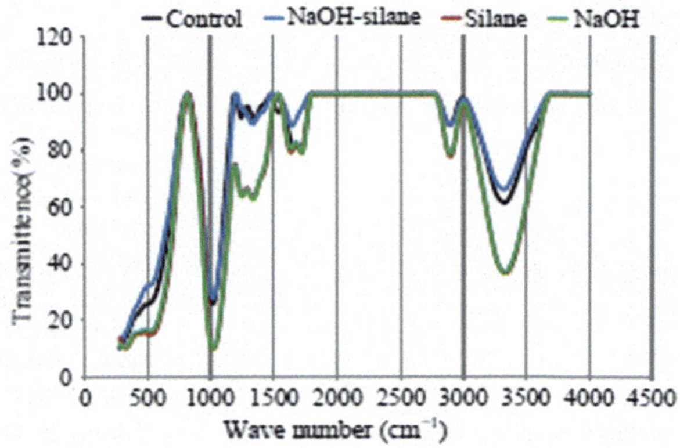

Fig. 1 FTIR spectra of untreated and treated PALF [4]

of CH_2 group of cellulose and hemicelluloses. A huge content of hydrogens (–OH group) are present in cellulosic structure, and it was showed through wide range peak from 3000 to 3700 cm^{-1}. Table 2 shows verity of functional groups at various wave numbers [4]. NaOH -treated PALF found absence of peaks at some wavelength such as 1200 and 1650 cm^{-1} leaching out of lignin partially and completely hemicellulose decomposition. Alkali treatment is more useful to remove hemicellulose rather than lignin and also showed that hemicellulose is easily removable structure [28]. PALF treated with silane also removed lignin and hemicelluloses but added a peak at 847 cm^{-1} duo to Si–C stretching bond, and showed presence of silane on the PALF surface [34]. Silane indication on PALF surface is positive sign of good interfacial bonding between fibre and matrix [16]. NaOH–silane-treated PALF indicated N–H and C–H bonding at vibration peaks at 1300 and 1675 cm^{-1} [24].

Another research based on fibre treatment studied by Threepopnatkul et al. [48], investigated both untreated and two types of silane-treated PALF. FTIR spectra revealed the chemical reaction between silane and cellulosic structure of PALF. Silane (Z-6011)-treated fibre showed peaks at 3379 cm^{-1} due to the C–NH_2 and transmission peaks at 1715 and 1597 cm^{-1} corresponding to NH_2.

Table 2 FTIR analysis through corresponding functional groups and vibrational types [25]

Wave number	Functional group	Vibrational type
3500–3700	– OH	O–H stretching
2912	– CH_2	C–H stretching
1715	R(CaO)OH	CaO stretching
1474	– CH_2	C–H bending, crystalline
1464	– CH_2	C–H bending, amorphous
908	R–CH_2CH_2	C–CH2 out of plane bending
650–1000	=CH	Alkenes

Silane (Z-6030)-treated fibres appeared peak at 1720 cm^{-1} corresponding to carbonyl group [20]. All treated fibres showed a peak around 3300 cm^{-1} corresponding to the –OH group [30]. Silane (Z-6011)-treated PALF revealed peak at 1715 cm^{-1} to ensure NH_2 means some silane reacted with PALF through direct condensation reactions between the hydrolyzed silane and hydroxyl groups of the pineapple leaf cellulose but silane (Z-6030) treatment did not show any confirmation of bonding [48].

Another study [24] on the FTIR spectroscopy characterization of surface-modified PALF investigated on several treatments such as NaOH and benzoyl peroxide (BPO). These treatments were used for surface modification and compared with untreated PALF. Untreated PALF showed –OH bonding of cellulose structure intense at ranging from 3100 to 3800 cm^{-1}. Another COH stretching vibration from OCH_2 functional group is found at 2917 cm^{-1}. The carbonyl group was detected at peak 1736 cm^{-1}, however, the peak at 1641 cm^{-1} was due to H_2O [39]. Some peaks at 1436 and 1254 cm^{-1} are due to the presence of lignin and hemicellulose structure, respectively [2, 28].

The NaOH treatment decomposed hemicellulose (ester type) and partially leached out lignin structure. Mwaikambo and Ansell [28] found that alkali treatment removed hemicellulose of natural fibres such as sisal and jute fibres rather than lignin. However, the treatment of BPO revealed new peaks of phenyl nucleus in BPO at about 1657 and698 cm^{-1} [38]. OOH group found in BPO-treated fibres, resulted broader absorption at 3300 cm^{-1} and peaks at 1735–1736 cm^{-1}.

5 Effects of Treatment on Interfacial Bonding

Interfacial bonding between natural fibres and matrix affects mechanical, thermal and physical properties of composites. Interfacial shear strength test provides data of bonding strength between fibres and matrix. To determine the interfacial shear strength, the analysis was done using the below given Eq. 1, schematics of the micro-droplet test have explained the procedure in Fig. 2 and also showed the before and after testing samples in Fig. 3.

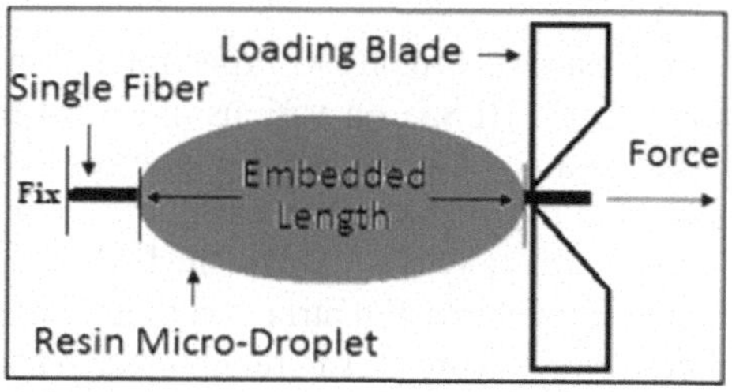

Fig. 2 Schematics of micro-droplet test [45]. Adopted

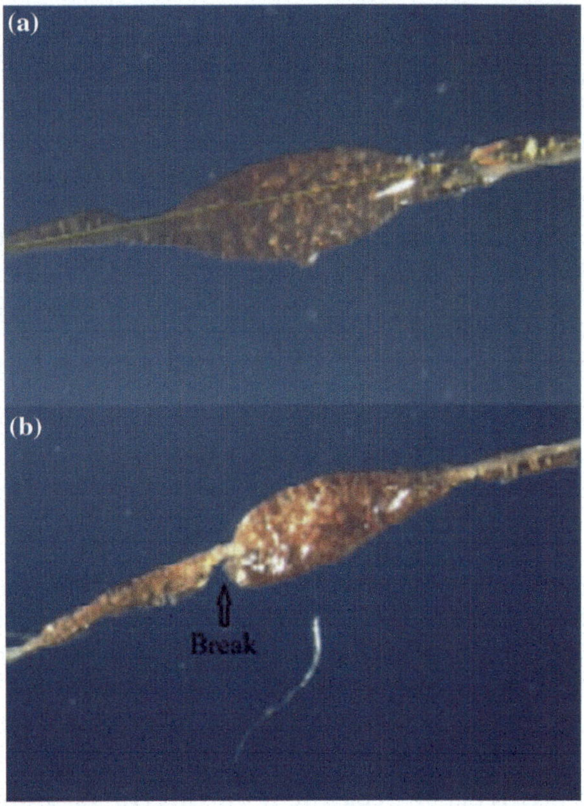

Fig. 3 Droplet test of PALF/phenolic rein: **a** before **b** after

$$\tau = \frac{F}{\pi DL} \tag{1}$$

where

τ indicates interfacial shear strength (MPa)
F indicates load at maximum stress (N)
D indicates single fibre diameter (m)
L indicates embedded length (m).

A research based on PALF/phenolic resin composite, investigated interfacial shear strength test (IFSS) on various chemical treatments and found that silane-treated PALF obtained highest IFSS properties [4]. Figure 4a showed a drop of matrix on fibres and Fig. 4b showed the failure of matrix.

Silane treatments improved the PALF and matrix bonding near about 40% higher than untreated PALF/matrix bonding. Other treatment such as NaOH did not show effectiveness with PALF though NaOH-treated kenaf fibre with phenolic resin

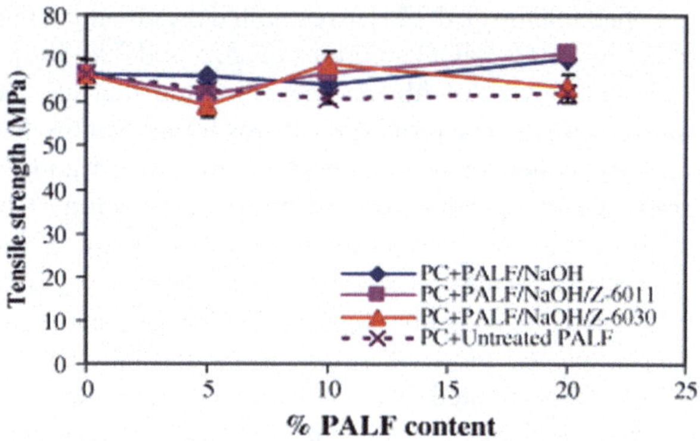

Fig. 4 Comparison of tensile strength of treated and untreated PALF/PC composites [48] (with permission)

showed very good bonding near about 130% better than untreated kenaf fibres. Nirmal et al. found alkali-treated improved bonding between fibres and polymers by 115% [31]. In comparative study of treated natural fibres, treated PALF showed less IFSS properties than kenaf fibres. Combine treatment of NaOH and silane of PALF improved 11.35%, however, kenaf fibres increased by 13.47%. Silane-treated PALF and kenaf fibres showed better IFSS than other chemical-treated PALF and Kenaf fibres such as NaOH and NaOH + silane.

6 Effects of Treatment on Mechanical Properties

Among various factors, chemical treatment ranked quite high on affecting mechanical properties of PALF, one of which is chemical treatment. So, several research studies based on chemical treatments of natural were carried out in order to achieve desired qualities [25, 34]. Chemical treatments remove lignin and hemicelluloses in fibres thus enhanced the tensile properties [4]. Siakeng et al. [41] treated PALF with three different chemicals such as 6% NaOH, 2% Silane and 6% $Ca(OH)_2$. They reported that tensile strength is higher in chemically treated PALF compared to the untreated PALF which had 281.6 MPa tensile strength. According to their result, highest tensile strength (455.8 MPa) was found in silane-treated PALF followed by 424.6 MPa in NaOH-treated fibres. And lastly, PALF treated by $Ca(OH)_2$ showed 303.4 MPa tensile strength. In addition, chemical-treated PALF showed better mechanical interfacial bonding between matrix and fibres due to lignin and hemicelluloses removal. However, elevated chemical concentration may decrease fibre's tensile strength [25]. Tensile modulus was increased in NaOH- and $Ca(OH)_2$-treated PALF by approximately 30 and 35%, whereas up to 70% increment of tensile modulus was found in silane-treated PALF compared to untreated fibre. Similar phenomenon was reported by Asim et al. [4] in which silane and NaOH treatments of PALF enhanced its tensile strength and modulus. In their study, they also suggests that the soaking time and concentration of chemicals can affect the tensile strength.

Lopattananon et al. [24] investigated the chemical modification of PALF in natural rubber by using NaOH solutions (1, 3, 5 and 7%, w/v) and benzoyl peroxide (BPO) (1, 3 and 5 wt% of fibre). They reported that all chemical treatments enhanced adhesion and tensile properties. It was found that 5% NaOH and 1% BPO treatments yielded excellent enhancement of tensile strength and all of the NaOH-treated PALF-reinforced composites showed higher elongation at break compares to untreated PALF composites. By comparisons of differential NaOH solution concentrations, they reported that 5% NaOH exhibit the best mechanical properties. In related study, Panyasart et al. [33] investigated PALF treatment in regard to the properties of the composites and the results are shown in Table 3 with code such as untreated (R-PALF), NaOH treated (Na-PALF) and silane-treated PALF (Si-PALF). Enhancement of mechanical properties of the composites after treatments was observed. Tensile strength and modulus were improved while elongation at break reduced by both treatments.

Another research based on fibre treatment by Threepopnatkul et al. [48] compared the effect of silane and NaOH treatment on performance of PALF in polycarbonate (PC) composites. Various mechanical properties like, tensile strength, Young's modulus, impact strength and elongation at break (%), etc., were considered in order to examine the reinforcement effect of treated PALF by silane and NaOH chemicals. It was found that tensile modulus of the composites increased significantly for NaOH-treated PALF composites owing to better adhesion between PALF and matrix which makes it possible for stress transfer from the matrix to the PALF, thus enhancing the composites strength. Silane-treated PALF composites exhibit lower modulus than NaOH-treated composites. The plausible reason given was the possible development of flexible polysiloxane (Si–O–Si) from self-condensation reaction of silane [47]. On the other hand, silane-treated fibre (PALF/Z-6011) composites exhibit highest impact strength as a result of the contact between the functional group of the treated PALF and the PC matrix. Figures 5 and 6 illustrate the tensile and impact

Table 3 Mechanical properties of polyamide 6 and different PALF/polyamide 6 composites [8] (with permission)

Sample	Young's modulus (MPa)	Tensile strength (MPa)	Elongation at break (%)
PA 6 (neat)	1007.7 ± 51.4	62.3 ± 2.1	48.6 ± 4.1
R-PALF 20	1420.0 ± 41.0	67.6 ± 1.8	10.3 ± 0.4
R-PALF 30	1600.0 ± 27.7	70.9 ± 0.9	8.2 ± 0.3
R-PALF 40	1962.5 ± 50.1	68.6 ± 2.2	5.2 ± 0.3
Na-PALF 20	1540.0 ± 36.0	73.7 ± 1.5	9.5 ± 0.3
Na-PALF 30	1820.0 ± 49.1	78.2 ± 2.1	8.0 ± 0.3
Na-PALF 40	2040.8 ± 40.1	73.5 ± 2.2	5.6 ± 0.3
Si-PALF 20	1570.0 ± 35.9	75.3 ± 1.4	10.8 ± 0.3
Si-PALF 30	1848.5 ± 48.1	81.8 ± 1.3	7.9 ± 0.2
Si-PALF 40	2091.3 ± 49.6	79.6 ± 2.7	6.0 ± 0.3

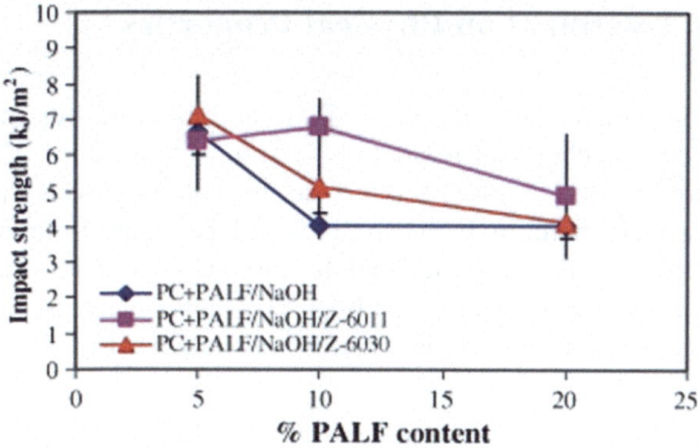

Fig. 5 Impact strength of treated and untreated PALF/PC composites [48] (with permission)

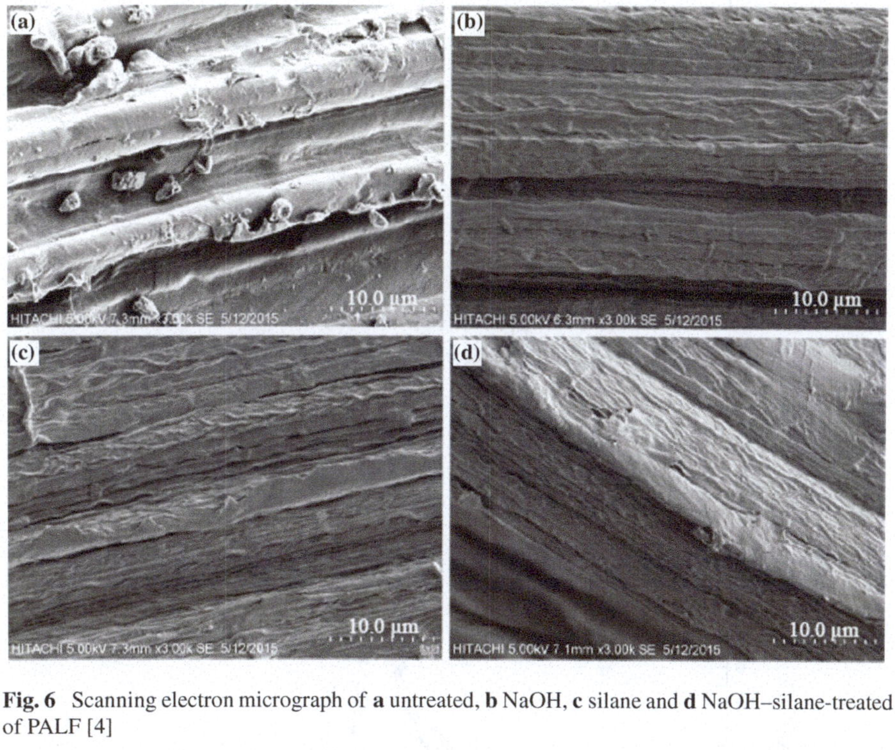

Fig. 6 Scanning electron micrograph of **a** untreated, **b** NaOH, **c** silane and **d** NaOH–silane-treated of PALF [4]

strength of the PALF-reinforced PC with the alkali treated and two different silanized PALF/PC composites in comparison with untreated composites.

7 Effects of Treatment on Physical Properties

Asim et al. [4] investigate the effects of silane and alkaline treatments on PALF and reported that these chemical treatments lead to separation of PALF into small fraction. These treatments increase fibre surface area and roughness which ultimately improves the interfacial bonding between fibres and the matrix ultimately enhancing the composites stiffness. They reported that the untreated fibres diameters were larger compared to alkaline- and silane-treated fibres. The chemical treatment washed away the dirt and other materials from the fibre surfaces as shown in Fig. 7. Alkali and silane

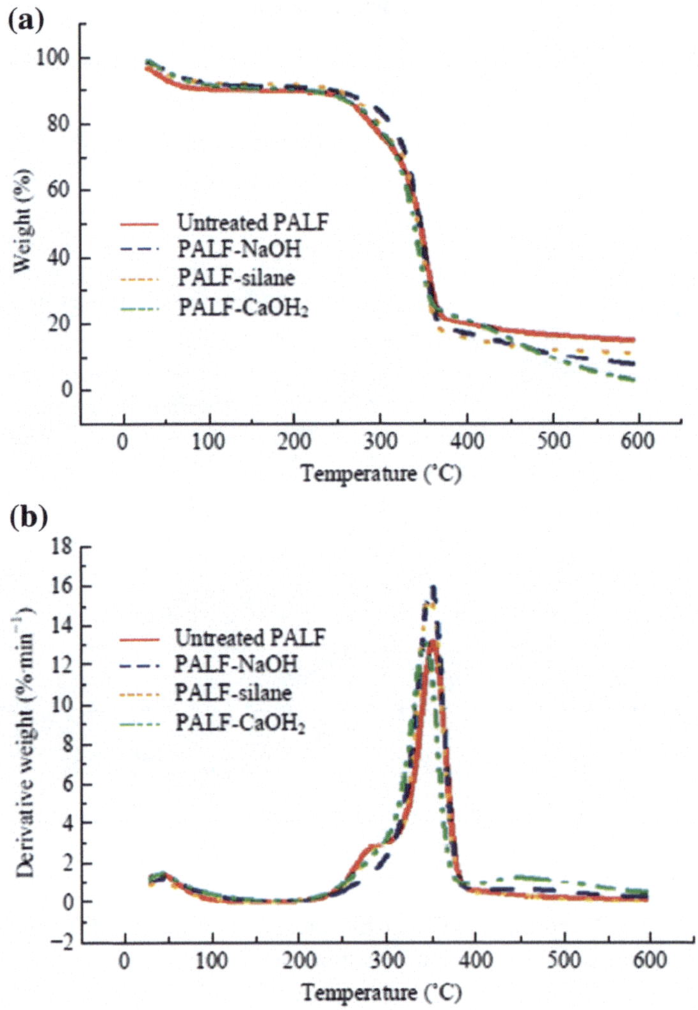

Fig. 7 **a** TGA, **b** DTG of untreated and treated PALF [41]

attacked on fibre surfaces which destroys lignin and hemicelluloses causing fibre separation from the bundles [18]. In addition, diameters were more homogeneous and smaller in treated fires compare to raw fibres. Reductions in diameters were observed in all treated fibres; however, the influences of different treatments on fibre diameters are slightly varied. Silane treatment was more effective compared to alkali treatment in this study.

In related study, Siakeng et al. [41] reported that chemical treatment of PALF was very effective to remove surface impurities. Untreated fibres exhibit largest average diameter compared to treated fibres such as 79.4, 50.6, 47.8 and 62.78 um diameter measurements in untreated and treated with NaOH, silane and $Ca(OH)_2$, respectively. This was due to removal of dirt and impurities from the fibre by treatments [18, 40, 49]. Silane treatment was reported the most effective one followed by NaOH treatment. In addition, higher chemical concentrations and longer soaking time may alter the effects of treatment in natural fibres. Predominantly, in case of NaOH, it could make the fibres more brittle [14] whereas $Ca(OH)_2$ treatment tends to cause calcium ions depositions on the surface of the fibre [22].

8 Effects of Treatment on Thermal Properties

Threepopnatkul et al. [48] studied thermogravimetric analysis of treated PALF and its composites. PALF showed its degradation in three stages: first degradation stage was indicated near 40–130 °C due to presence of moisture content in fibres; second stage was revealed at near 195–360 °C due to the decomposition of fibre constituents such as hemicellulose and cellulosic content. The final stage of thermal degradation was showed at 360–470 °C because of decomposition of lignin [15].

Siakeng et al. [41] studied thermal degradation of untreated and treated PALF and coir fibres and found two-stage thermal degradation of fibres; however, fibres treated with calcium hydroxide showed three distinct peaks. Initial weight loss was about 3–6% and detected at the temperature range from 30 to 150 °C is due to the presence of hydroxyl group [6]. The second degradation showed highest degradation among all three stages. Figure 7 showed thermal degradation as well as weight loss of all samples.

The thermal decomposition of materials determined form the points where the slopes of the TG curves begin to change drastically. The temperature of thermal degradation of untreated, alkali-treated, silane- and calcium hydroxide-treated coir fibre were 266.71 °C, 269.37 °C, 302.01 °C and 298.06 °C, respectively. The temperature of thermal degradation of untreated, alkali treated, silane- and calcium hydroxide-treated PALF illustrated at 319.31 °C, 323.58 °C, 319.47 °C and 312.32 °C, respectively. PALF showed more thermally stable than coir fibres [41]. Calcium hydroxide-treated fibres showed third peak in between 265 and 330 °C in accordance with the degradation of cellulose and lignin [37, 42]. The silane and calcium hydroxide treatment of coir fibre improved the thermal stability due to hydrolyzed substances and hemicellulose removal that degrade earlier than major cellulose and lignin. On the other hand, alkali treatment improved the thermal stability of PALF.

9 Conclusion

Natural fibres are considered as potential replacement for man-made fibres in composite materials due to their obvious advantages of being lightweight, low cost, renewable and environmentally safe. But these natural fibres like PALF do have a downside. The main problem of natural fibres is their inherent incapability to adhere with hydrophobic polymer matrices due to their strong polar character. To partially solve this adhesion problem, chemical treatments, are potentially able to overcome this incompatibility problem. Chemical treatment of natural fibre is essential to enhance compatibility between the hydrophilic natural fibres and hydrophobic matrices and also to reduce water absorption. Formation of covalent bonds seems to be the most promising approach in chemical treatments. Most research studies reviewed the effect of chemical treatments on the resulting fibre composites. Many of the chemical treatments decreased the strength properties because of disintegration of the non-cellulosic materials an also due to bond structure breakage. Alkali and silane treatments improved the modulus of the fibres.

Evidently, alkali treatment is one of the most common and efficient methods of chemical modification of PALF and it has been used to treat almost all other natural fibres with positive results. Silane treatment follows alkali in the popularity chart showing mixed results in the final composites and the fibre itself. Chemical treatment methods discussed in this paper have different effectiveness in causing adhesion between the fibre and matrix. Thus far, chemical treatments have achieved various levels of accomplishment in improving fibre and composites strength, fibre–matrix adhesion in natural fibre-based composites and overall fibre's performance. Further research must be conducted to develop novel methods of chemical treatments of PALF and examine the exact means of reactions taking place at the fibre–matrix interface.

References

1. Asim M, Abdan K, Jawaid M, Nasir M, Dashtizadeh Z, Ishak M, et al (2015) A review on pineapple leaves fibre and its composites. Int J Polym Sci
2. Asim M, Jawaid M, Abdan K, Ishak M (2017) Effect of pineapple leaf fibre and kenaf fibre treatment on mechanical performance of phenolic hybrid composites. Fibers Polym 18(5):940–947
3. Asim M, Jawaid M, Abdan K, Ishak M, Alothman O (2018) Effect of hybridization on the mechanical properties of pineapple leaf fiber/kenaf phenolic hybrid composites. J Renew Mater 6(1):38–46

4. Asim M, Jawaid M, Abdan K, Ishak MR (2016) Effect of alkali and silane treatments on mechanical and fibre-matrix bond strength of kenaf and pineapple leaf fibres. J Bionic Eng 13(3):426–435
5. Asim M, Jawaid M, Nasir M, Saba N (2018) Effect of fiber loadings and treatment on dynamic mechanical, thermal and flammability properties of pineapple leaf fiber and kenaf phenolic composites. J Renew Mater 6(4):383–393
6. Asim M, Paridah M, Saba N, Jawaid M, Alothman OY, Nasir M et al (2018) Thermal, physical properties and flammability of silane treated kenaf/pineapple leaf fibres phenolic hybrid composites. Compos Struct 202:1330–1338
7. Belgacem MN, Gandini A (2008) Surface modification of cellulose fibres. Monomers, polymers and composites from renewable resources (pp. 385–400). Elsevier
8. Cherian BM, Leão AL, de Souza SF, Costa LMM, de Olyveira GM, Kottaisamy M et al (2011) Cellulose nanocomposites with nanofibres isolated from pineapple leaf fibers for medical applications. Carbohyd Polym 86(4):1790–1798
9. Cherian BM, Leão AL, De Souza SF, Thomas S, Pothan LA, Kottaisamy M (2010) Isolation of nanocellulose from pineapple leaf fibres by steam explosion. Carbohyd Polym 81(3):720–725
10. Dey S, Bhattacharyya G, Bhattacharyya S (2005) Magic yarns from ramie and pineapple—a new dimension in 21st century. In: Proceedings of the 20th Indian Engineering Congress, p 69
11. Essabir H, Achaby ME, Hilali EM, Bouhfid R, Qaiss A (2015) Morphological, structural, thermal and tensile properties of high density polyethylene composites reinforced with treated argan nut shell particles. J Bionic Eng 12(1):129–141
12. George J, Sreekala M, Thomas S (2001) A review on interface modification and characterization of natural fiber reinforced plastic composites. Polym Eng Sci 41(9):1471–1485
13. Gurunathan T, Mohanty S, Nayak SK (2015) A review of the recent developments in biocomposites based on natural fibres and their application perspectives. Compos A Appl Sci Manuf 77:1–25
14. Hossain MK, Dewan MW, Hosur M, Jeelani S (2011) Mechanical performances of surface modified jute fiber reinforced biopol nanophased green composites. Compos B Eng 42(6):1701–1707
15. Huda M, Drzal L, Mohanty A, Misra M (2007) The effect of silane treated-and untreated-talc on the mechanical and physico-mechanical properties of poly (lactic acid)/newspaper fibers/talc hybrid composites. Compos B Eng 38(3):367–379
16. Huda MS, Drzal LT, Mohanty AK, Misra M (2008) Effect of chemical modifications of the pineapple leaf fiber surfaces on the interfacial and mechanical properties of laminated biocomposites. Compos Interfaces 15(2–3):169–191
17. Kabir M, Wang H, Lau K, Cardona F (2012) Chemical treatments on plant-based natural fibre reinforced polymer composites: an overview. Compos B Eng 43(7):2883–2892
18. Kabir M, Wang H, Lau K, Cardona F (2013) Tensile properties of chemically treated hemp fibres as reinforcement for composites. Compos B Eng 53:362–368. https://doi.org/10.1016/j.compositesb.2013.05.048
19. Kalia S, Kaith B, Kaur I (2009) Pretreatments of natural fibers and their application as reinforcing material in polymer composites—a review. Polym Eng Sci 49(7):1253–1272
20. Khan MA, Mina F, Drzal L (2000) Influence of silane coupling agents of different functionalities on the performance of jute-polycarbonate composite. In: 3rd International Wood and Natural Fibre Composite Symposium, pp 1–8
21. Ku H, Wang H, Pattarachaiyakoop N, Trada M (2011) A review on the tensile properties of natural fiber reinforced polymer composites. Compos B Eng 42(4):856–873. https://doi.org/10.1016/j.compositesb.2011.01.010
22. Le Troedec M, Sedan D, Peyratout C, Bonnet JP, Smith A, Guinebretiere R et al (2008) Influence of various chemical treatments on the composition and structure of hemp fibres. Compos A Appl Sci Manuf 39(3):514–522. https://doi.org/10.1016/j.compositesa.2007.12.001
23. Lee KY, Delille A, Bismarck A (2011) Greener surface treatments of natural fibres for the production of renewable composite materials. Cellulose fibers: bio-and nano-polymer composites (pp. 155–178). Springer

24. Lopattananon N, Panawarangkul K, Sahakaro K, Ellis B (2006) Performance of pineapple leaf fiber–natural rubber composites: the effect of fiber surface treatments. J Appl Polym Sci 102(2):1974–1984. https://doi.org/10.1002/app.24584
25. Meon MS, Othman MF, Husain H, Remeli MF, Syawal MSM (2012) Improving tensile properties of kenaf fibers treated with sodium hydroxide. Procedia Eng 41:1587–1592
26. Mishra S, Mohanty AK, Drzal LT, Misra M, Hinrichsen G (2004) A review on pineapple leaf fibers, sisal fibers and their biocomposites. Macromol Mater Eng 289(11):955–974
27. Mohanty A, Misra MA, Hinrichsen G (2000) Biofibres, biodegradable polymers and biocomposites: an overview. Macromol Mater Eng 276(1):1–24
28. Mwaikambo LY, Ansell MP (2002) Chemical modification of hemp, sisal, jute, and kapok fibers by alkalization. J Appl Polym Sci 84(12):2222–2234
29. Nam TH, Ogihara S, Tung NH, Kobayashi S (2011) Effect of alkali treatment on interfacial and mechanical properties of coir fiber reinforced poly (butylene succinate) biodegradable composites. Compos B Eng 42(6):1648–1656
30. Nasir M, Khali D, Jawaid M, Tahir P, Siakeng R, Asim M et al (2019) Recent development in binderless fiber-board fabrication from agricultural residues: a review. Constr Build Mater 211:502–516
31. Nirmal U, Singh N, Hashim J, Lau ST, Jamil N (2011) On the effect of different polymer matrix and fibre treatment on single fibre pullout test using betelnut fibres. Mater Des 32(5):2717–2726
32. Omojasola PF, Jilani OP, Ibiyemi S (2008) Cellulase production by some fungi cultured on pineapple waste. Nat Sci 6(2):64–79
33. Panyasart K, Chaiyut N, Amornsakchai T, Santawitee O (2014) Effect of surface treatment on the properties of pineapple leaf fibers reinforced polyamide 6 composites. Energy procedia 56:406–413
34. Puglia D, Monti M, Santulli C, Sarasini F, De Rosa IM, Kenny JM (2013) Effect of alkali and silane treatments on mechanical and thermal behavior of phormium tenax fibers. Fibers Polym 14(3):423–427. https://doi.org/10.1007/s12221-013-0423-x
35. Rahman MA (2011) Study on modified pineapple leaf fiber. J Text Appar, Technol Manag 7(2)
36. Ray D, Sarkar B (2001) Characterization of alkali-treated jute fibers for physical and mechanical properties. J Appl Polym Sci 80(7):1013–1020
37. Rosa MF, Chiou BS, Medeiros ES, Wood DF, Mattoso LH, Orts WJ et al (2009) Biodegradable composites based on starch/EVOH/glycerol blends and coconut fibers. J Appl Polym Sci 111(2):612–618
38. Samal R, Acharya S, Mohanty M, Ray M (2001) FTIR spectra and physico-chemical behavior of vinyl ester participated transesterification and curing of jute. J Appl Polym Sci 79(4):575–581
39. Samal R, Ray MC (1997) Effect of chemical modifications on FTIR spectra. II Physicochemical behavior of pineapple leaf fiber (PALF). J Appl Polym Sci 64(11):2119–2125
40. Sawpan MA, Pickering KL, Fernyhough A (2011) Effect of fibre treatments on interfacial shear strength of hemp fibre reinforced polylactide and unsaturated polyester composites. Compos A Appl Sci Manuf 42(9):1189–1196
41. Siakeng R, Jawaid M, Ariffin H, Salit MS (2018) Effects of surface treatments on tensile, thermal and fibre-matrix bond strength of coir and pineapple leaf fibres with poly lactic acid. J Bionic Eng 15(6):1035–1046
42. Siakeng R, Jawaid M, Ariffin H, Sapuan S (2018b) Thermal properties of coir and pineapple leaf fibre reinforced polylactic acid hybrid composites. In: IOP Conference Series: Materials Science and Engineering, vol 368, no 1, p 012019
43. Siakeng R, Jawaid M, Ariffin H, Sapuan S (2019) Mechanical, dynamic, and thermomechanical properties of coir/pineapple leaf fiber reinforced polylactic acid hybrid biocomposites. Polym Compos 40(5):2000–2011
44. Siakeng R, Jawaid M, Ariffin H, Sapuan S, Asim M, Saba N (2019) Natural fiber reinforced polylactic acid composites: a review. Polym Compos 40(2):446–463
45. Sreenivasan S, Ibraheem SA, Sulaiman S, Baharudin B, Ariffin MK, Abdan K (2014) Evaluation of combined treatments of natural fibers: kenaf, abaca and oil palm fibers using micromechanical and SEM methods. In: Advanced Materials Research, vol 912, pp 1932–1939. Trans Tech Publication

46. Terpakova E, Kidalova L, Eštoková A, Čigášová J, Števulová N (2012) Chemical modification of hemp shives and their characterization. Procedia Eng 42:931–941
47. Thongsang S, Sombatsompop N (2006) Effect of NaOH and Si 69 treatments on the properties of fly ash/natural rubber composites. Polym Compos 27(1):30–40
48. Threepopnatkul P, Kaerkitcha N, Athipongarporn N (2009) Effect of surface treatment on performance of pineapple leaf fiber–polycarbonate composites. Compos B Eng 40(7):628–632
49. Yan L, Chouw N, Yuan X (2012) Improving the mechanical properties of natural fibre fabric reinforced epoxy composites by alkali treatment. J Reinf Plast Compos 31(6):425–437. https://doi.org/10.1177/0731684412439494
50. Yusof Y, Yahya SA, Adam A (2015) Novel technology for sustainable pineapple leaf fibers productions. Procedia CIRP 26:756–760

Chemical, Physical and Biological Treatments of Pineapple Leaf Fibres

F. N. M. Padzil, Z. M. A. Ainun, Naziratulasikin Abu Kassim, S. H. Lee, C. H. Lee, Hidayah Ariffin and Edi Syams Zainudin

Abstract Pineapple leaves are known as organic wastes that left behind after pineapple fruit have been harvested. In Malaysia, waste management of these leaves is improving time to time, whereby the leaves are collected and consigned for research and industry utilization. Based on studies, pineapple leaf contains an amount of 2.5–3.5% of fibres that layered by hydrophobic waxy substances. The fibres of pineapple leaf (PALF) are extracted and beneficial in textile industry since eighteenth century. In order to optimised the usage of the PALF in high technology application which not only in textile industry, thus, numerous chemicals, physicals and biological or even combination of fibre treatments are applied by researchers and industrial players. For instance, the PALF is recognized as suitable candidates as reinforcing agent in polymeric matrices due to its high specific strength and sustainability. It is proved that attributable to inexpensive, abundant and good mechanical strength obtained by controlling the treatment methods has positioned the PALF as popular fibres in the development of functionalized smart and intelligent products.

Keywords Biocomposite · Biological treatment · Chemical treatment · Pineapple leaf fibres · Physical treatment · Sustainability

---

F. N. M. Padzil · Z. M. A. Ainun (✉) · N. Abu Kassim · S. H. Lee · C. H. Lee · H. Ariffin
Laboratory of Biopolymer and Derivatives, Institute of Tropical Forestry and Forest Products (INTROP), Universiti Putra Malaysia, 43400 UPM Serdang, Selangor, Malaysia
e-mail: ainunzuriyati@upm.edu.my

F. N. M. Padzil
e-mail: farahnadia@upm.edu.my

E. S. Zainudin
Department of Mechanical and Manufacturing Engineering, Faculty of Engineering, Universiti Putra Malaysia, 43400 Serdang, Selangor, Malaysia

M. Jawaid et al. (eds.), *Pineapple Leaf Fibers*, Green Energy and Technology,
https://doi.org/10.1007/978-981-15-1416-6_5

1 Pineapple Leaves Fibre (PALF)

In recent days, the field of research has expanded in line with the development of technology in various fields globally which more focus towards eco-friendly technology and products as well as environmental preservation. Thus, a lot of research involved with natural-based material. The utilization of natural plant fibre is very common in producing versatile and sustainable high value-added products. The natural fibres mostly offer several advantages and successfully attract many researchers, including industries from various fields like textile, automobile, building, medical, pulp and paper sectors. The natural fibre is a common term which covers a wide range of animal, mineral and vegetable fibres, while for composite industry; they are referred as agro-based and wood fibre seed, bast, leaf and stem fibres. Few examples of natural fibre used as raw materials are cotton, kenaf, rice husk, oil palm, banana, sisal, bamboo, coir, pineapple and others. Most of them are abundantly available, sustainable, biodegradable, renewable, low cost and eco-friendly compared to synthetic fibres which are normally vice versa [49]. Even though natural fibres usually portray amazing mechanical and physical properties, it may vary depends on the plant species, source, geography, breed and so on.

Pineapple plant (*Ananas comosus* (L.) Merr.) from Bromeliaceae family is originated from tropical South America and also widely grown in other parts of the world that include all tropical and subtropical areas. The pineapple is a monocotyledonous, xerophytic, perennial plant, with tough and spiny-tipped leaves which includes fragile dusty bloom on the underside as well as the waxy upper surface. Being the third most important tropical fruit in the world after citrus and banana, the pineapple has a great potential in many applications. In year 1960s and 1970s, Malaysia was one of the top players in the pineapple industry among three main pineapple producing countries after the Philippines and Costa Rica. However, in years afterward, Malaysia was not included as the main producer anymore and has been replaced by other countries. The top ten pineapple producers supplying approximately 70.5% pineapple yield around the world have been entitling towards Colombia, Mexico, China, Nigeria, India, Indonesia, Thailand, Philippines, Brazil and Costa Rica [1]. Pineapple in Malaysia is considered as a commodity and industrial top grown crop with high prospective for international commercialization [40]. In 2017, it is planted in a large area of around 12,898 ha with approximately 340,722 metric tons estimated production which takes up almost 13–15 months to grow with around 1–2 m height and 1–2 m width. The pineapple production has been decreased up to 6.3% for 2017 from a year before due to the limitation of plantation area. The largest growth areas are in Johor, Sabah and Sarawak with 84% from the whole state in Malaysia and mainly produced in Johor [15].

There are seven component structures of the pineapples mainly are the peduncle, the stem, the multiple fruit, the crown, the shoots, the roots and the leaves as shown in Fig. 1.

PALF is one of the enormously available discarded materials in most countries including Malaysia and has various potential applications that still can be explored

Fig. 1 Morphological structures of pineapple

and utilized [6]. One of the PALF good features is that it does not wilt and low water loss during carbon dioxide absorption due to stomata opening at night instead of during the day in which decarboxylated process took place by the accumulated malic acid. This happens because of crassulacean acid metabolism (CAM) feature of PALF. The interesting fact is that PALF has almost similar cellulose content with cotton about 70–82% and it is higher than other natural resources like coir, banana as well as oil palm frond stem fibres. Therefore, in terms of mechanical and physical strength of matrix reinforced biocomposites application can be improved due to the high cellulose content and its lower microfibrillar angle [5]. Besides that, other compositions of PALF are 5–12% lignin and 1.1% ash [53].

There are several benefits of PALF which are low lignin content, high aspect ratio, high fineness index, biodegradable, environmentally friendly, renewable, high stiffness, specific strength and low mass type [29, 39]. PALF also attributed as a waste for many farmers after harvesting season. However, it also brings a new source of income for their utilization. Prior to any application using PALF, the extraction method plays an important role in order to obtain its long fibre which commonly used in the manufacture of non-woven mats, yarn, woven knitted and woven fabrics. This is also including handmade products as well as animal feed and pharmaceuticals products from its mucilage as their by-product from extraction [28].

2 Chemical Composition

Based on Ghosh et al. [22], PALF is comprised of cellulose, hemicellulose, lignin and other extractives like wax, fat, pectin, colouring matter, pentosan, uronic anhydride and inorganic substances. Table 1 shows the main chemical composition of several natural fibres. PALF consists of high cellulose with low lignin content which provides more strength to the fibre as it is a main fibre structural component [44].

Table 1 Chemical composition of natural fibres

Fibre	Cellulose (%)	Hemicellulose (%)	Lignin (%)
Pineapple	80–81	6–19	4.6–12
Banana	60–65	6–19	5–10
Cotton	82–96	2–6	0.5–1
Jute	51–84	12–20	5–13
Ramie	68–76	13–15	0.6–1
Hemp	70–92	18–22	3–5
Flax	72–84	16–18	0.6–5

Source Mwaikambo [36]

The PALF also contains hemicellulose that functions as filler between lignin and cellulose that consists of sugars including glucose, mannose, xylose, arabinose and galactose [23]. Despite acts as a filler, hemicellulose does not give significant contribution on the strength and stiffness of fibres or individual cells but only bound to the fibre through some ester bonding between lignin and cellulose [20]. Lignin acts as a binder agent between the fibrils and individual fibre cells forming the cell wall of highly cross-linked molecular complex with amorphous structure [33]. Lignin is a very important chemical composition of a plant since it provides protection against biological attacks, acts as water-holding capacity and strengthens the stem against gravity forces and wind. The arrangement of molecules in PALF is similar to cotton cellulosic fibres due to its high cellulose content [32].

3 Importance of Fibre Treatment

The compensations of PALF over non-fibrous materials are mainly due to acceptable properties such as its biodegradability, low density, low cost, low energy consumption, high specific strength, high filling level and non-abrasive compared to synthetic fibres [32]. However, one of the deficient factors of using lignocellulosic fibre such as PALF is related to the poor compatibility between fibres and matrixes, and intrinsic moisture absorption. One of the major chemical components in lignocellulosic fibres like PALF is cellulose that contains elementary unit anhydro-D-glucose with three hydroxyls. The hydroxyls may form hydrogen bonds via its inner and outer macromolecules that represent the hydrophilic properties of the lignocellulosic fibres. Appropriate treatments are crucial in order to minimize the issue of moisture absorption and wettability. Therefore, dimensional changes and ability to transfer stress from matrix to the fibre are possible because the internal adhesion is improved by fibre surface topology modification [50]. In addition, insufficient adhesion between polymer matrix and fibres in a composite can be reduced.

In general, the type of fibre treatment can be categorized into four which itemized as chemical, physical, biological and their combinations. Almost 60% of the treatments involved combination treatments such as chemical–physical, chemical–biological or chemical–physical–biological approaches. It is well known that conventional treatment of PALF principally occupied physical treatment such as fibre scrapping, combing, scratching, macerating, retting, milling, steaming and more. The intention covers on removing the softy surfaces of the pineapple leaves and gaining the fibres. The fibre produced from each treatment is absolutely differs in terms of chemical and strength properties. Products can be obtained by blending PALF with or without physical–chemical modifications or treatments. The PALF has a textile fibre like properties which consists of long fibre and good tensile strength [26].

4 Chemical Treatment of Pineapple Leaf Fibre

There are several treatments on the natural fibres mainly to modify fibre parameters, improve processing results, assist the process and achieve desired properties of the specific applications. For PALF, there is a particular treatment need to be done prior any process which is degumming. The fibres are bunched together by gummy substances even after being extracted from the leaves. Basically, the content of the gum is the pectineus matter associated with the fibre which will become an inhibitor further fibre wet processing. Therefore, it is important to remove the gum that usually through retting process or chemical/biochemical methods namely as degumming. The degumming could be done either by enzymes, microorganisms or chemicals. The results after the treatment are enhancement in mechanical and physical properties [26].

4.1 Chemical Degumming

The most prominent chemical treatment for natural fibres is alkaline treatment. The treatment using alkali resulted in fibre swelling which native cellulose I crystalline structure transforms into cellulose II. The transformation into cellulose II can be attributed by temperature, treatment time, concentration and alkali type, as well as fibres nature [25]. The main difference in both celluloses is its alignment. Cellulose I is in parallel direction, whereas cellulose II is in antiparallel direction [14]. The crucial part in alkaline degumming of PALF is NaOH (caustic soda). It is used to remove the gum content in the PALF. According to Munawar et al. [35], 2–4 w/v% of NaOH solution is used to treat extracted fibres at 95 °C for 2 h. The treated fibres were washed with water and neutralization with mild acid. Certain amounts of hemicelluloses, lignin, wax and oils from the fibre will be removed through this process. The hemicellulose removal from the decorticated fibre can be seen as the peak decline. Beside tensile modulus, strength and absorbency of the treated PALF are increased over untreated fibres. Other than degumming process, acids like oxalic and sulphuric acids in combination with detergent can also be used for fibre extraction

[26]. Both processes are cost effective and simple, but less eco-friendly. Another drawback is that if the process is not conducted in proper manner, cellulose contents of the fibre may adversely affected.

4.2 Chemical Modification of PALF

Fibre modification may lead an enhancement in fibre properties. There are several treatments to modify the fibre like physical, chemical and thermal treatments in order to improve the fibres mechanical properties. Common chemical treatment on PALF is alkaline treatment that assists in fibre swelling and lead to increment of moisture absorption. This treatment will remove lignin, hemicellulose and pectin components. The untreated and chemically treated PALF have multicellular structure and ultimate fibre structures which can be observed via morphological observation [42]. Fibre strength becoming weak as cellulose has contact with strong acids. At this point, the cellulose is called hydrocellulose. However, not much effect occurs if the cellulose has contact with weak acids. There are several methods to remove lignin. One of the chemical methods which can be carried out is treatment with sodium chlorite at pH 4 [51]. Previous studies proved that high delignification will not split the fibre into ultimate cells. In addition, surface features of the fibres can progressively improve with the removal of waxes, pectins and gums [26]. Depend on targeted application, some features of the fibres required modification. In instance, the moisture absorption capacity of the PALF should be reduced particularly for the making of composite materials. This can be done by applying chemical treatments like acetylation. Surface modification can be conducted using grafting method. Grafting will enhance the mechanical properties and thermal stability of the fibres. The main purpose of fibre surface modification is to improve the bond strength between the matrix and fibres, especially in composite application [45].

PALF is a natural fibre which has a hydrophilic nature consists of abundant hydroxyl group. It needs chemical modification in order to enhance interfacial properties between resin and fibre by reducing the hydrophilic tendency of the reinforcing fibre (Zin et al. 2018; Rajesh Bapu et al. 2015). The common chemical treatments that have been done previously are alkali and silane treatments on the fibre which will provide the best result for the composite production. There are several previous studies that have been done using alkali treatment which shown in Table 2.

5 Physical Treatment of PALF

Physical treatment of PALF starts from fresh leaves and utilize mechanical action to separate long fibres from the soft covering materials. Conventional methods of physical treatment usually provide low yield, coarse fibre bundles and hard to scale up. These deliverables are mostly influenced by the manual activities along the flow

Table 2 Previous studies on alkali treatment of natural fibre for reinforcement materials

Composite			
PALF/PLA	5% NaOH at RT for 1 h	Superior mechanical properties – Flexural strength: 107.53 MPa – Flexural modulus: 0.32 GPa	Ramli et al. [48]
PALF/polyamide 6	5% NaOH at RT for 5 h	– Tensile strength: 73.5–78.2 MPa – Young modulus: 1540–2040 MPa	Panyasart et al. [43]
Kenaf/PLA	5% NaOH at RT for 2 h	– Flexural strength: 8.3 GPa (34% increment compared to untreated sample) – Flexural modulus: 56.3 MPa (48% increment compared to untreated sample)	Huda et al.
Sugar palm-epoxy	2.5–5% NaOH for 1–8 h	– Tensile strength: 16.4% increment compared to untreated sample – Tensile modulus: 20.4% increment	Bachtiar et al. [7]
Flax-epoxy	0.5–10% NaOH at RT	– Tensile strength and modulus: 30% increment compared to untreated sample	Li et al. [31]

chain in extracting the PALF. Banik et al. [9] applied scraping-macerating-retting technique in extracting PALF. They developed a machine that comprised with metal knife scrapper and serrated rollers that can scrap waxy layers. The leaf surfaces were then macerated and broken into form that acceptable for easy retting. Finally, the PALF was properly extracted by retting the macerated fibres in water. In another study by Das et al. [16], peeling-scrutching-scrapping-retting-decortication technique was used to obtain fibres from pineapple leaves which began with peeling, scrutching and scrapping of green leaves. Later, the fibres have undergone retting in water and mechanical decortication. In order to possible the decortication process, a handy mechanical scratcher of sisal decortication was altered. Therefore, an amount of 1500 kg of green leaves can be transformed into PALF in a day. Kengkhetkit and Amornsakchai [27] introduced mechanical milling by employing mechanical force to crush the leaves into paste. The soft covering material was broken down into fines while the fibres still resilient to the force. It is claimed as a simple method and easy to be scaled up for larger scale of production mainly short PALF. An interesting physical treatment was presented by Cherian et al. [13] whom applied steam explosion process in extracting cellulose nanofibrils from PALF which effectively depolymerized and defibrillated the fibres. The nanocellulose of extracted from PALF could be applied in tissue engineering, medical implants, wound dressings and drug delivery of biomedical and biotechnological applications [13].

6 Biological Treatment of PALF

Biological treatment is becoming an important technology due to its advantages in environmentally friendly impacts and economic wise. It is applied in various industries such as textile, pulp, paper, composites and more. In instance, enzyme is produced from living organisms by conducting a role as biocatalysts. The process is known as biochemical represents two parts of protein substances known as enzyme carrier and active part. Noteworthy properties of enzymes are its catalytic power, specificity scope and choice of substrates. Enzymes act as accelerator in changing substrate into targeted product. An enzyme particle interacts with other particles in substrate which velocity of changes or activity depends on process conditions. Enzymes can be obtained from various microorganisms and bacteria. *Aspergillus niger, Botrytis cinereal, Erwinia carotovora* and *Rhizopus artocarpi* are among microorganisms used as enzymes source. Besides that, enzymes can be sourced from bacteria such as *Clostridium felsineum*. Related to its biodegradability in effluents therefore it complied the need of green chemistry.

6.1 Textile Industry

Enzyme was gradually employed in the textiles industry for certain functions explicitly for desizing, scouring, polishing, washing, degumming, bleaching and decolouring of dyehouse wastewater since decades. Application of enzymes highpoints the lessening in water and energy consumption. In addition, toxic chemical agents can be replaced and wastewater can be recovered to be reused which satisfy environmental requirement and economic aspects.

In textile industry, scoring process aimed to remove dirt and impurities to ease the following processes but can also be carried out at the end stage to increase fabric wettability. Two popular enzymes used in textile industry are cellulase and pectinase can be employed either separately or combination showing adequate absorbency of fibres with short treatment times [30]. Other enzymes used in fibre biological treatment such as pectinase, lipase and protease enzymes may improve water wettability and strength retention [24]. In improving water absorbency of the treated fibres, pectinase worked better than lipase or protease [11]. Cellulase can remove accessible fibrils which resulting smoother and softer fabric of textile [12].

Sricharussin et al. [56] scoured pineapple leaves with pectinase and cellulase before dying the fibres using natural dyes. In his study, metal salts mordants were replaced with natural dye garments because of environmental concerns and satisfaction demand. Series of samples comprised conventional and enzymatic scouring were prepared. Test listed as morphological observation, tensile strength, weight losses, dye uptake, colour measurements and opacity were carried out. It was proved that enzymes progressed to higher dye exhaustion and tensile strength.

Enzymes are also employed in almost all steps and purposes in such industry. The enzymes such as amylases are used in desizing, cellulases for denim finishing, laccases for decolourization of textile effluents and bleaching, and proteases aid in detergent formulations. New developments in enzymology that are more environmentally friendly are being developed time to time [3].

6.2 *Delignification in Pulp and Papermaking*

Biological treatment is a bioremediation method in which microorganisms are utilized to degrade and metabolize chemical substances. It is known as cheap, sustainable and benign treatment to the environment that divided into biostimulation and bioaugmentation [63]. Applying biological treatment of PALF may soften the pineapple leaves by using enzyme, fungal or microbe extracts. In other words, the aim of conducting enzymatic treatment is to alter the physical properties of PALF by removing the cementing layers of lignin in the fibres. Sedelnik [52] explained the usage of biological fibre treatment is to separate technical fibre bundles into smaller units and transform them to becoming softer. The fibre treatments can be subjected to period of enzyme, culture and anaerobic treatments. Therefore, the mechanical properties such as tenacity (g/tex) and average peak elongation can be improved as well as surface smoothness. However, enzymatic treatment was investigated to be less performed compare to alkali treatment in removing the lignin content besides hemicellulose and waxes [63].

Depletion of wood sources to be used in pulp and papermaking attract many attentions included industry players. An alternative of fibre sources arises from abundance of agro-based lignocellulosic materials such as PALF. The PALF displayed outstanding and excellent paper properties [37] due to its high cellulose content (70–82%) [10]. Biological pulping or biopulping is another popular pulping method compared to conventional chemical pulping. Lignocellulosic materials consist of lignin, hemicelluloses and cellulose are exposed to natural wood decay fungi attack such as white-rot fungi prior to thermochemical pulping.

Singhal et al. [55] examined that biopulping degraded holocellulose and depolymerized lignin significantly. By removing the lignin or delignification, PALF converted into pulps. Nayan et al. [38] proved that biopulping of PALF can be efficiently carried out by applying *Ceriporiopsis subvermispora*. The produced paper achieved desirable tensile and tear indices with uniform morphological fibre structures at 0.3% of fungus treatment. In addition, Abd Razak et al. [2] enhanced PALF sheets with a biopolymer called poly (lactic acid) (PLA) for better mechanical strength and hydrophobicity as a green packaging material.

Moya et al. [34] conducted a study on PALF by subjecting the fibres to colonization of *Pleurotus ostreatus* and *Trametes versicolor* in order to produce biopulp. By expanding the exposure period to fungal colonies from 2–6 weeks to 4–8 weeks, yield of biopulps increased 35–50% to 55–70% accordingly. Longer colonization period simultaneously increased holocellulose and reduced lignin and extractives content. Moreover, colonization by *Pleurotus ostreatus* and *Trametes versicolor* led the biopulp to change into whiteness based on the reading of CIELAB. Lightness (parameter L*) arising greater while leaving the decrement in green (parameter a*) and yellow (parameter b*).

6.3 Reducing Sugar in Biofuel Production

Due to ease obtainability, low price and high holocellulose content, lignocellulosic material is well known as potential biomass for biofuel production. Oil palm frond [61], poplar [46], switch grass [59] and barley straw [41] have been examined their performances in producing bioenergy. Another lignocellulosic waste is pineapple leaf which recognized as one of worldwide pineapple production abundance excess. Currently, the leaf is used to make rope and paper. Pineapple leaf is also found as having excellent potential for bioethanol production. Banerjee et al. [8] applied laccase to increase the enzymatic digestibility of the substrate. This was carried out in order to reduce sugar content in the PALF. They investigated that maximum delignification of 78.57% (w/w) resulted in reducing sugar of 492.33 ± 3.1 mg/g within 5.30 h. Laccase mediated delignification of pineapple leaf waste is found as cleaner sustainable process which sugar is reduced so that it can be potentially meet the requirement for biofuels.

7 Application of Treated Pineapple Leaf Fibre

In practice, after harvest pineapple fruit, the PALF will be burned which usually involve open burning since they are in bulk quantity. This activity imposes a negative impact on the environment due to smoke and haze. Many approaches have been done to cater PALF, the post-harvest agricultural waste into high-end products. There are several good products from PALF which have been commercialized globally in various industries such as pulp, biocomposite, paper, and textiles as well as butanol production mainly due to high cellulose, hemicellulose and other carbohydrates content.

7.1 Filler in Food Packaging

Since few decades, the food packaging issue successfully attracts attention of many researchers or stakeholders due to environmental problem that emerge around the world particularly on marine debris issue, non-biodegradable packaging or non-renewable resource materials. The global focus has shifted to natural renewable resources material that can sustain for long period of time. Therefore, numerous researches have been conducted using natural resources or agricultural waste to develop safe, versatile, sophisticated high-end food packaging products as shown in Fig. 2. The well-known conventional source of plastic-making, petroleum is able to be replaced by bio-nanocomposite material that was derived from tapioca starch with plasticizer glycerol as a matrix and PALF as filler for food packaging. This is a crucial part which uses other alternatives that is more environmentally friendly and due to limited availability of petroleum nowadays. Besides, it would reduce the level of CO_2 that pollute air and the PALF is easier to be decomposed by bacterial (biodegradable). Moreover, its mechanical properties are better if we compared with fibre glass. PALF can be used as reinforcement material for various types of polymer up to composite material as known as biocomposites prior several pretreatments [].

By utilizing the PALF as a filler bio-nanocomposite packaging, it assists in making an alternative product for synthetic plastics packaging as well as reduce the agricultural waste by transforming it into useful innovative products. Moreover, PALF is a great replacement for petroleum-based food packaging which needs longer time to renew their resources besides being an environmentally friendly as well as widely available raw material that similar to starch, chitin and chitosan [39].

Fig. 2 Versatile biodegradable food packaging

7.2 PALF/ABS Composite

Natural fibre is also well known in composites area as biocomposite. The characteristics of the fibre which improve mechanical properties of the composite, cost-saving and productivity improvement have gathered a lot of interest among researchers in various applications. For instance, acrylonitrile-butadiene-styrene terpolymer (ABS) plastics possess several prominent properties such as strength, rigidity and toughness. ABS is resistance to chemicals and endurance provision for wide ranges of temperature usage. There will be different effects on the ABS properties from each component of terpolymer. Heat and chemical resistance are attributed from acrylonitrile. While impact properties obtained from butadiene due to its rubbery phase. Therefore, the outstanding properties of ABS plastics could be tailored-made by adjusting the proportion of its components [57]. The ABS composite with natural resources still lacking in terms of in-depth research and development, mostly been carried out using wood sawdust as reinforcements. However, as mentioned earlier, natural fibre reinforced composites are getting more demands nowadays. Enormous number of natural fibres is available globally such as PALF which portrays excellent mechanical properties because of its high cellulose content of more than 70% [58]. This composite can be applied in automobile sector as compartments of the vehicles as shown in Fig. 3.

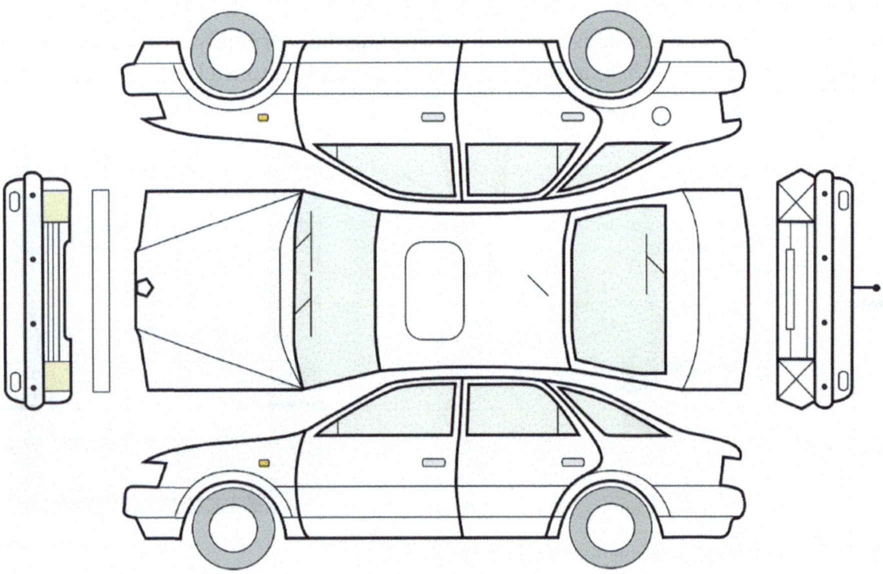

Fig. 3 Vehicle compartments from biocomposite

Fig. 4 Paper productions from lignocellulosic materials

7.3 Paper Production Using PALF

Figure 4 shows paper products which are widely used everywhere, anywhere and all-time favourite thing that can be seen in many sectors, especially in working or educational environments that need a lot of records during the process. The fast-growing technology nowadays has reduced the paper-used, but it is still crucial for certain sectors like cleaning, educational and packaging. Paper is mainly produced from cellulosic fibres and possesses an array of specialized function. Cellulosic fibres commonly obtained from wood resources, as the main raw material in global pulp and paper production. This is tremendously contributed to depletion of forest resources, thus resulted in negative impact on environment [4]. Recently, around 50% of industrial timber production is used in Europe pulp production industry. While approximately 38% of the global production is domineering by the main pulp and paper manufacturer, USA, followed by Canada with 15%, Japan 7.5% and Sweden 6% [54].

For the past few decades, the wood raw material has been transferred to abundance non-wood lignocellulosic fibres with less devastated environmental impact compared to wood for pulp resources. Numerous non-wood lignocellulosic fibre or even agricultural waste like banana fibre, carpet grass, kenaf and so on have previously been commercialized as substitute [60]. PALF is still considerably new and lacking in terms of research in pulp and paper industry. However, PALF consists of high cellulose content almost like cotton. Pineapples cultivated under selective breeding were researched to have higher cellulose content than wood fibre, and hence indicated that such non-timber fibre makes good substitutes. Thus, there will be less depletion of forest and other advantages of PALF are low lignin content which contributes in low chemical and energy usage during pulping process. Moreover, pineapple is cultivated in a short growth cycle [4].

7.4 PALF Turns into Valuable Textiles

In textile sector, PALF is not a new resource anymore. It is widely exploited in various kinds of textiles. Even in late 1970s, Ghosh and Sinha [21] understood and assessed the textile value of PALF. They studied and did a lot of research in detail. One of their studies is about the possibility of spinning and further weaving of PALF in jute/flax spinning system which targeting technical textile material development and fashion. However, some efforts have been made in recent days to utilize PALF for different commercial values. As fashionable textile fibre, PALF is graded in between jute and ramie or jute and cotton. All textile properties are portrayed by PALF and well blended with cotton, ramie, jute and some other synthetic fibres [17]. Thus, PALF is considered as potential commercial-grade textile fibre among natural fibres; however, there is need of its assured supply to textile processing industry in adequate quantities.

As previously mentioned, technical textile material development and fashion have developed plain and twill woven cloths from the pineapple-jute-blended yarn and pineapple yarn. These fabrics are also known for its lightweight properties which suitable for curtain, furnishing fabrics as well as fashion bag. Therefore, they reach the agreement that concludes these PALF or PALF-blended products have vast potential for sustainable development of fashion fabrics. Ghosh et al. [22] have made a significant amount of research on processing PALF in cotton machinery. They studied and compared the physical and mechanical properties of PALF, jute and cotton prior processing in cotton spinning system. The results showed that PALF is not 100% possible to spin into yarn in cotton spinning machinery.

The yarn tenacity declines significantly while increasing the proportion of PALF more than 50% in the cotton blend during blending process. There are also other attempts to study acrylic fibre and PALF blending in jute spinning system performance in terms of their similarities [18]. In regard to various research papers, Doraiswami and Chellamani [19] concluded that PALF has immense potential to develop luxury textiles for the sustainable development of natural fibre-based products. The whole process is covered starting from fibre extraction to the final yarn and fabric from PALF and its blends. The final product might be costly since the eco-friendly in nature and by-product utilization point must be taken into consideration.

PALF that is used as fashionable textiles is commercialize high-end products which have huge demand in the global market. Recent days, unlike previous decades, the socio-economic sound people globally are giving more preferences in the natural fibre-based fashionable textile apparels for which then can afford even much higher price. This led to numerous fashion shows and other promotional advertisements about sustainable fashion textile products from natural fibres for instance PALF. Nevertheless, apart from fancy and fashion apparels, there is big demand market on shopping bag, curtains and furnishing materials can be developed from this environmentally friendly sustainable PALF. Besides that, the PALF-blended products also have equally demand which can minimize the dependency on the man-made or synthetic fibres. Blended even the waste or by-product generated during the extraction of PALF has enough potential in the area of agricultural manure sustainable [17].

8 Conclusion

PALF has outstanding physical and mechanical properties as well as high strength despite having large diameter size which possible to convert into high value-added products. Moreover, PALF is considered to be economic, recyclable, biodegradable and environmentally friendly material, especially as reinforcement material in bio-composite or biopolymer. However, due to the hydroscopic nature of PALF, it would be quite challenging to utilize it as a filler by its own form. Therefore, there are several surface modifications or fibre treatments which should be done before being proceeded with the whole process. The properties of modified PALF are able to surpass or partially substitute the synthetic fibres for certain applications. There are tremendous applications related to composite nowadays compared to grand textiles sector which is a good step in expanding the utilization of PALF. In light with that, numerous in-depth studies need to be conducted to improve the different functional and properties in order to suit with specific application. Apart from previous holistic study, a lot of concerted effort is required to commercialize the utilization of PALF and corporation of it in other potential products in different sectors like compartment for vehicles or others. Thus, it will lead to increment in terms of country's socioeconomy.

References

1. Ab Halim N (2016) Policy intervention for the development of the pineapple industry in Malaysia. Agricult Mark Policy. http://ap.fftc.agnet.org/ap_db.php?id=716&print=1 Accessed 13 Feb 2019
2. Abd Razak SI, Ahmad Sharif NF, Mat Nayan NH et al (2015) Impregnation of poly (lactic acid) on biologically pulped pineapple leaf fiber for packaging materials. BioRes 10(3):4350–4359
3. Araujo R, Casal M, Cavaco-Paulo (2008) Application of enzymes for textile fibres processing. Biocatal Biotransform 26(5):332–349
4. Aremu MO, Rafiu MA, Adedeji KK (2015) Pulp and paper production from nigerian pineapple leaves and corn straw as substitute to wood source. Int Res J Eng Technol 2(4)
5. Asim M, Jawaid M, Abdan K et al (2018) Effect of alkali treatment on physical and mechanical strength of pineapple leaf fibres. IOP Conf Ser: Mater Sci Eng 290:012030
6. Asim M, Khalina Abdan M, Jawaid M et al (2015) A review on pineapple leaves fibre and its composites. Int J Polym Sci 1–16
7. Bachtiar D, Sapuan SM, Hamdan MM (2008) The effect of alkaline treatment on tensile properties of sugar palm fibre reinforced epoxy composites. Mater Des 29(7):1285–1290
8. Banerjee R, Chintagunta AD, Ray S (2019) Laccase mediated delignification of pineapple leaf waste: an ecofriendly sustainable attempt towards valorization. BMC Chem 13:58
9. Banik S, Nag D, Debnath S (2011) Utilization of pineapple leaf agro-waste for extraction of fibre and the residual biomass for vermicomposting. Ind J Fibre Text Res 36:172–177
10. Buana MSAS, Pasbaskhsh P, Goh KL et al (2013) Elasticity, microstructure and thermal stability of foliage and fruit fibres from four tropical crops. Fibers Polym 14(4):623–629
11. Buchert J, Pere J, Puolakka A et al (2000) Scouring of cotton with pectinases, proteases, and lipases. Textile Chem Colorist Am Dyestuff Report 32(5)
12. Buschle-Diller G, Fanter C, Loth F (1999) Structural changes in hemp fibers as a result of enzymatic hydrolysis with mixed enzyme systems. Text Res J 69:244–251

13. Cherian BM, Leão AL, de Souza SF et al (2010) Isolation of nanocellulose from pineapple leaf fibres by steam explosion. Carbohydr Polym 81(3):720–725
14. Cordeiro N, Gouveia C, John MJ (2011) Investigation of surface properties of physico-chemically modified natural fibres using inverse gas chromatography. Indust Crops Prod 33:108–115
15. DOA (2017) Fruit crops statistics 2017. Department of Agriculture Malaysia
16. Das PK, Nag D, Debnath S et al (2010) Machinery for extraction and traditional spinning of plant fibres. Indian J Tradit Knowl 9(2):386–393
17. Debnath S (2016) Pineapple leaf fibre—a sustainable luxury and industrial textiles. In: Muthu S, Gardetti M (eds) Handbook of sustainable luxury textiles and fashion. Environmental footprints and eco-design of products and processes. Springer, Singapore
18. Dey SK, Nag D, Das PK (2009) New dimensions of pineapple leaf fibre—an agrowaste for textile application. In: Shukla JP (ed) New technologies for rural development having potential of commercialisation. Allied Publishers Private Limited, Delhi, pp 115–127
19. Doraiswami I, Chellamani P (1993) Pineapple leaf fibres. Text Prog 24(1):1–37
20. Dutta S, Bhattacharyya D (2013) Enzymatic, antimicrobial and toxicity studies of the aqueous extract of *Ananascomosus* (pineapple) crown leaf. J Ethnopharmacol 150:451–457
21. Ghosh SK, Sinha MK (1977) Assessing textile values of pineapple fibre. Indian Text J 88(111–115):8
22. Ghosh SK, Sinha MK, Dey SK et al (1982) Processing of pineapple leaf fibre (PALF) in cotton machinery. Text Trends 24:49–53
23. Hansen CM, Björkman A (1998) The ultra-structure of wood from a solubility parameter point of view. Holzforschung 52:335–344
24. Hartzell MM, Hsieh YL (1998) Enzymatic scouring to improve cotton fabric wettability. Text Res J 68(4):233–241
25. John JM, Thomas S (2008) Biofibers and biocomposites. Carbohydr Polym 71:343–364
26. Jose S, Salim R, Ammayappa L (2016) An overview on production, properties, and value addition of pineapple leaf fibres (PALF). J Nat Fibres 13(3):362–373
27. Kengkhetkit N, Amornsakchai T (2012) Utilisation of pineapple leaf waste for plastic reinforcement: 1. A novel extraction method for short pineapple leaf fiber. Indust Crops Prod 40:55–61
28. Leao AL, Souza SF, Cherian BM et al (2010) Pineapple leaf fibres for composites and cellulose. Mol Cryst Liq Crystals 522(1):36–41
29. Leao AL, Cherian BM, Narine S et al (2015) The use of pineapple leaf fibres (PALSs) as reinforcement in composites. Biofibre Reinf Compos Mater 211–235
30. Li Y, Hardin ZR (1997) Enzymatic scouring of cotton: effects on structure and properties. Cellulose 94(88):96
31. Li X, Tabil LG, Panigrahi S (2007) Chemical treatments of natural fibre for use in natural fibre-reinforced composites: a review. J Polym Environ 15(1):25–33
32. Misra M, Hinrichsen G (2004) A review on pineapple leaf fibers, sisal fibers and their biocomposites. Macromol Mater Eng 289:955–974
33. Mohanty AK, Misra M, Hinrichsen G (2000) Biofibres biodegradable polymers and biocomposites: an overview. Macromol Mater Eng 276–277(1):1–24
34. Moya R, Berrocal A, Rodríguez-Zúñiga A et al (2016) Biopulp from pineapple leaf fiber produced by colonization with two white-rot fungi: *Trametes versicolor* and *Pleurotus ostreatus*. BioRes 11(4):8756–8776
35. Munawar SS, Umemura K, Tanaka F et al (2008) Effects of alkali, mild steam, and chitosan treatments on the properties of pineapple, ramie, and sansevieria fiber bundles. J Wood Sci 54:28–35
36. Mwaikambo LY (2006) Review of the history, properties and application of plant fibres. African J Sci Technol Sci Eng Ser 7:120–133
37. Nayan NHM, Razak SIA, Rahman WAWA et al (2013) Effects of mercerization on the properties of paper produced from Malaysian pineapple leaf fiber. Inter J Eng Technol 13(4):1–6

38. Nayan NHM, Razak SIA, Rahman WAWA (2014) Biopulping by *Ceriporiopsis subvermispora* towards pineapple leaf fiber (PALF) paper properties. Adv Mater Res 1043:180183
39. Nikmatin S, Rudwiyanti JR, Prasetyo KW et al (2015) Mechanical and optical characterization of bio-nanocomposite from pineapple leaf fibre material for food packaging. In: International seminar on photonics, optics and its applications (ISPhOA2014), vol 9444, pp 1–6
40. Othman MH, Buang L, Mohd Khairuzamri MS (2011) Rejuvenating the Malaysia pineapple industry. Acta Horti 902
41. Panagiotopoulos IA, Lignos GD, Bakker RR et al (2012) Effect of low severity dilute-acid pretreatment of barley straw and decreased enzyme loading hydrolysis on the production of fermentable substrates and the release of inhibitory compounds. J Clean Prod 32:45–51
42. Pandey SN (2007) Ramie fibre: Part II. Physical fibre properties. A critical appreciation of recent developments. Text Prog 39:189–268
43. Panyasart K, Chaiyut N, Amornsakchai T et al (2014) Effect of surface treatment on the properties of pineapple leaf fibers reinforced polyamide 6 composites. Energy Procedia 56:406–413
44. Paster M, Pellegrino JL, Carole TM (2003) Industrial bioproducts: today and tomorrow. Report prepared for the US Department of Energy, Washington, DC
45. Payae Y, Lopattananon N (2009) Adhesion of pineapple-leaf fiber to epoxy matrix: the role of surface treatments. Songklanakarin J Sci Technol 31:189–194
46. Porzi GF, Prussi M, Chiaramonti D et al (2012) Modelling lignocellulosic bioethanol from poplar: estimation of the level of process integration, yield and potential for co-products. J Clean Prod 34:66–75
47. Rajesh Bapu TN, Kumaragurubaran SB, Venkataramanan R et al (2015) A review on effect of chemical treatment of natural fibres on mechanical properties. Inter J of Applied Engin Research 10(19):14703–14714
48. Ramli SNR, Fadzullah S, Mustafa Z (2016) Mechanical performance of pineapple leaf fiber reinforced poly lactic acid (PLA) biocomposites. Proc Mech Eng Res 131–132
49. Rana S, Pichandi S, Parveen S et al (2014) Natural plant fibres: production, processing, properties and their sustainability parameters (Chap. 1). In: Roadmap to sustainable textiles and clothing. Textile science and clothing technology. Springer, Berlin, pp 1–35
50. Rout J, Misra M, Tripathy SS et al (2001) The influence of fibre treatment on the performance of coir-polyester composites. Compos Sci Technol 61(9):1303–1310
51. Sarkar PB, Chatterjee H (1948) The bleaching of jute with chlorite. J Text Inst 39:74–81
52. Sedelnik N (2004) Properties of hemp fibre cottonised by biological modification of hemp hackling noils. Fibres Text East Eur 12(1):58–60
53. Siakeng R, Jawaid M, Ariffin H et al (2018) Physical properties of coir and pineapple leaf fibre reinforced polylactic acid hybrid composites. IOP Conf Ser Mater Sci Eng 290:012031
54. Sibaly S, Jeetah P (2017) Production of paper from pineapple leaves. J Environ Chem Eng 5:5978–5986
55. Singhal A, Jaiswal PK, Thakur IS (2015) Biopulping of bagasse by *Cryptococcus albidus* under partially sterilized conditions. Inter Biodeter Biodegrad 97:143–150
56. Sricharussin W, Ree-iam P, Phanomchoeng W et al (2009) Effect of enzymatic treatment on the dyeing of pineapple leaf fibres with natural dyes. Sci Asia 35(1):31–36
57. Threepopnatkul P, Teppinta W, Sombatsompop N (2011) Effect of co-monomer ratio in ABS and wood content on processing and properties in wood/ABS composites. Fibers Polym 12:1007–1013
58. Threepopnatkul P, Krachang T, Teerawattanan W et al (2012) Study of surface treatment of pineapple leaf fiber (PALF) on performance of PALF/ABS composites. In: 15th European conference on composite materials, Venice, Italy, pp 1–7
59. Yang Y, Sharma-Shivappa R, Burns JC, Cheng JJ (2009) Dilute acid pretreatment of oven-dried switchgrass germplasms for bioethanol production. Energy Fuels 23:3759–3766
60. Yusof Y, Ahmad MR, Wahab MS et al (2012) Producing paper using pineapple leaf fiber. Adv Mater Res 383:3382–3386

61. Zahari MAKM, Abdullah SSS, Roslan AM et al (2014) Efficient utilization of oil palm frond for bio-based products and biorefinery. J Clean Prod 65:252–260
62. Zin MH, Abdan K, Mazlan N et al (2018) The effect of alkali treatment on the mechanical and chemical properties of pineapple leaf fibres (PALF) and adhesion to epoxy resin. IOP Conf Ser: Mater Sci and Engin 368:012035
63. Zwane PE, Masarirambi MT, Thwala JM et al (2014) Enzymatic processing of plant fibres for diversified uses. In: RUFORUM fourth biennial conference, Maputo, Mozambique, 19–25 July 2014, pp 391–392

Physical, Morphological, Structural, Thermal and Mechanical Properties of Pineapple Leaf Fibers

C. H. Lee, A. Khalina, S. H. Lee, F. N. M. Padzil and Z. M. A. Ainun

Abstract Natural fibers have drawn significant attention globally for its adverse effect on the environment, lower cost and superior performance. Leaf or hard fibers are tough plant fibers, extracted from leaves of a monocotyledonous plant which has parallel-veined leaves. Pineapple leaf fibers (PALFs) are usually disposed of with an extremely low value due to lack of adequate skills. With a suitable platform, it can be fully utilized. PALF was found to be very high in cellulose contents which contribute to high strength performance. However, various factors make it perform differently. The changes in density and diameter of PALF had been found closely related to its strength. Apart from this, surface morphology of PALF reviewed that the location of leaf fiber and surface conditions provided various interlocking quality and optimum applications. On the other hand, PALF treatment observed better strength properties with evidence under infrared spectroscopy. The nanofibrils PALF from acid hydrolysis treatment provided better adhesion force and higher crystallinity index but high hydrophilicity verified by high moisture absorptions. Higher crystallinity index provided the fiber a good strength performance and an excellent spinnability, which allows it to be used in yarn and textile industries. On the contrary, high cellulose content of PALF has a promising fire-retardant behavior. PALF has a high potential for advanced material substitutions. Unfortunately, underutilized PALF is only disposed of as landfills and low-cost feedstock. The development and utilization of PALF could be the solution for the disposal problem as well as to increase the national income of a country.

Keywords Pineapple leaf fibers · Mechanical properties · Physical properties · Thermal properties · Morphological properties

---

C. H. Lee (✉) · S. H. Lee · F. N. M. Padzil · Z. M. A. Ainun
Institute of Tropical Forestry and Forest Products (INTROP), Universiti Putra Malaysia, 43300 Serdang, Selangor, Malaysia
e-mail: leechinghao@upm.edu.my

A. Khalina
Engineering Faculty, Universiti Putra Malaysia, 43300 Serdang, Selangor, Malaysia

M. Jawaid et al. (eds.), *Pineapple Leaf Fibers*, Green Energy and Technology,
https://doi.org/10.1007/978-981-15-1416-6_6

1 Introduction

Nowadays, environmental conservation and maximizing the potential of raw materials are becoming an obligation for sustainable product development. For this reason, the application of natural fibers has been the prior objective for industries and academic research. The reduction of environmental impact is achieved using natural fibers to avoid the increasing number of waste volume while reducing the production cost [15, 20, 30, 32, 31]. The natural fibers can be obtained from bast stem, leaf and fruit which naturally exist as fiber bundles. Retting and decortication are some most common methods to extract fibers from the leaf of plants.

The leaf fibers or "hard" fibers are extracted from leaves of the monocotyledonous plant which has parallel-veined leaves [45]. They are the toughest of the plant fibers which is the most likely to be used to produce rope and twine. Pineapple is one of the most popular tropical fruits in Malaysia. However, the main priority for pineapple agriculture focuses on the quality of the fruits. The other parts from the plant (roots, stems and leaves) are considered as disposable waste with an extremely low value due to lack of adequate skills and methods [34]. These harvest residues will mainly be burned or composted, yet this practice is not effective in yield improvement.

A warm and humid surrounding is the best climate to grow pineapple. The optimal temperature ranges from 20 to 30 °C or 23 to 24 °C if specifically [47, 58]. Pineapple is a herbaceous plant with the root system that may develop up to 1–2 m wide and 0.85 m depth [7]. A matured pineapple plant grows about 68–82 leaves in well-arranged structure [23]. Pineapple leaf is an agricultural waste, which is very low in value and biodegradable. With a suitable platform, pineapple leaves can generate huge profits. PALF is rich and famous in high cellulose content among the natural fibers due to its low microfibrillar angle (14°) [13]. The chemical composition and physic properties of some of the leaf's fibers are shown in Tables 1 and 2, respectively. Superior strength-to-weight ratio, abundantly available and relatively inexpensive, creates a good opportunity for PALF to be used in fiber-reinforced polymer composite, paper making, yarn production and acoustic barriers.

Compared to other natural fibers (jute, hemp, kenaf fibers), little research project is found on PALF utilization. As the outcome, PALF still contains a huge potential in additional value. The northeast region of Brazil supplies about 7% of the world's pineapple demands [23]. A large portion of Kenya's economic income depends on

Table 1 Chemical composition of some leaf's fibers [45]

	Cellulose	Hemicellulose	Lignin	Pectin
Abaca	61–64	21	12	0.8
Banana	60–65	6–10	5–10	3–5
Phormium	67	30	11	–
Pineapple	80–81	16–19	4.6–12	2–3
Sisal	43–78	10–13	4–12	0.8–2

Table 2 Physical properties of some leaf's fibers [45]

	Diameter, μm	Length, mm	Aspect ratio, *l/d*	Microfibril angle, θ	Bulk density, kg/m^3	Moisture regain, %
Abaca	17.0–21.4	4.6–5.2	257	–	1500	14
Banana	–	2–3.8	–	11–12	1300–1350	–
Phormium	15.4–16.4	5.0–5.7	337	–	–	–
Pineapple	20–80	–	–	6–14	1520–1560	–
Sisal	18.3–23.7	1.8–3.1	115	10–22	1300–1500	14

pineapple production and 90% of production from large-scale producers. Value-added PALF products are expected to raise up the country's overall income [46].

To increase the value of abundantly available biodegradable PALF, a lot of research and developments must be reexamined and reevaluated. Starting with the fundamental of enhancement, the properties of PALF must be figured out before making any modifications. Several questions should arise in this stage: How is the strength related to the surface condition of fiber? What are the contributions for each component in the fiber? How do the thermal properties of single fiber affect the thermal behavior of product structure?

Characterization methods and testing standards have been revised frequently to adapt to the latest materials. In this chapter, we aim to explain the working principle of major testing methods (physical, morphologies, mechanical and thermal) involved in PALF characterization.

2 Physical Properties

2.1 Density Measurement

To measure the density of PALF, the fiber is required to be dried for 48 h by calcium chloride in a non-hygroscopic desiccator [80]. Then, the fiber is impregnated with toluene for 2 h to eliminate microbubbles in the fibers and kept in a pycnometer [69] with Eq. 1 of natural fiber density,

$$\rho = \left(\frac{m_2 - m_1}{(m_3 - m_1)(m_4 - m_2)}\right)\rho_T \tag{1}$$

where m_1, m_2, m_3, m_4 are the mass of the empty pycnometer, pycnometer filled with chopped fibers, pycnometer filled with toluene and pycnometer filled with chopped fibers and toluene solution, respectively [61], and ρ_{T} is the density of toluene liquid solution (0.867 g/cm^3). Table 3 shows the density value of PALF recorded by previous studies.

Table 3 Physical and mechanical strength of PALF [3]

Study	Density, g/cm^3	Tensile strength, MPa	Young's modulus, GPa	Elongation at break, %	Diameter, μm	Microfibril angle, °
1	1.52	413–1627	34.5–82.51	1.6–3	20–80	–
2	1.526	170	62.10	3	–	–
3	1.44	413–1627	34.5–82.51	1.6	–	–
4	1.526	413	62.10	1.6	50	–
5	1.07	126.60	4.405	2.2	–	–
6		413–1627	34.5–82.5	1.6	20–80	14
7	1.526	170	6.26	3	–	–
8	1.52	170	6.21	3	–	–
9	1.07	126.60	4.405	2.2	–	–
10	1.526	414	6.5	1.6	30–60	–
11	1.526	170	62.10	3	–	–
12	1.44	413–1627	34.5–82.51	–	20–80	8–14
13	1.44	413–1627	34.5–82.51	1.6	20–80	–
14	1.44	170	6.26	1.6	5–30	12
15	1.526	413	4.2	3.0–4.0	50	14
16	1.44	–	–	–	1.56–4.5	8–15
17	–	293.08	18.934	1.41	150–300	–

On the other hand, Oliveira Glória et al. [49] used Weibull's statistic method to relate the density PALF to its diameter. The range of fiber diameters varies from 0.10 to 0.28, with an average diameter of 0.20 mm. By considering the cylindrical volume of PALF, the density of each fiber was obtained by using Eq. 2,

$$\rho = \frac{4m}{\pi d^2 l} \tag{2}$$

where m is the mass, d is the fiber diameter, and l is the length of the fiber.

More than 100 PALF have been tested and concluded an inverse dependence between the density and fiber diameter, whereby the larger the diameter, the lower the density. In PALF, large standard deviations (Fig. 1) were found on the average fiber density because of the high probability of defects and flaws on the fibers especially thicker fibers (proven in Sect. 3.1).

The low density of PALF provides favorable information; higher specific stiffness can be achieved for PALF-reinforced polymer composites than glass fiber (density of E-glass, 2.5 g/cm^3)-reinforced polymer composites [51]. George et al. [18] studied the effect of fiber loading on low-density polyethylene (LDPE). Insertion of 30 wt% of PALF only increases 14% of density value (1.03 g/cm^3), which resulted in 175 and 746% increment on tensile strength and Young's modulus, respectively. This

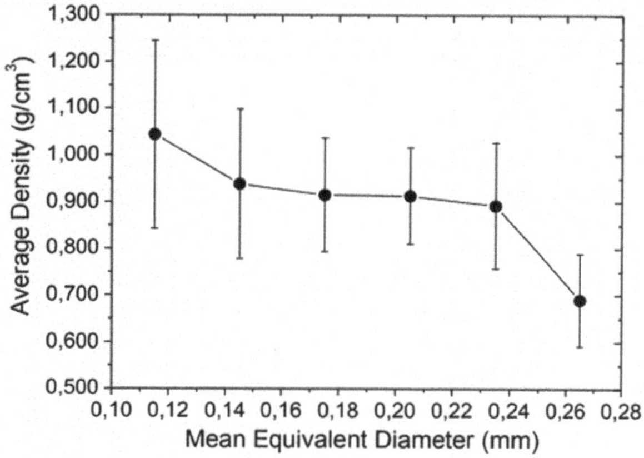

Fig. 1 Diameters of the PALF and their respective average densities [49]

shows that PALF is an efficient reinforcement in the LDPE matrix. Besides, good recyclability of PALF composites was observed as properties remain constant up to the third recycle by using 120 °C extrusion temperature. Beyond that, the strength decreases due to thermal degradation of PALF.

2.2 *Diameter Measurement*

One of the major disadvantages of natural fiber from the selection is the inconsistency of every single fiber to another. Figure 2 shows the histogram for the distribution of PALF diameter in the study. The growth conditions, climate, location, altitude as well as retting method will affect the content of chemical components (cellulose, hemicellulose and lignin) in each natural fiber. Besides, the natural fiber consists of a tight bundle of natural fibers which makes the fiber grow in a non-circular cross section with varying thickness. However, the fiber is measured on several locations along its length by using an optical microscope to calculate the average diameter. Table 3 shows the average diameter of PALF obtained from the previous study.

Kaewpirom and Worrarat [27] measured the diameter of PALF by using an optical microscope on fiber obtained from the leaf tip, leaf center and leaf base. Three locations were inspected on each fiber to calculate the average value of fiber diameter. It is found that the leaf base fiber has the highest diameter (0.219 mm), which is due to different chemical component compositions. The thicker fiber has more defects as shown in the SEM micrograph displaying the lowest tensile value among others.

Asim et al. [4] have studied the diameter of PALF due to the effects of chemical surface treatments. It is found out that treatment has removed impurities effectively on the fiber surface to get a rougher structure for better interfacial bonding. Two

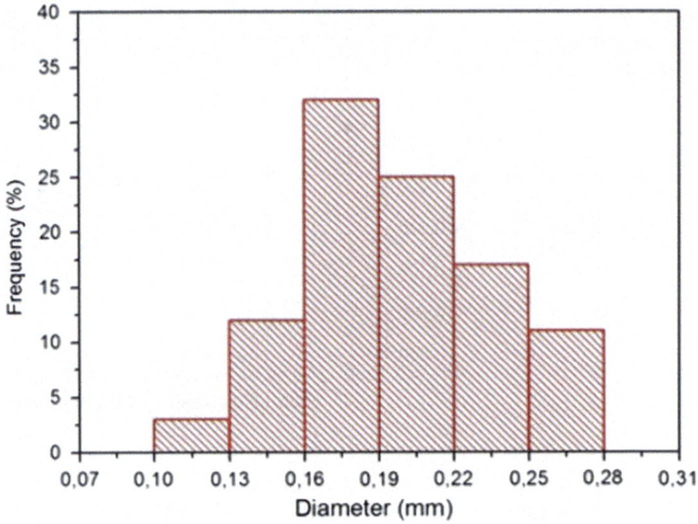

Fig. 2 Histogram of the frequency of the PALF of diameter interval [49]

percent of silane was the most effective treatment as the diameters of silane-treated PALF were reduced the most by removing the hemicellulose and lignin web of fiber bundles while still maintaining the strength performance. Figure 3 shows the diameter measurement of PALF by using an optical microscope [42].

Fig. 3 Diameter measurement of PALF by using optical microscope [42]

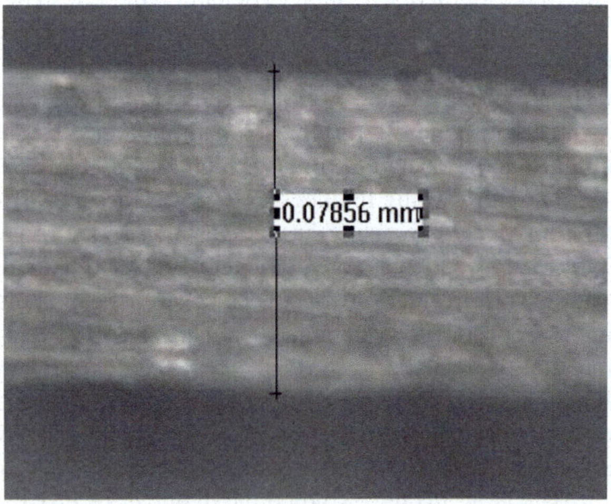

3 Surface Morphology on PALF

3.1 Scanning Electron Microscope (SEM)

A scanning electron microscope (SEM) produces images by scanning the specimen's surface with a focused electron beam. The electrons reflected and produce signals from the surface and composition of the specimen to form images. The energy spectrum of the electrons emitted is contributed from Auger electrons (AEs), backscattered electrons (BSEs) and secondary electrons (SEs). The SE is created when the electrons have excited and overcome the work function, thereby decelerating to the Fermi level, while backscattered electrons are caused by the deceleration of electrons that undergo several large-angle scatterings and energy losses. Besides, Auger electron is an alternative to characteristic X-ray emission after ionization, the energy released when an electron moves from the upper shell to the ionized shell. SE and AE penetrate only a few nanometers thick, leaving the surface via elastic or inelastic scattering. BSE penetrates into large parts of the specimen, creating a lower magnification with a better shadow effect.

Roque et al. [64] have studied the structure of pineapple leaf by using an SEM micrograph. The structure of the pineapple leaf is fabricated by more cell layers at the bottom part and lesser cell layers at the tip of the leaf (Fig. 4a). The upper epidermis of the whole pineapple leaf is constructed from two to three layers of thin-walled polygonal shape cells, allowing effective penetration of light for the photosynthesis process [2]. The lower epidermis of the leaf tip has similar epidermis structure but tapered radially oriented shape or elongated columnar of cells with a smaller diameter found at the bottom of the leaf (Fig. 4b).

The epidermis layer is reported to protect the leaf from regulating optimum temperature under intense sunlight, mechanical stress and predation by animals [28]. The hydrophobic nature of the epidermis cells also helps to reduce dehydration.

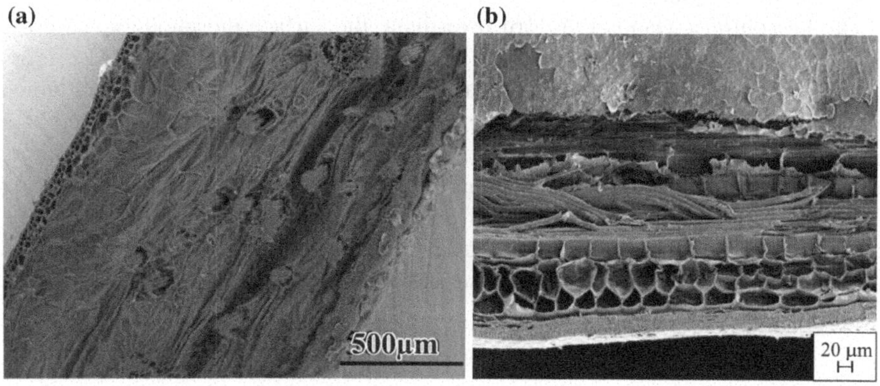

Fig. 4 SEM micrographs of **a** cross section [64] and **b** longitudinal section [2] of pineapple leaf

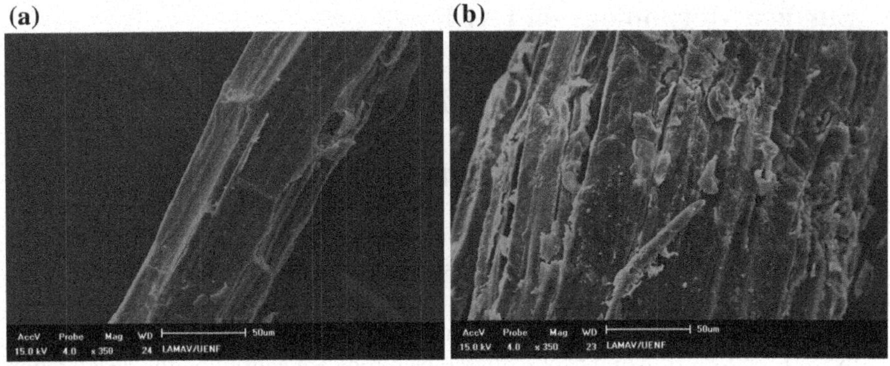

Fig. 5 SEM of PALF with **a** thinner diameter and **b** thicker diameter [79]

Besides, small stomata size and naturally protected heavy coat of waxy trichomes made pineapple leaf to have a very low rate of transpiration [7]. A rough pineapple leaf surface provides a large number of anchorage points and offers a good fiber/resin interface for its fiber. On the other hand, the elongated palisade parenchyma is located under the epidermis layer which consists of only a few layers and responsible as the main photosynthesis tissue of the leaf. The spongy mesophyll cells have an open and net structure and are involved in carbon dioxide, oxygen and water vapor exchanging processes.

Varied diameters were present in fiber bundles, with a bigger diameter in the bottom part and a decreasing diameter when approaching the upper part of the pineapple leaf. The inverse correlation relationship between the diameter of PALF and the tensile value was proven in SEM micrographic (Fig. 5). A thicker PALF displayed higher defects and porosities with an uneven rupture, and a thinner PALF observed a more uniform structure with simultaneous rupture of its microfibrils. A smoother fiber surface provides a better interlocking between fiber and matrix, fabricating a better composite quality.

On the contrary, Daud et al. [16] investigate the surface morphology and cross sections of PALF and other two fibers (corn stalk and Napier grass). PALF cross sections show a rougher structure compared to other fibers and the presence of lumen surrounding the cell wall. A closer micrograph (Fig. 6a) shows PALF formed by a lumen and polygonal cell wall [40]. Besides, Fig. 6b reveals that PALF has a more compact surface and fiber matrices than the Napier grass and corn stalk, showing higher fiber contents in pineapple leaf and producing papers with better quality. Therefore, the authors concluded that PALF is a potential substitute for wood fibers in paper production. The high cellulose and low lignin content of PALF produce better pulp and paper quality.

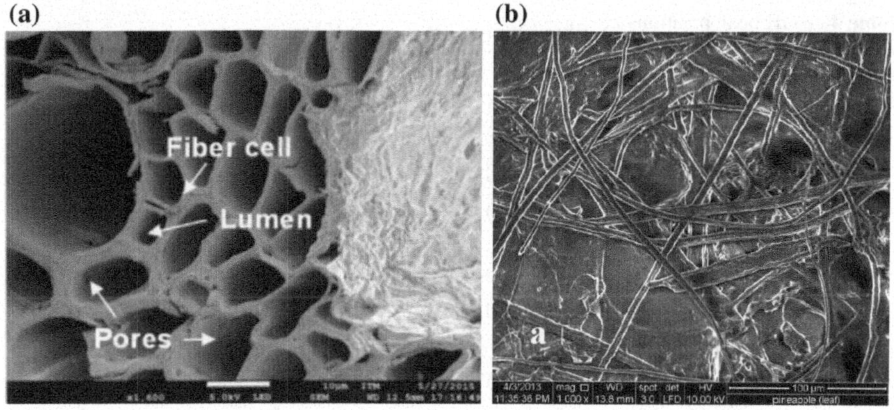

Fig. 6 SEM images for PALF from **a** cross section [40] and **b** surface morphology [16]

3.2 Fourier-Transform Infrared (FTIR) Spectroscopy

Infrared spectroscopy is a technique based on the vibrations of the atom in the specimen. An infrared spectrum is recorded after the specimen is passed through infrared radiation. The recorded spectrum determines the fraction of absorbed incident radiation by the specimen at a specific frequency. Vibrations can involve either from a change in bond angle (bending) or bond length (stretching) while some bonds can stretch in-phase (symmetrical stretching) or out-of-phase (asymmetric stretching).

Fourier-transform infrared (FTIR) spectroscopy is based on the idea of the interference of radiation between two beams to yield an interferogram, a signal produced as a function of the change of pathlength between two beams. The two domains of distance and frequency are changeable by the mathematic method of Fourier transformation. It is considered a nondestructive analysis that can provide the qualitative and quantitative outcomes of natural fibers. Some of the FTIR peak positions and the chemical stretching allocations happened on natural fiber as shown in Table 4.

Figure 7 shows the FTIR spectra for six types of PALF studied. PALF exhibited a broadband between 3200 and 3600 cm^{-1} which donated to O–H stretching of hydroxyl groups of cellulose structure and adsorbed water molecules [35]. This shows the appearance of free water on PALF and will thermally disappear below 100 °C as recorded in TGA testing (see Sect. 5.1). A small band around 2920 cm^{-1} was found in PALF due to C–H stretching in the methyl group ($-CH_2$) of cellulose and hemicellulose [43]. Besides, 1710 cm^{-1} observed in PALF IR spectra represents stretching vibrations of C=O in ketones and carboxylic acids in the xylan component of hemicellulose [77] while 1640 cm^{-1} was attributed to the water absorbed [37]. The functional groups of cellulose and lignin were recorded at the peak between 1500 and 1050 cm^{-1}. The bands at 1460, 1425 and 1220 cm^{-1} were reported for C–H and C–O bending or stretching vibrations of lignin and other carbohydrate components [75]. On the other hand, C–OR or C–OR–C stretching of cellulose

Table 4 FTIR peak positions and the chemical stretching allocations happened on natural fibers [67]

Wave number, cm^{-1}	Vibration mode
420–520	S–S stretching
735–770	C–C deformation
600–800	C–S stretching
750–815	N–O stretching
1020–1050	Symmetric C–OH stretching of lignin
1070–1150	C–O–C stretching
1260–1350	C–O stretching
1365	CH bending (deformation
1385–1395	t-butyl stretching
Below 1500	Single bonds and bending vibrations
Above 1500	C=O, NH_2, NH, C=C, C=N functional group stretching
1380, 1567	NO_2 stretching
1400–1450	CH_2 symmetric bending
1450, 1500, 1580, 1600	C=C stretching
1600	C=O stretching (Amide)
1630–1650	OH (Absorbed water)
1650–1680	N=O stretching
1600, 1655–1685	A, β—unsaturated stretching
1720–1740	C=O stretching of hemicelluloses
2540–2590	S–H stretching
2850–2960	C–H stretching
3000–3100	C=C–H stretching
3500	N–H stretching (amide)
3300–3500	N–H stretching (amide)
3500–3700	O–H stretching of α-cellulose

showed the bands of 1055 and 1162 cm^{-1}. The bands of 1430, 1375 and 1320 cm^{-1} assigned to C–H deformation of crystalline cellulose were associated with aromatic skeletal vibrations, C–H bending in cellulose and O–H vibrations of alcohol groups [11, 48, 55], respectively.

FTIR has been commonly used to characterize treatments on fiber like mercerization, grafting or coupling. With FTIR technique, researchers are able to observe the changes in the molecular bonding stage due to treatments or condition of fiber/matrix interface bonding. The appearance of 2245 cm^{-1} shows successful acrylonitrile grafted in PALF which is responsible for nitrile group vibrations [39]. The bands of 1200–1300 cm^{-1} (aromatic ring vibration of lignin) disappeared after various chemical and mechanical surface treatment, showing a smoother PALF surface by

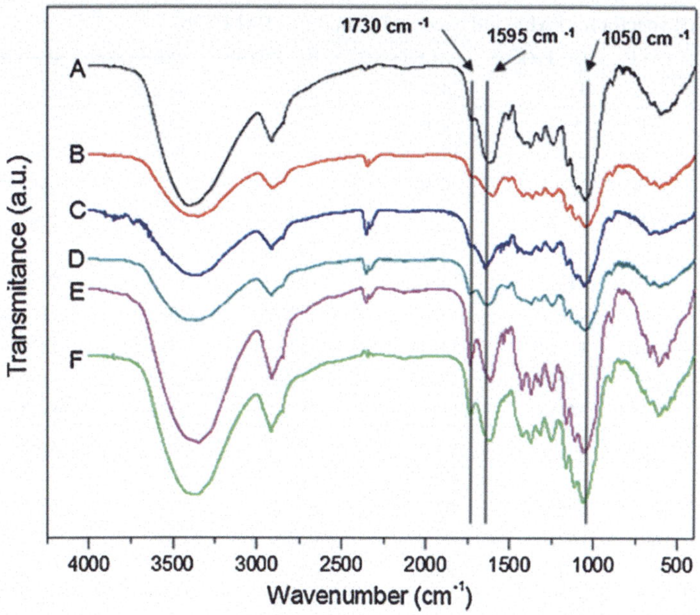

Fig. 7 FTIR of six types of PALF (A) *Bromelia* sp. (B) *Ananas comosus* var. *comosus* (C) *Bilbergis* sp. (D) *Ananas comosus* var. *bracteatus* (E) *Ananas comosus* var. *erectifolius* (F) *Ananas macrodontes* × *Primavera* [73]

the removal of lignin [10, 11, 14]. Saha et al. [65] have observed the disappearance of 1730 cm^{-1} band (vibrations of C=O for xylan component in hemicellulose), and a weakening band of 1195 cm^{-1} after alkali treatment on PALF as NaOH removes amorphous components and makes a smoother surface for a better interface. Table 5 lists the FTIR spectra for PALF and chemically modified PALF, showing the disappearance of original bands and appearance of new bands after treatment on PALF.

Besides, FTIR can be used to differentiate the various types of natural fibers. Prado and Spinacé [57] have concluded that cotton textile residues have a wider IR band at 3200–3600 cm^{-1} since cotton textile residue fibers contain more hydrogen bonds from a stable cellulose II structure while the intense band at 1047 cm^{-1} for rich husk fiber related to Si–O (silicon-oxygen tetrahedrons) stretching vibrations. Mittal and Chaudhary [41] have noticed a great difference in IR spectra between PALF and coconut husk fiber (COIR) (Fig. 8). The narrower bank peak in the region of 3200–3500 cm^{-1} for COIR shows a less amount of hydroxyl group (–OH) in COIR, giving lesser active sites for effective fiber/matrix adhesion. Less lignin amount in PALF than COIR also reduces the peak at 1244 cm^{-1}.

Table 5 FTIR spectra for PALF and chemically modified PALF [66]

FTIR	PALF	PALF-ONa[a]	PALF-O-NO_2^b	PALF-DNP[c]	PALF-OBz[d]	PALF-OBz-OAc[e]
spectra absorbance, cm^{-1}	3349.9	3347.3	3411.5	3405.4	3355.0	3413.7
	2903.8	2902.3	2916.2	2903.5	2915.8	2904.7
	1737.4	–	1753.2	–	1737.2	1736.7
	1634.9	1637.6	1639.1	–	–	1631.5
	1608.3	1610.0	–	1605.4	1603.4	1604.6
	–	–	1530.5	1528.7	–	–
	1429.4	1430.8	1431.0	1430.7	1429.4	1430.5
	1374.2	1372.1	1376.6	1373.0	1373.4	1374.6
	1321.1	1320.0	1324.6	1322.3	1325.4	1320.7
	1163.3	1164.6	1163.2	1164.7	1163.5	1163.6
	1112.1	1116.6	1113.4	1112.6	1112.4	1113.8
	1058.6	1059.9	–	1059.6	1055.9	1059.5
	–	–	1024.5	–	–	–
	897.9	898.9	896.7	896.8	897.2	897.6
	668.1	667.2	–	667.2	666.6	667.1
	613.0	607.5	606.3	611.2	–	609.0
	561.5	559.3	560.6	559.4	559.3	559.7
	434.3	438.0	434.8	434.4	435.1	436.4

[a] Alkali treated
[b] Nitration treated
[c] Dinitrophenylation treated
[d] Benzoylation treated
[e] Benzoylation–acetylation treated

3.3 Atomic Force Microscopy (AFM)

Atomic force microscopy (AFM) is a method to see the shape of a surface in 3D detail without any complicated preparations. It can image all materials in all state of matters including biological structures, regardless of conductivity or opaqueness. The surface morphology is not perceived in the usual way, but by using touch (force probe) to image a surface and measure the repulsive and attractive force between the probe tip and the surface. The color gradient is used to display the height of measurements and provide precise information about the dimensions of fibers especially nanoscale PALF fibers [74].

A cellulose nanofibrils fiber fabricated from acid hydrolysis of PALF gave an effective reduction in fiber size into the submicron level (5–15 nm) [12]. The PALF nanocellulose is claimed to have the desired properties for biomedical applications due to its biocompatibility and durability. A longer nanocellulose sisal fiber (30.9 nm) was observed by AFM technique in the previous study compared to PALF nanocellulose [44]. Besides, AFM images from Fig. 9 showed micro-sized PALF and some needle-like nanoparticles from successful acid hydrolysis for PALF [68]. However, a short treatment period led to the observation of agglomeration in the specimen.

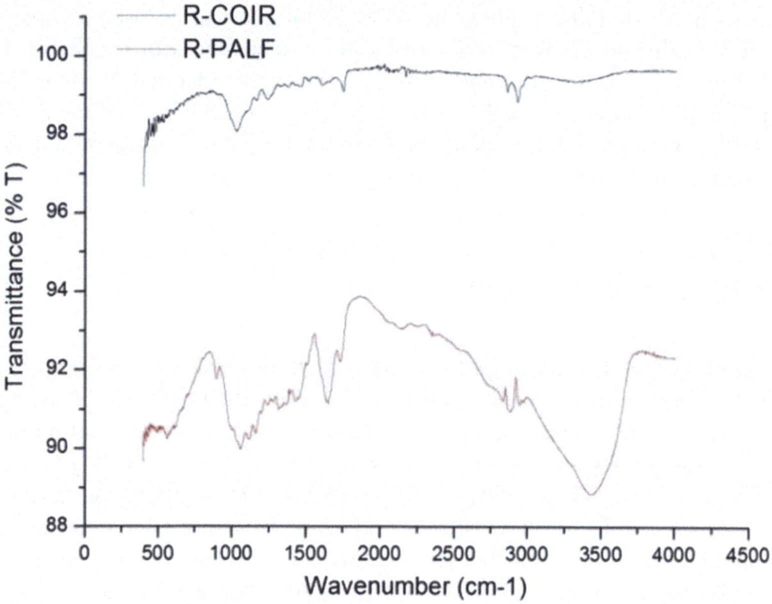

Fig. 8 IR spectra of COIR (coconut husk fiber) and PALF [41]

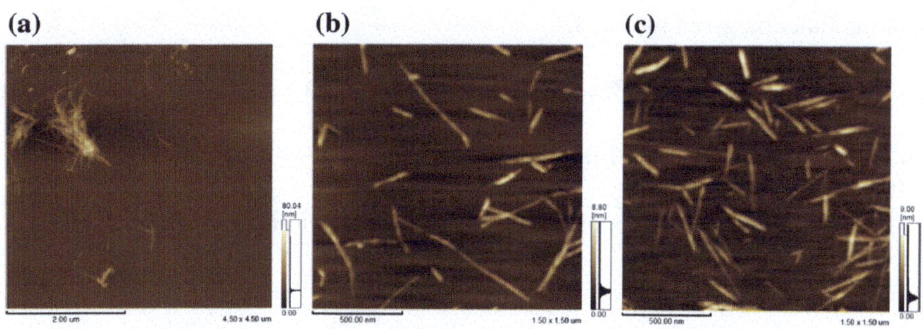

Fig. 9 AFM images for cellulose nanocrystals from PALF for **a** 5 min, **b** 30 min and **c** 60 min acid hydrolysis treatment period [68]

Balakrishnan et al. [5] used powdered PALF, undergoing alkali treatment and acid hydrolysis treatment to fabricate cellulose nanofibers and cellulose nanocrystals, respectively. PALF nanofibers were observed to have a long slender continuous fiber structure while nanocrystals showed needle-like structure. Both reinforcements enhanced composite effectively. PALF nanofibers have better reinforcing efficiency due to higher aspect ratio and network structure. On the other hand, they reported a rod-like PALF nanocrystal with 73% crystallinity index with high hydrophilicity verified by high moisture absorptions.

Other than the size of the fiber, the AFM technique can be used to investigate the surface conditions of fiber to determine the effectiveness of treatment. Pietak et al. [52] displayed a smooth, highly aligned and well-separated fibrils indicating the structure of the fiber's secondary cell wall. The removal of the primary cell wall from the fiber was successfully done by steam and alkaline treatment and thereby increases adhesion force.

3.4 X-ray Diffraction (XRD)

XRD is a nondestructive and rapid analytical technique used to identify chemical composition and crystallographic structure of natural fibers. X-ray diffractograms were collected using a sample holder mounted on a diffractometer with monochromatic CuKα radiation ($\lambda = 0.15418$ nm) and operated under high voltage. The most measured intensities were in the range of $0° < 2\theta < 90°$, normally having a scan steps of 0.05° with 1.5°/min. Peak separations are done by using Gaussian deconvolution to calculate and compare under few parameters. The *d*-spacings were calculated using the Bragg equation while *Z*-discriminant function developed by Wada and Okano [82] to determine cellulose crystalline structure (triclinic or monoclinic). By employing discriminant analysis, it is possible to categorize cellulose as Type I_α or I_β by using *Z*-value (I_α and I_β represent by positive and negative *Z*-value, respectively) with the function given by Eq. 3:

$$Z = 1693d1 - 902d2 - 594 \tag{3}$$

where $d1$ is the *d*-spacing of the peak (1–10), and $d2$ is the *d*-spacing of the peak (110).

The crystalline index proposed by Hermans and Weidinger [21] is presented in Eq. 4:

$$\text{Crystalline index} = \frac{A_{\text{cryst}}}{A_{\text{total}}} \tag{4}$$

where A_{total} is the total area under the diffractograms, A_{cryst} is the sum of crystalline band areas, and CI is the crystalline index

On the other hand, Eq. 5 was the empirical method proposed by Segal et al. [70] to looking for crystalline index, which is:

$$\text{CI} = 1 - \frac{I_{\text{AM}}}{I_{200}} \times 100\% \tag{5}$$

where I_{200} and I_{AM} are the intensity of is the maximum intensity of the (200) lattice diffraction and the amorphous phase, respectively.

The crystallite size (CS) can be calculated by using the Scherrer equation [71],

$$CS = \frac{0.89\lambda}{\beta \cos \theta} \quad (6)$$

where β is the full width at half maximum of the peak and θ is the Bragg angle.

XRD spectra of raw PALF are shown in Fig. 10. Three 2θ values peaks have been observed at 16.862°, 22.136° and 28.4882° in the work of Jain et al. [25]. Amorphous region of cellulose and hemicellulose rings in PALF attributed to the less intensity angle of 16.862° and 28.4882°, whereas at 22.136°, sharp and intense diffraction pattern shows the highly crystalline nature of the PALF. The crystal size is also determined at these three different peaks (16.862°, 22.136° and 28.4882°) are 11.9, 29.9 and 8.9 nm, respectively. The crystallinity index of untreated PALF is reported at 37.67% in the previous study. Table 6 shows the crystallinity index of raw PALF found by previous studies.

Cellulose nanofiber is expected to have a higher crystallinity index; cellulose I content preserved in the specimen is able to be observed by using XRD spectra which indicated at the peak of 22.7° [83]. PALF founded a slight shoulder at a scattering angle of 22.7°, showing the presence of cellulose I in raw fiber. Crystalline cellulose I was found stable to chemical attack. Therefore, acid-treated PALF observed the highest increment of crystallinity compared to steam treated and bleached fiber, by evidence of the highest peak at 22.7° in XRD scanning (Fig. 11) [13]. Only amorphous content of PALF was hydrolyzed leaving a larger semicrystalline cellulose fraction [14]. The crystallinity of raw PALF and acid-treated nanofiber is 42.01 and 75.38%, respectively. This is a comparable high value of crystallinity index to that of cassava bagasse (54.10%), wood fibers (69.34%) and wheat straw (79.87%) [9, 30, 81]; Kaushik et al. [9, 30, 81]. On the other hand, a lower crystalline index for mechanical treatment was due to the damage to the crystalline structure of cellulose

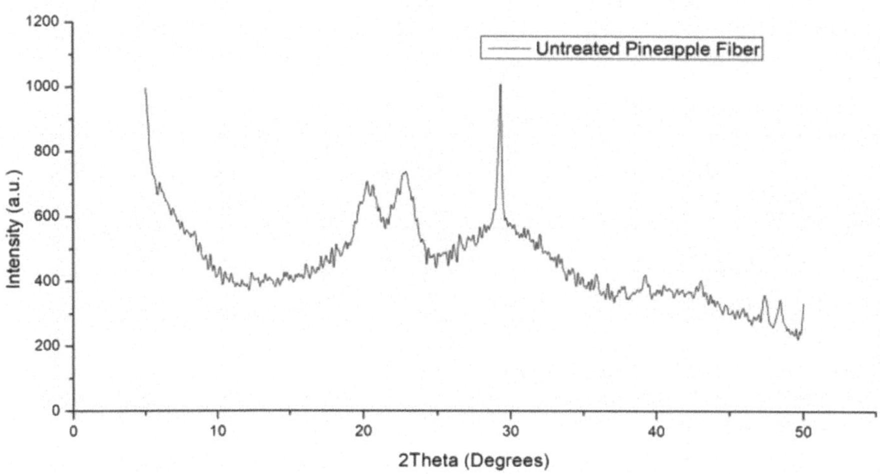

Fig. 10 XRD spectra of raw PALF [25]

Table 6 Crystallinity index of raw PALF

Source	Crystallinity, %	Calculation method	Reference
Argo-industrial residues by a pineapple-in-syrup processing plant at Puebla, Mexico	32	Segal	Ramos Cassellis et al. (60)
	42	Segal	Sipiao et al. (76)
Local plantation in Pekan Nanas, Pontian, Johor, Malaysia	64	Segal	Ain Ibrahim et al. (1)
Pusat Teknologi Tanaman Nanas Alor Bukit, Pontian, Johor, Malaysia	72[a], 63[b] and 58[c]	Segal	Sarah et al. (2018)
Procured from Go Green Products, Chennai, Tamil Nadu, India	37.67	Herman	Jain et al. [25]
Collected from neighboring fields around Kottayam District, Kerala, India	42.01	Segal	Balakrishnan et al. (6)
Supplied by the company Superpolpa from Iaras—SP, Brazil	35.9	–	Cherian et al. [13]
–	60.82	Herman	Hong Dong et al. [22]
Collected from an agricultural plant in Subang, West Java, Indonesia	41.5	Segal	Mahardika et al. [37]
Obtained from residues after harvesting pineapple in the São Mateus farm (Comendador Gomes, Minas Gerais, Brazil).	49	Segal	Santos et al. [68]
Received from Embrapa Cassava and Tropical Fruits (CNPMF) activebank of pineapple germplasm (APGB) (Cruz das Almas, BA, Brazil)	61.0–78.7	Herman	Sena Neto et al. [72]

[a]Semi-mechanical PALF extraction method
[b]Chemical PALF extraction method
[c]Mechanical PALF extraction method

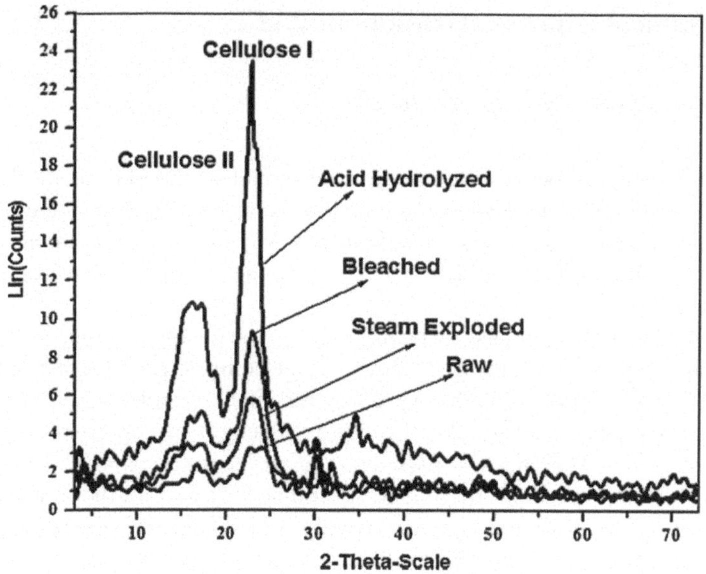

Fig. 11 X-ray diffractometry for raw and treated PALF fibers [13]

upon broken intermolecular hydrogen bonds between cellulose chains [86]. Table 7 shows crystallinity index for some natural fibers.

Table 7 Crystallinity of some natural fibers [54]

Natural fibers	Crystalline index by Segal method, %
Eucalyptus grandis	49.3
Pinus elliottii	43.4
Dipteryx odorata	55.7
Mezilaurus itauba	52.7
Curaua	43.5
Jute	34.3
Kenaf	34.9
Ramie	34.8
Sisal	57.3
Buriti	71.2

4 Mechanical Characterization on PALF

4.1 Single-Fiber Tensile Testing

Test results of single-fiber characterization are reliant on the fiber itself. Therefore, placing the single fiber as the specimen and in the testing machine should be done with extra care, securely without misalignment. Any failure in specimen mounting could cause bending stresses on the fiber at the grips and errors in measurement could be the consequence.

Some of the fiber specimens can be mounted on the testing machine with sticking adhesive tape at both ends of the fiber. However, the common practice of mounting the individual fibers on cardboard tabs to produce repeatable results ensures consistency and reduces systematic errors. A 25 mm gauge length of cardboard is normally applied in research. The fiber was pasted on the cardboard by using a drop of quick-drying epoxy to minimize misalignment. After that, both ends of the cardboard were cut out according to the dotted line shown in Fig. 12 to avoid extra load capacity from the epoxy adhesive glue. Next, the cardboard is then gripped on the testing machine and the central cutout is cut off before starting the test, ensuring that only fiber is loaded during the test, as illustrated in Fig. 13.

Fibers are the main load-carrying component in the composite. The strength of the fiber influences the effective properties of the composite. However, the inconsistency of the natural fibers' quality is a major drawback. This is because natural fibers often exist in a bundle of fibers, which causes an irregular shape of every single fiber depending on the way they are packed together. Besides, the growth conditions, local climate and retting method will decide the fiber's chemical components (cellulose, hemicellulose, lignin and pectin). Apart from this, physical or chemical treatment will be deployed to gain a greater interfacial bonding between fiber and matrix. Therefore, reliable single fiber properties are crucial during material selection in order to get the most optimum fiber and matrix composite match.

The strength of the fiber is attributed to cellulose component. Repeated unit of cellulose bonded with hydroxyl groups resulted in crystalline arrangements [50]. Crystallinity index refers to the degree of structural order in fiber. Six cultivars of fibers of different pineapple varieties have been studied to investigate the properties of PALF [73]. It is observed that the cross-sectional area of the fiber, Young's modulus, tensile strength and elongation at the break strongly depended on crystallinity index. Another study of Sena Neto et al. [72] has concluded the correlation between the tensile strength of PALF with diameter, crystallinity index, lignin, cellulose and

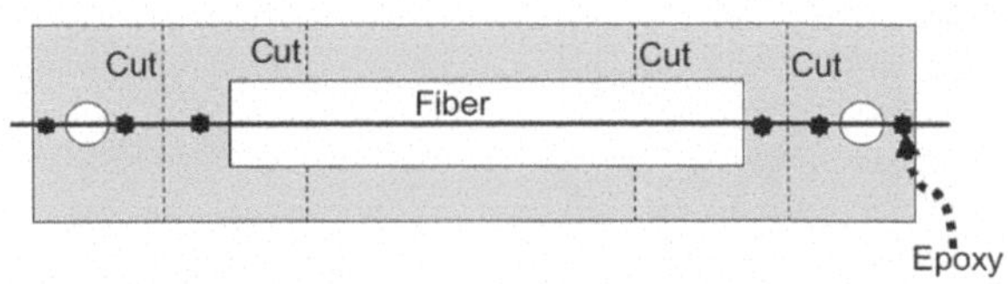

Fig. 12 Cardboard mounting specimen [8]

Fig. 13 Cutting line on single-fiber mounted specimen [8]

holocellulose contents for 12 types of PALF. Obvious correlations were observed for all findings, and the authors recommended that cellulose content and crystallinity index may be used as the criteria for fiber selection. Table 8 shows the cellulose contents, crystallinity index and tensile strength of 12 types of PALF.

Besides that, all lignocellulosic fibers including PALF present heterogenous properties due to inconsistent fiber diameters and chemical components. An adverse effect of average fiber tensile strength to the PALF diameter was reported by Teles et al. [79] as shown in Fig. 14. SEM micrographic observed a more uniform structure with simultaneous rupture of thin fiber's microfibrils while thicker fibers found more defects and microstructural porosity, resulting in deteriorated strength properties. In other words, a thinner fiber has closer packed of microfibrils which enhances fiber resistance, hence higher in strength value.

High crystallinity content (60.82%) of PALF provided the fiber a good strength performance and excellent spinnability, which allows it to be used in yarn and textile

Table 8 Cellulose contents, crystallinity index and tensile strength of 12 types of PALF [72]

PALF type	Cellulose contents, %	Crystallinity index, I_c	Tensile strength, MPa
1	57	73.0	629 ± 312
2	55	73.5	1049 ± 382
3	63	70.3	1092 ± 414
4	55	74.7	1071 ± 444
5	63	78.7	1231 ± 508
6	60	77.6	1221 ± 348
7	62	61.0	1100 ± 366
8	58	73.6	1309 ± 473
9	51	62.9	891 ± 387
10	52	74.7	832 ± 354
11	49	75.3	988 ± 344
12	56	70.8	828 ± 317

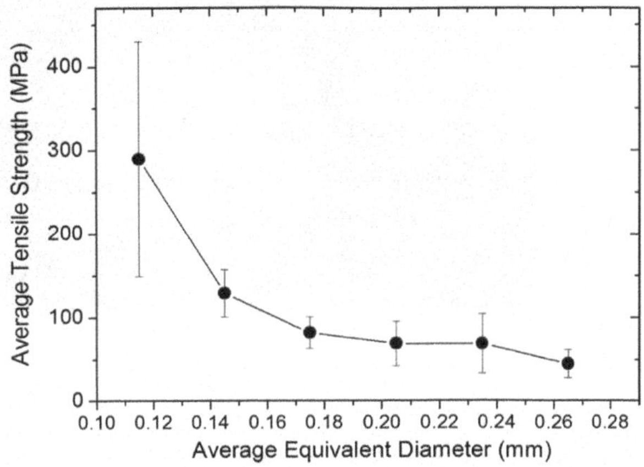

Fig. 14 Average tensile strength of PALF corresponding to its diameter [49]

industries [22]. One thing worth mentioning in the study is the similarities of performance observed in both dry and wet states. This is important as force is applied when washing the textiles. Yusof et al. [84] also agreed that NaOH-treated PALFs (single and twisted fibers) recorded a promising tensile strength (84.67–331.53 MPa) and comparable to cotton yarn.

4.2 *Single-Fiber Droplet Test*

Interfacial shear strength (IFSS) between fibers and matrix plays an important role in high-strength composite materials. A good adhesive strength between fibers and matrix represents a stronger composite. Droplet test is used to study the interfacial bonding strength [29].

IFSS properties are strongly dependent on the type of matrix. Thermoset resin specimen is normally applied on fiber and cure at ambient temperature overnight. Thermoplastic resin can apply heat to melt when placed on the fiber; however, excessive heating often degrades the PALF [62].

To produce a reliable calculation of interfacial shear strength of composite, Kelly and Tyson model is the main approach [85]. "Kelly and Tyson" assuming matrix is completely plastic and only transmits shear stress, $\tau(x)$. During axial tension, shear stress, $\sigma(x)$, is found on the interface, stressing transmission from the matrix to fiber. Thus, tensile stress is gained in the fiber. By considering fiber is in a perfect cylinder shape,

$$\tau(x)2\pi r\mathrm{d}x = \pi r^2[\sigma(x+\mathrm{d}x) - \sigma(x)] \tag{7}$$

hence, rearranging Eq. 7 to get,

$$\tau(x) = \frac{r\mathrm{d}\sigma}{2\mathrm{d}x} \tag{8}$$

or,

$$\sigma(x) = \frac{2}{r}\int_0^{\pi} \tau(\delta)\mathrm{d}\delta \tag{9}$$

Equation 8 shows that the shear stress is zero, but tensile stress reaches the maximum in the middle part of the fiber. On the other hand, Eq. 9 tells that tensile stress is zero, but shear stress is at a maximum at the ends of the fiber.

A plastic displacement region located at the ends of the fiber which the shear stress, τ, is constant in Kelly and Tyson model. This has produced a linear increment in tensile stress at the ends of the fiber while constant tensile value, σ_{o}, at the middle part of the fiber. When reaches ultimate tensile strength of the fiber is reached, σ_{c}, the fiber breaks at the segment coordinate, $[-(1/2) + x_{\mathrm{c}}, (1/2 - x_{\mathrm{c}})$, where x_{c} is the width of fiber displaced. This fragmentation process cycle repeats upon further stressing and continues until all the fragments are less than $2x_{\mathrm{c}}$. The value of fiber critical length, $l_{\mathrm{c}} = 2x_{\mathrm{c}}$. Since τ is constant, in Kelly and Tyson model,

$$\int_0^{\pi} \tau(\delta)\mathrm{d}\delta = \tau(x) \tag{10}$$

Hence,

$$\sigma_{\mathrm{c}} = \frac{2\tau x_{\mathrm{c}}}{r} \tag{11}$$

And substitute l_{c} into the equation (11) and rearranging to become,

$$\tau = \frac{\sigma_c d}{2 l_c} \quad (12)$$

where τ is the interfacial shear strength, σ_c is ultimate fiber strength, d is the fiber diameter, and l_c is the fiber critical length.

IFSS improvement in PALF was found to be very less compared to other treated natural fiber. One study reported that silane-treated PALF has a 40% higher interfacial shear strength, IFSS, compared to untreated PALF (IFSS = 1.70 MPa) while alkaline-treated and alkaline-silane-treated PALF has improved 7% and 111.35%, respectively [4]. The improvement was because of impurity removal, creating voids on the fiber surface, which help to make an anchor between fiber and matrix while the coupling agent penetrated the fiber and deposited at interfibrillar regions. A similar finding was done by Zin et al. [87], IFSS between 6% alkaline-treated PALF for one-hour soaking period, and epoxy resin has improved 106% to 42.67 MPa compared to untreated PALF [87]. The promising results show comparability to IFSS of carbon fiber/epoxy [53]. An effective fiber wetting by a higher number of possible reactions sites is due to a rougher fiber surface.

A plot of IFSS versus embedded PALF area is shown in Fig. 15. Mean IFSS value from 50 PALF/PHBV specimens is 8.23 MPa with 17.75% of variations [36]. This low result is generally because the large number irregularity of the PALF surface and high viscosity of PHBV resin preclude much mechanical bonding, making the fiber debonding to happen. A good IFSS value will observe fiber failure (break of fiber) before debonding. SEM photographic of the micro-droplet part shown in Fig. 16 observed a clear fiber surface with no attached resin, and no fibrils pulled out after the testing, respectively, indicating a failed adhesion of PHBV resin on PALF.

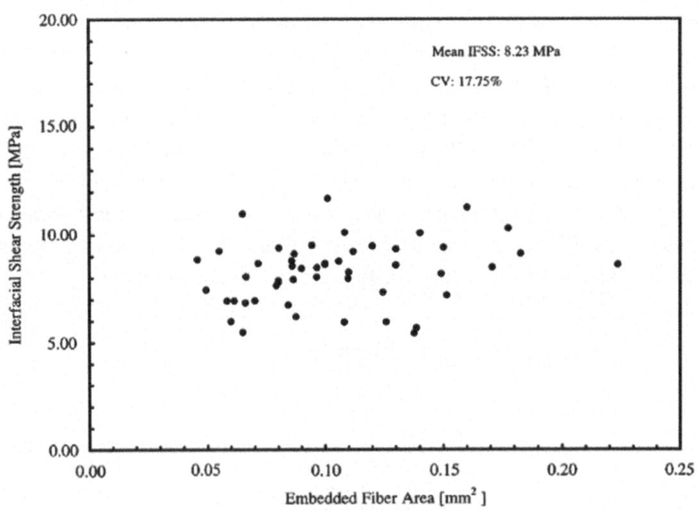

Fig. 15 A plot of IFSS versus embedded PALF area [36]

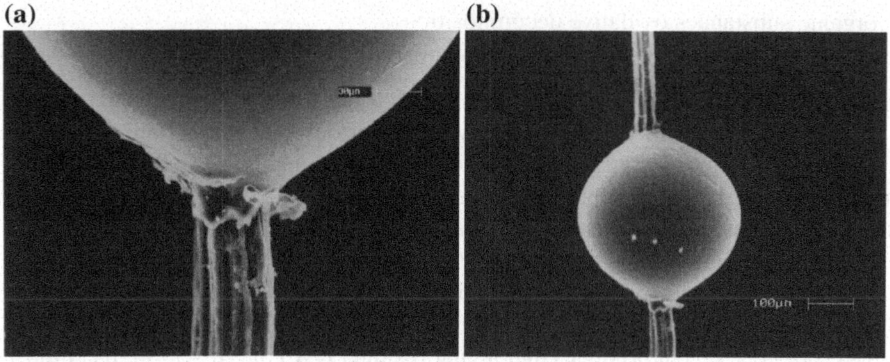

Fig. 16 SEM photographic of PHBV resin micro-droplet on PALF. **a** Before testing, **b** after testing [36]

5 Thermal Characterization of PALF

5.1 TGA

For the last 30 years, glass, carbon and aramid fibers have dominated the composite industry because of their high stability of thermomechanical properties [56]. Low thermal stability nature of PALF has limited it from the hazardous environmental application. Hence, thermal characterization of a single fiber is important, to select the most optimum natural fiber for a specific application. PALF is non-thermoplastic and pyrolyzed before glass transition or melting temperature. It generally has poorer fire stability than wood fibers. One report found that thermal degradation behavior of sisal and jute fiber was similar, but flax fiber degrades at a higher temperature [38].

Thermogravimetric analysis (TGA) is an experimental technique that measures the mass of a specimen as a function of temperature or time. The specimen is subjected to a constant temperature (isothermal heating) or heated with a constant heating rate (dynamic heating). Some of TGA machine also supports nonlinear heating to the sample, to illustrate specific environment condition. On the other hand, few types of gases are available in the TGA experiment, depending on the user to use either inert or reactive gas.

TGA can be used to investigate the thermal stability of a material. In the desired temperature range, no mass loss will be observed from a high thermal stable material. TGA also gives the limited working temperature of a material. Beyond the working temperature, the material will begin to degrade. Oxidation mass loss is the most common thermal degradation loss in TGA. Some of the reactions can cause a specimen to have changes in the mass, such as

- drying of specimen
- loss of water
- metal oxidation

- organic substances oxidative decomposition
- gaseous production from thermal decomposition
- heterogeneous chemical reactions.

Slight differences in thermal stability from various plant fibers are due to the fiber chemical compositions. Cellulose, hemicellulose and lignin are three main constituents in the fiber. Decomposition of cellulose happens at 260–350 °C, producing flammable gases, noncombustible gases and char. Higher cellulose content tends to increase the flammability of the fiber. On the other hand, hemicellulose starts decomposing at temperatures of 200–260 °C and produces more noncombustible gases than thermal decomposition of cellulose. Lignin decomposes at a wide range of temperatures (160–400 °C) but produces major char residue. PALF has a very high cellulose content (80% from Table 1), and its fire-retardant behavior is expected to be better than other natural fibers. Figure 17 shows the TGA curves of untreated PALF and alkali plus bleached treated PALF. All PALF specimens show three thermal degradation peaks, representing moisture removal at first peak, hemicellulose and lignin thermal decompositions at the second peak and degradation of cellulose at the third peak.

The second thermal degradation peak is the most important in TGA testing, to analyze the thermal stability of PALF. It was responsible for lignin, hemicellulose, pectin degradation and reported a higher onset temperature compared to other natural fibers, representing a greater thermally stable fiber and higher working temperature limits. PALF has a very high cellulose content (Table 1) and therefore is expected to have better thermal resistance behavior. The onset temperature was found proportional to the cellulose content by Sena Neto as closely packed cellulose component is contributed to the fiber's strength as well as thermally stable [59, 72]. Santos et al. [68] have proven that cellulose nanocrystals using PALF have higher thermal stability due to the poor thermal stability of hemicellulose and lignin constituent removal

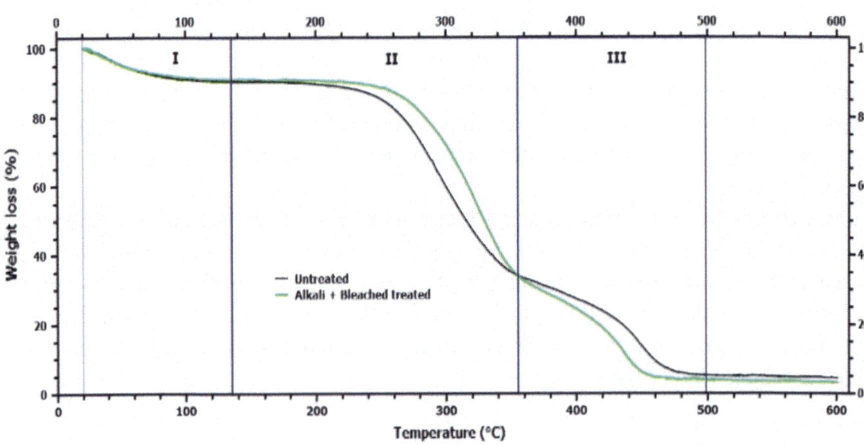

Fig. 17 TGA curves of untreated PALF and alkali + bleached treated PALF [19]

during the acid hydrolysis process. A longer processing period has observed the worst outcome where incorporation of sulfate groups (from sulfuric acid) on the cellulose surface has a catalytic effect in thermal degradation reactions [63]. Only 1.2–6.7% of ash contents remain after the testing (600 °C) was observed under a synthetic atmosphere (20% oxygen and 80% nitrogen) [73]. Pyrolysis of PALF in inert environment reported a higher final mass residual (17% at 800 °C) which is a carbonaceous product of cellulose degradation [24, 78].

5.2 DSC

A DSC measures the energy changes observed in a specimen during heating, cooling or held isothermally. The energy changes enable the user to locate melting processes, glass transitions and/or complex reactions via endothermic or exothermic. Melting is an endothermic process while glass transition is the temperature where the material changes from glassy material to leathery in nature.

The main property measure in DSC is the heat flow as a function of temperature. The energy flows in or out of the specimen representing certain processes. Therefore, encapsulation of the specimen is very important to make sure the specimen has good thermal contact with the furnace. A result with pan burst will record unwanted peaks in the DSC curve.

Typically, the mass used in DSC experiments is about 5–10 mg. However, a single-fiber specimen has a mass less than the standard weight. Therefore, an extremely high heating rate of power compensation DSC (100–400 °C/min) should be used to provide high sensitivity to the low mass materials. The faster the scan rate, the bigger the size of the peak that will be measured. This allows for small transitions to be observed easily, increasing accuracy for the specific heat of materials.

The heat flow PALF is relatively simpler than the heat flow of composite, as shown in Fig. 18. The PALF moisture evaporation temperature was recorded at 75 °C. This was agreed by other previous studies on DSC of PALF, summarizing a broad endothermic peak found due to the absorption of heat for moisture evaporation [17, 22]. Besides, this moisture removal temperature was found to synchronize with DSC of jute fiber [26]. Higher evaporation temperature will determine for treated fiber as lesser moisture content.

6 Conclusion and Future Perspective

In conclusion, there has been an increasing interest in the manufacturing and use of environmentally friendly PALF since decades due to a high strength-to-low weight ratio, relatively low cost, abundant availability and environmental advantages. Fibers can be obtained from bast stem, leaf and fruit which naturally exist as fiber bundles. PALF is famous in high cellulose content among natural fibers due to its low

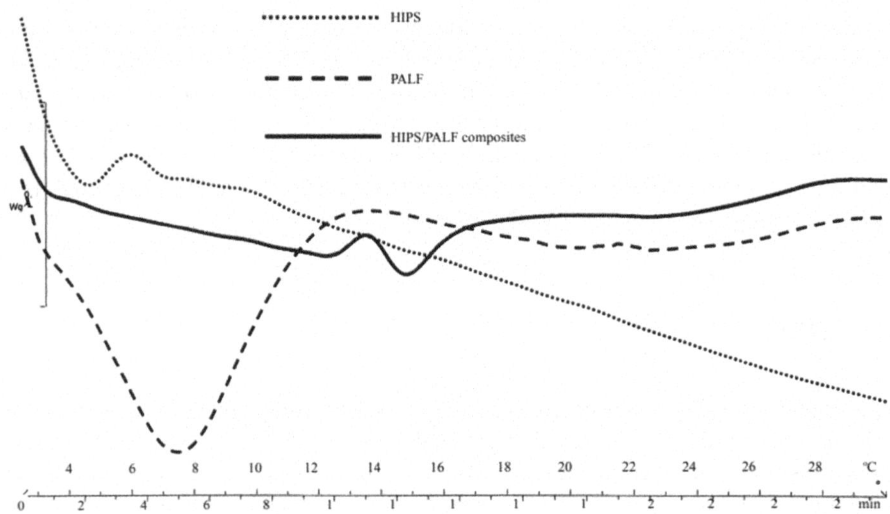

Fig. 18 DSC analysis of HIPS, PALF and HIPS/PALF composites [78]

microfibrillar angle. Superior strength-to-weight ratio, abundantly available and relatively inexpensive, creates a good opportunity for PALF to be used in fiber-reinforced polymer composite, paper making, yarn production and acoustic barriers. However, growth conditions, climate, location, altitude as well as retting method will affect the content of chemical components (cellulose, hemicellulose and lignin) in each natural fiber. Besides, natural fiber bundles tightly a bundle of natural fibers together, which makes the fiber grow in a non-circular cross section, varying in thickness.

The structure of the pineapple leaf is fabricated by thicker cell layers at the bottom part and thinner cell layers at the tip of the leaf which have different functions for every single part. Therefore, varied sizes of fiber diameters were recorded, inversely correlating with the PALF tensile value. FTIR spectra of PALF observed natural fiber components as other natural fibers, yet a great difference between PALF and coconut husk fiber on 3200–3500 cm^{-1} and 1244 cm^{-1} showed a varying amount of hydroxyl group and lignin contents, respectively. Besides, AFM analysis observed the success of cellulose nanofibrils fiber fabricated from acid hydrolysis of PALF. Higher crystallinity index and smaller in size of nanofibers have been reported as compared to raw PALF (32–78.7% crystallinity index). On the other hand, PALF shows good fire-retardant behavior due to higher crystallinity cellulose content. Three thermal degradation peaks were found on PALF by TGA, representing moisture removal at the first peak, hemicellulose and lignin thermal decompositions at the second peak and degradation of cellulose at the third peak. DSC reported a broad endothermic peak due to absorption of heat for moisture evaporation and found to synchronize with other natural fibers. PALF has great potential to become an advanced material substitution due to its superior properties. Unfortunately, the majority of underutilized PALF is currently only disposed of as landfills and low-cost feedstock. The development and utilization of PALF could be a solution for the disposal problem as well as increasing the national income of a country.

References

1. Ain Ibrahim N, Azraaie N, Aimi Mohd Zainul Abidin N, mamat razali NA, Abdul Aziz F, Zakaria S (2014) Preparation and characterization of alpha cellulose of pineapple (Ananas comosus) leaf fibres (PALF). Adv Mat Res, 895, 147–150. https://doi.org/10.4028/www.scientific.net/AMR.895.147
2. Alwani MS, Khalil A, Islam MN, Nadirah WW, Dungani R (2014) Fundamental approaches for the application of pineapple leaf fiber in high performance reinforced composites. Polimery 59. https://doi.org/10.14314/polimery.2014.798
3. Asim M, Abdan K, Jawaid M, Nasir M, Dashtizadeh Z, Ishak MR, Hoque ME (2015) A review on pineapple leaves fibre and its composites. Int J Polym Sci 2015:16. https://doi.org/10.1155/2015/950567
4. Asim M, Jawaid M, Abdan K, Ishak MR (2016) Effect of alkali and silane treatments on mechanical and fibre-matrix bond strength of kenaf and pineapple leaf fibres. J Bionic Eng 13(3):426–435. https://doi.org/10.1016/S1672-6529(16)60315-3
5. Balakrishnan P, Gopi S, Geethamma VG, Kalarikkal N, Thomas S (2018) Cellulose nanofiber vs nanocrystals from pineapple leaf fiber: a comparative studies on reinforcing efficiency on starch nanocomposites. Macromol Symposia 380(1):1800102. https://doi.org/10.1002/masy.201800102
6. Balakrishnan P, Sreekala MS, Kunaver M, Huskic M, Thomas S (2017) Morphology, transport characteristics and viscoelastic polymer chain confinement in nanocomposites based on thermoplastic potato starch and cellulose nanofibers from pineapple leaf. Carbohydr Polym, 169, 176–188. https://doi.org/10.1016/j.carbpol.2017.04.017
7. Bartholomew D, Paull R, Rohrbach DP (2003) The pineapple: botany, production and uses. CABI Publishing, Wallingford
8. Bunsell AR, Joannès S, Marcellan A (2018) 2—Testing and characterization of fibers. In: Bunsell AR (ed) Handbook of properties of textile and technical fibres, 2nd ed. Woodhead Publishing, pp 21–55. https://doi.org/10.1016/B978-0-08-101272-7.00002-X
9. Chen W, Yu H, Liu Y, Chen P, Zhang M, Hai Y (2011) Individualization of cellulose nanofibers from wood using high-intensity ultrasonication combined with chemical pretreatments. Carbohyd Polym 83(4):1804–1811. https://doi.org/10.1016/j.carbpol.2010.10.040
10. Chen W, Yu H, Liu Y, Hai Y, Zhang M, Chen P (2011) Isolation and characterization of cellulose nanofibers from four plant cellulose fibers using a chemical-ultrasonic process. Cellulose 18(2):433–442. https://doi.org/10.1007/s10570-011-9497-z
11. Chen X, Yu J, Zhang Z, Lu C (2011) Study on structure and thermal stability properties of cellulose fibers from rice straw. Carbohyd Polym 85(1):245–250. https://doi.org/10.1016/j.carbpol.2011.02.022
12. Cherian BM, Leão AL, de Souza SF, Costa LMM, de Olyveira GM, Kottaisamy M, Nagarajan ER, Thomas S (2011) Cellulose nanocomposites with nanofibres isolated from pineapple leaf fibers for medical applications. Carbohyd Polym 86(4):1790–1798. https://doi.org/10.1016/j.carbpol.2011.07.009
13. Cherian BM, Leão AL, de Souza SF, Thomas S, Pothan LA, Kottaisamy M (2010) Isolation of nanocellulose from pineapple leaf fibres by steam explosion. Carbohyd Polym 81(3):720–725. https://doi.org/10.1016/j.carbpol.2010.03.046
14. Chirayil CJ, Joy J, Mathew L, Mozetic M, Koetz J, Thomas S (2014) Isolation and characterization of cellulose nanofibrils from *Helicteres isora* plant. Ind Crops Prod 59:27–34. https://doi.org/10.1016/j.indcrop.2014.04.020
15. Dashtizadeh Z, Khalina A, Cardona F, Lee CH (2019) Mechanical characteristics of green composites of short kenaf bast fiber reinforced in cardanol. Adv Mater Sci Eng 2019:6. https://doi.org/10.1155/2019/8645429
16. Daud Z, Zainuri Mohd Hatta M, Kassim A, Awang H, Mohd Aripin A (2013) Exploring of agro waste (pineapple leaf, corn stalk, and napier grass) by chemical composition and morphological study. BioResources 9. https://doi.org/10.15376/biores.9.1.872-880

17. Gebino G, Muhammed N (2018) Extraction and characterization of ethiopian pineapple leaf fiber. Curr Trends Fash Technol Text Eng 4(4):1–7. https://doi.org/10.19080/CTFTTE.2018.04.555644
18. George J, Bhagawan S, Prabhakaran N, Thomas S (1995) Short pineapple-leaf-fiber-reinforced low-density polyethylene composites. J Appl Polym Sci 57. https://doi.org/10.1002/app.1995.070570708
19. Grumo J, Jabber LJ, Patricio J, Magdadaro MR, Lubguban A, Alguno A (2017) Alkali and bleach treatment of the extracted cellulose from pineapple (*Ananas comosus*) leaves. J Fundam Appl Sci 9. https://doi.org/10.4314/jfas.v9i7s.13
20. Hao LC, Sapuan SM, Hassan MR, Sheltami RM (2018) 2—Natural fiber reinforced vinyl polymer composites. In: Sapuan SM, Ismail H, Zainudin ES (eds) Natural fibre reinforced vinyl ester and vinyl polymer composites. Woodhead Publishing, pp 27–70. https://doi.org/10.1016/B978-0-08-102160-6.00002-0
21. Hermans PH, Weidinger A (1961) Quantitative investigation of the X-ray diffraction picture of some typical rayon specimens: Part I. Text Res J 31(6):558–571. https://doi.org/10.1177/004051756103100607
22. Hong Dong C, Lv Z, Zhang L, Jie Shen H, Na Li N, Zhu P (2014) Structure and characteristics of pineapple leaf fibers obtained from pineapple leaves. Adv Mater Res 998–999. www.scientific.net/AMR.998-999.316
23. Hossain F (2016) World pineapple production: an overview. Afr J Food Agric Nutri Dev 16. https://doi.org/10.18697/ajfand.76.15620
24. Huda MS, Drzal LT, Mohanty AK, Misra M (2008) Effect of chemical modifications of the pineapple leaf fiber surfaces on the interfacial and mechanical properties of laminated biocomposites. Compos Interfaces 15(2–3):169–191. https://doi.org/10.1163/156855408783810920
25. Jain J, Jain S, Sinha S (2018) Characterization and thermal kinetic analysis of pineapple leaf fibers and their reinforcement in epoxy. J Elastom Plast 1–20. https://doi.org/10.1177/0095244318783024
26. Kabir M, Islam M, Wang HAO (2013) Mechanical and thermal properties of jute fibre reinforced composites, vol 1. J Multifunct Compos. https://doi.org/10.12783/issn.2168-4286/1.1/islam
27. Kaewpirom S, Worrarat C (2014) Preparation and properties of pineapple leaf fiber reinforced poly(lactic acid) green composites. Fibers Polym 15. https://doi.org/10.1007/s12221-014-1469-0
28. Kang J, Dengler N (2004) Vein pattern development in adult leaves of *Arabidopsis thaliana*. Int J Plant Sci 165(2):231–242. https://doi.org/10.1086/382794
29. Kang S-K, Lee D-B, Choi N-S (2009) Fiber/epoxy interfacial shear strength measured by the microdroplet test. Compos Sci Technol 69(2):245–251. https://doi.org/10.1016/j.compscitech.2008.10.016
30. Kaushik A, Singh M, Verma G (2010) Green nanocomposites based on thermoplastic starch and steam exploded cellulose nanofibrils from wheat straw. Carbohyd Polym 82(2):337–345. https://doi.org/10.1016/j.carbpol.2010.04.063
31. Lee CH, Sapuan SM, Hassan MR (2018) Thermal analysis of kenaf fiber reinforced floreon biocomposites with magnesium hydroxide flame retardant filler. Polym Compos 39(3):869–875. https://doi.org/10.1002/pc.24010
32. Lee CH, Sapuan SM, Lee JH, Hassan MR (2016) Melt volume flow rate and melt flow rate of kenaf fibre reinforced floreon/magnesium hydroxide biocomposites. SpringerPlus 5(1):1680. https://doi.org/10.1186/s40064-016-3044-1
33. Lee CH, Salit MS, Hassan MR (2014) A review of the flammability factors of kenaf and allied fibre reinforced polymer composites. Adv Mater Sci Eng 1–8. https://doi.org/10.1155/2014/514036
34. Leão AL, Cherian BM, Narine S, Souza SF, Sain M, Thomas S (2015) 7—The use of pineapple leaf fibers (PALFs) as reinforcements in composites. In: Faruk O, Sain M (eds) Biofiber reinforcements in composite materials. Woodhead Publishing, pp 211–235. https://doi.org/10.1533/9781782421276.2.211

35. Lim WC, Srinivasakannan C, Al Shoaibi A (2015) Cleaner production of porous carbon from palm shells through recovery and reuse of phosphoric acid. J Clean Prod 102:501–511. https://doi.org/10.1016/j.jclepro.2015.04.100
36. Luo S, Netravali AN (1999) Interfacial and mechanical properties of environment-friendly "green" composites made from pineapple fibers and poly(hydroxybutyrate-co-valerate) resin. J Mater Sci 34(15):3709–3719. https://doi.org/10.1023/a:1004659507231
37. Mahardika M, Abral H, Kasim A, Arief S, Asrofi M (2018) Production of nanocellulose from pineapple leaf fibers via high-shear homogenization and ultrasonication. Fibers 6(2):28
38. Manfredi LB, Rodríguez ES, Wladyka-Przybylak M, Vázquez A (2006) Thermal degradation and fire resistance of unsaturated polyester, modified acrylic resins and their composites with natural fibres. Polym Degrad Stab 91(2):255–261. https://doi.org/10.1016/j.polymdegradstab.2005.05.003
39. Maniruzzaman M, Rahman MA, Gafur MA, Fabritius H, Raabe D (2012) Modification of pineapple leaf fibers and graft copolymerization of acrylonitrile onto modified fibers. J Compos Mater 46(1):79–90. https://doi.org/10.1177/0021998311410486
40. Mecánica I, Jaramillo N, Hoyos D, Santa J (2016) Composites with pineapple-leaf fibers manufactured by layered compression molding. Ingeniería y competitividad 18
41. Mittal M, Chaudhary R (2018) Experimental study on the water absorption and surface characteristics of alkali treated pineapple leaf fibre and coconut husk fibre. Int J Appl Eng Res 13.15 (2018): 12237–12243
42. Mohamed AR, Sapuan SM, Shahjahan M, Khalina A (2009) Characterization of pineapple leaf fibers from selected Malaysian cultivars. J Food Agric Environ 7(1):235–240
43. Mopoung S, Amornsakchai P (2016) Microporous activated carbon fiber from pineapple leaf fiber by H_3PO_4 activation. Asian J Sci Res 9:24–33. https://doi.org/10.3923/ajsr.2016.24.33
44. Morán JI, Alvarez VA, Cyras VP, Vázquez A (2008) Extraction of cellulose and preparation of nanocellulose from sisal fibers. Cellulose 15(1):149–159. https://doi.org/10.1007/s10570-007-9145-9
45. Mwaikambo L (2006) Review of the history, properties and application of plant fibres. Afr J Sci Technol 7(2):120–133
46. Ndungu S (2014) A report on conventional pineapple production in Kenya Swedish Society for Nature Conservation (SSNC), Sweden
47. Neild RE, Boshell F (1976) An agroclimatic procedure and survey of the pineapple production potential of Colombia. Agric Meteorol 17(2):81–92. https://doi.org/10.1016/0002-1571(76)90024-8
48. Nelson ML, O'Connor RT (1964) Relation of certain infrared bands to cellulose crystallinity and crystal latticed type. Part I. Spectra of lattice types I, II, III and of amorphous cellulose. J Appl Polym Sci 8(3):1311–1324. https://doi.org/10.1002/app.1964.070080322
49. Oliveira Glória G, Altoe G, Amoy Netto P, Margem F, de Oliveira Braga F, Neves Monteiro S (2016) Density Weibull analysis of pineapple leaf fibers (PALF) with different diameters. Mater Sci Forum 869. www.scientific.net/MSF.869.384
50. Park S, Baker JO, Himmel ME, Parilla PA, Johnson DK (2010) Cellulose crystallinity index: measurement techniques and their impact on interpreting cellulase performance. Biotechnol Biofuels 3:10. https://doi.org/10.1186/1754-6834-3-10
51. Pickering KL, Efendy MGA, Le TM (2016) A review of recent developments in natural fibre composites and their mechanical performance. Compos A Appl Sci Manuf 83:98–112. https://doi.org/10.1016/j.compositesa.2015.08.038
52. Pietak A, Korte-Kerzel S, Tan E, Downard A, Staiger M (2007) Atomic force microscopy characterization of the surface wettability of natural fibres. Appl Surf Sci 253. https://doi.org/10.1016/j.apsusc.2006.07.082
53. Pitkethly MJ, Favre JP, Gaur U, Jakubowski J, Mudrich SF, Caldwell DL, Drzal LT, Nardin M, Wagner HD, Di Landro L, Hampe A, Armistead JP, Desaeger M, Verpoest I (1993) A round-robin programme on interfacial test methods. Compos Sci Technol 48(1):205–214. https://doi.org/10.1016/0266-3538(93)90138-7

54. Poletto M, Ornaghi HL, Zattera AJ (2014) Native cellulose: structure, characterization and thermal properties. Mater (Basel) 7(9):6105–6119. https://doi.org/10.3390/ma7096105
55. Poletto M, Zattera AJ, Santana RMC (2012) Structural differences between wood species: evidence from chemical composition, FTIR spectroscopy, and thermogravimetric analysis. J Appl Polym Sci 126(S1):E337–E344. https://doi.org/10.1002/app.36991
56. Pornwannachai W, Ebdon JR, Kandola BK (2018) Fire-resistant natural fibre-reinforced composites from flame retarded textiles. Polym Degrad Stab 154:115–123. https://doi.org/10.1016/j.polymdegradstab.2018.05.019
57. Prado KdSd, Spinacé MAdS (2015) Characterization of fibers from pineapple's crown, rice husks and cotton textile residues. Mater Res 18:530–537
58. Py C, Lacoeuilhe JJ, Teisson C (1987) The pineapple: cultivation and uses. Maisonneuve & Larose, Paris
59. Rachini A, Le Troedec M, Peyratout C, Smith A (2009) Comparison of the thermal degradation of natural, alkali-treated and silane-treated hemp fibers under air and an inert atmosphere. J Appl Polym Sci 112(1):226–234. https://doi.org/10.1002/app.29412
60. Ramos Cassellis M, Sánchez-Pardo M, López MR, Mora-Escobedo R (2014) Structural, physicochemical and functional properties of industrial residues of pineapple (Ananas comosus). Cell Chem Technol, 48, 633–641
61. Rao KMM, Rao KM (2007) Extraction and tensile properties of natural fibers: Vakka, date and bamboo. Compos Struct 77(3):288–295. https://doi.org/10.1016/j.compstruct.2005.07.023
62. Revol BP, Thomassey M, Ruch F, Bouquey M, Nardin M (2017) Single fibre model composite: interfacial shear strength measurements between reactive polyamide-6 and cellulosic or glass fibres by microdroplet pullout test. Compos Sci Technol 148:9–19. https://doi.org/10.1016/j.compscitech.2017.05.018
63. Roman M, Winter W (2004) Effect of sulfate groups from sulfuric acid hydrolysis on the thermal degradation behavior of bacterial cellulose. Biomacromolecules 5. https://doi.org/10.1021/bm034519+
64. Roque R, Muñoz-Acosta F, Roy Soto F, Mata-Segreda J (2013) An anatomical comparison between bunch and fruit of oil palm with pineapple leaf and three woods from plantations in costa rica. J Oil Palm Res 25
65. Saha SC, Das BK, Ray PK, Pandey SN, Goswami K (1991) Infrared spectra of raw and chemically modified pineapple leaf fiber (*Annanus comosus*). J Appl Polym Sci 43(10):1885–1890. https://doi.org/10.1002/app.1991.070431013
66. Samal RK, Ray MC (1997) Effect of chemical modifications on FTIR spectra. II. Physicochemical behavior of pineapple leaf fiber (PALF). J Appl Polym Sci 64(11):2119–2125
67. Sanjay MR, Siengchin S, Parameswaranpillai J, Jawaid M, Pruncu CI, Khan A (2019) A comprehensive review of techniques for natural fibers as reinforcement in composites: preparation, processing and characterization. Carbohydr Polym 207:108–121. https://doi.org/10.1016/j.carbpol.2018.11.083
68. Santos RMd, Flauzino Neto WP, Silvério HA, Martins DF, Dantas NO, Pasquini D (2013) Cellulose nanocrystals from pineapple leaf, a new approach for the reuse of this agro-waste. Ind Crops Prod 50:707–714. https://doi.org/10.1016/j.indcrop.2013.08.049
69. Sathishkumar TP, Navaneethakrishnan P, Shankar S, Rajasekar R (2013) Characterization of new cellulose sansevieria ehrenbergii fibers for polymer composites. Compos Interfaces 20(8):575–593. https://doi.org/10.1080/15685543.2013.816652
70. Segal L, Creely JJ, Martin AE, Conrad CM (1959) An empirical method for estimating the degree of crystallinity of native cellulose using the X-ray diffractometer. Text Res J 29(10):786–794. https://doi.org/10.1177/004051755902901003
71. Seki Y, Sarikanat M, Sever K, Durmuşkahya C (2013) Extraction and properties of Ferula communis (chakshir) fibers as novel reinforcement for composites materials. Compos B Eng 44(1):517–523. https://doi.org/10.1016/j.compositesb.2012.03.013
72. Sena Neto AR, Araujo MAM, Barboza RMP, Fonseca AS, Tonoli GHD, Souza FVD, Mattoso LHC, Marconcini JM (2015) Comparative study of 12 pineapple leaf fiber varieties for use as mechanical reinforcement in polymer composites. Ind Crops Prod 64:68–78. https://doi.org/10.1016/j.indcrop.2014.10.042

73. Sena Neto AR, Araujo MAM, Souza FVD, Mattoso LHC, Marconcini JM (2013) Characterization and comparative evaluation of thermal, structural, chemical, mechanical and morphological properties of six pineapple leaf fiber varieties for use in composites. Ind Crops Prod 43:529–537. https://doi.org/10.1016/j.indcrop.2012.08.001
74. Senthamaraikannan P, Saravanakumar SS, Arthanarieswaran VP, Sugumaran P (2015) Physicochemical properties of new cellulosic fibers from bark of *Acacia planifrons*. Int J Polym Anal Charact 21. https://doi.org/10.1080/1023666x.2016.1133138
75. Silva MC, Lopes OR, Colodette JL, Porto AO, Rieumont J, Chaussy D, Belgacem MN, Silva GG (2008) Characterization of three non-product materials from a bleached eucalyptus kraft pulp mill, in view of valorising them as a source of cellulose fibres. Ind Crops Prod 27(3):288–295. https://doi.org/10.1016/j.indcrop.2007.11.005
76. Sipiao BLS, Paiva RLM, Goulart SAS, Mulinari DR (2011) Effect of chemical modification on mechanical behaviour of polypropylene reinforced pineapple crown fibers composites. Procedia Engineer, 10, 2028–2033. https://doi.org/10.1016/j.proeng.2011.04.336
77. Siqueira G, Bras J, Dufresne A (2010) *Luffa cylindrica* as a lignocellulosic source of fiber, microfibrillated cellulose and cellulose nanocrystals. BioResources 5(2)
78. Siregar J, Sapuan S, Ab Rahman MZ, Mohd Dahlan KZH (2011) Thermogravimetric analysis (TGA) and differential scanning calometric (DSC) analysis of pineapple leaf fibre (PALF) reinforced high impact polystyrene (HIPS) composites. Pertanika J Sci Technol 19(1):161–170
79. Teles MCA, Glória GO, Altoé GR, Amoy Netto P, Margem FM, Braga FO, Monteiro SN (2015) Evaluation of the diameter influence on the tensile strength of pineapple leaf fibers (PALF) by Weibull method. Mater Res 18:185–192
80. Varma I, Anantha Krishnan SR, Krishnamoorthy S (1989) Effect of chemical treatment on density and crystallinity of jute fibers. Textile Res J 59. https://doi.org/10.1177/004051758905900609
81. Versino F, García MA (2014) Cassava (*Manihot esculenta*) starch films reinforced with natural fibrous filler. Ind Crops Prod 58:305–314. https://doi.org/10.1016/j.indcrop.2014.04.040
82. Wada M, Okano T (2001) Localization of Iα and Iβ phases in algal cellulose revealed by acid treatments. Cellulose 8(3):183–188. https://doi.org/10.1023/a:1013196220602
83. Wada M, Sugiyama J, Okano T (1993) Native celluloses on the basis of two crystalline phase (Iα/Iβ) system. J Appl Polym Sci 49(8):1491–1496. https://doi.org/10.1002/app.1993.070490817
84. Yusof Y, bin Mat Nawi N, Bin Alias MSH (2016) Pineapple leaf fiber and pineapple peduncle fiber analyzing and characterization for yarn production. ARPN J Eng Appl Sci 11
85. Zhandarov SF, Pisanova EV (1996) Two interfacial shear strength calculations based on the single fiber composite test. Mech Compos Mater 31(4):325–336. https://doi.org/10.1007/BF00632619
86. Zhao J, Zhang W, Zhang X, Zhang X, Lu C, Deng Y (2013) Extraction of cellulose nanofibrils from dry softwood pulp using high shear homogenization. Carbohyd Polym 97(2):695–702. https://doi.org/10.1016/j.carbpol.2013.05.050
87. Zin MH, Abdan K, Mazlan N, Zainudin ES, Liew KE (2018) The effects of alkali treatment on the mechanical and chemical properties of pineapple leaf fibres (PALF) and adhesion to epoxy resin. IOP Conf Ser Mater Sci Eng 368(1):012035

Improving Flame Retardancy of Pineapple Leaf Fibers

S. H. Lee, C. H. Lee, Z. M. A. Ainun, F. N. M. Padzil, Wei Chen Lum and Zakiah Ahmad

Abstract Pineapple leaf fibers (PALF) are very suitable to act as reinforcing composite matrixes. Nevertheless, PALF is highly susceptible to the risk of fire hazard. Therefore, priority is often being placed in order to improve the fire retardancy of the PALF and its composite products. This chapter discusses the behavior of natural fibers in fire and various fire properties testing methods that can evaluate the fire performance of natural fibers. Different conventional fire retardant additives and its effects to the PALF fibers and its resultant composites are also been reviewed. Aluminum trihydroxide is the most popular flame retardant in the world. However, due to the prohibition of halogenated retardants, phosphorus-based flame retardants are expected to witness a gratifying market gains in the next few years. Flame retardants that are commonly used in improving flame retardancy of a material could be divided into reactive retardants, active fillers, and inert fillers. It also can be categorized based on their chemical nature, namely phosphorus-, halogen-, silicon-, and mineral-based flame retardants as well as nanometric particles. Different types of flame retardants have different mode of action and, therefore, is also functioned differently, where the mode of action of a flame retardant can be conveniently classified into physical action and chemical action.

Keywords Pineapple leaf fibers · Fire test · Flame retardants · Fire behavior · Composite

S. H. Lee (✉) · C. H. Lee · Z. M. A. Ainun · F. N. M. Padzil
Institute of Tropical Forestry and Forest Products, Universiti Putra Malaysia (UPM), 43400 UPM Serdang, Selangor, Malaysia
e-mail: lee_seng@upm.edu.my

W. C. Lum · Z. Ahmad
Institute of Infrastructure Engineering and Sustainability Management, Universiti Teknologi MARA, 40450 Shah Alam, Selangor, Malaysia
e-mail: lumweichen@outlook.com

M. Jawaid et al. (eds.), *Pineapple Leaf Fibers*, Green Energy and Technology,
https://doi.org/10.1007/978-981-15-1416-6_7

1 Introduction

Pineapple (*Ananas comosus*) is a common fruit that exists in abundance in tropical regions. According to Arib et al. [4], pineapple is the most essential tropical fruits globally, after banana and citrus fruits. As a waste after the collection of pineapple fruits, the pineapple leaves could be used for the extraction of fibers. Pineapple leaf fiber (PALF) is a type of leaf fibers, one of the six main types of plant fibers other than bast fibers, seed fibers, straw fibers, bass fibers, and wood fibers [45]. A hectare of pineapple plantation is able to generate approximately 15 tonnes of PALF annually as reported by Leao et al. [31]. However, lack of relevant technologies and awareness of the farmers have caused these PALF always been regarding as waste and left for rotting and composting in the field. What has been missed in PALF is a thriving business opportunity. In contrast with the softwood fibers that cost an astronomical high price of USD250 per tonne, PALF cost only as low as USD15 per tonne [30]. Effective utilization of these raw materials not only could bring great profits to the manufacturers but also are able to generate extra incomes for the farmers.

Like many other natural fibers, PALF is a potential candidate to be utilized as reinforcement in polymer composite owing to their ease of extraction from its leaves, environmentally friendly nature, and high tensile strength [1]. PALF has been used in various applications such as textile thread for making clothing, paper, and nanocellulose extraction [22]. PALF has the best modulus among leaf fibers and comparable to that of the synthetic fibers such as aramid and glass. PALF also bestowed with the highest tensile strength among the leaf fibers [45]. Apart from that, PALF possesses a high fineness index, more superior compared to the other vegetal fibers. This special feature made PALF very suitable for various industrial applications, especially yarn and woven fabric [30]. Application of natural fibers offers several advantages as they are greener, cheaper, and lighter but stronger. Nevertheless, due to their hydrophilic nature, natural fibers were restricted only to be used in indoor environment for construction sector. In addition, natural fibers have poor fire resistance which inhibited them to be used effectively in aerospace and marine industries [20]. Comparing to synthetic fibers, natural fibers reinforced polymer composites have a main disadvantage in terms of fire resistance. Chai et al. [13] compared the flammability of epoxy composite reinforced with flax fiber and glass fiber. Based on the results obtained, it was observed that epoxy composite reinforced with flax fibers exhibited lower fire resistance as it ignite faster and release higher amount of heat compared to that of the glass fiber reinforced epoxy composite.

Flammability is a vital parameter for a material was it to install in an area that poses possible fire threat. Flammability of a material, or so-called reaction-to-fire, describes the way a material responds to heating or fire. The reactions mentioned here include ignitability, heat release rate, spread rate of the fire, and the emittance of smoke or other toxic substances. Fire endurance, on the other hand, often called fire retardancy, is the ability of a material to prevent ignition or to inhibit the spread of fire if ignition did occur [8]. Materials can be protected from fire hazard by applying a variety of methods. Bourbigot and Duquesne [9] have identified three main methods to reduce the flammability of a material in their review: (i) application of inherently flame retarded polymers; (ii) conduct chemical modification on the polymer; and (iii) application of flame retardants. Flame retardants are chemicals that are used or applied to slow or prevent the start or growth of fire. Normally, better fire resistance could be attained by adding the chemical substances into the materials [19]. In recent years, fire retarding coatings have become increasingly popular owing to their effectiveness and efficiency [60]. In this chapter, the application of flame retardants in improving the fire resistance of PALF and its composites were focused. The types and modes of action of different types of flame retardants were also highlighted.

2 World Consumption of Flame Retardants

Flame retardants are chemicals that are used or applied to slow or prevent the start or growth of fire. Flame retardants are often used in products such as furnishings, electronics and electrical devices, building and construction materials, and transportation products. Due to their vital role in preventing fire hazards, the global consumption of flame retardants has seen a steady growth during 2016–2021, recorded an average annual growing rate of about 3%. According to The Freedonia Group, a leading international business research company, 2.8 million metric tons or $7.0 billion worth of flame retardants are on demand globally in the year of 2018. Figure 1 illustrates the

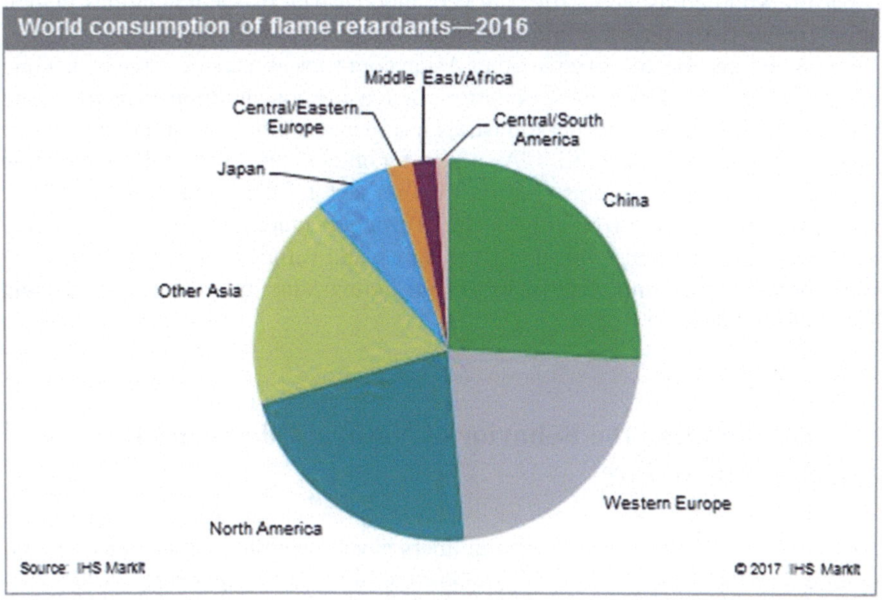

Fig. 1 World consumption of flame retardants in 2016 [25]

global consumption of flame retardants by region in the year of 2016. Over 50% of flame retardant consumptions are from the Asia/Pacific region in 2018 where China, India, and Thailand are the top three nations in consuming flame retardants [52].

Aluminum trihydroxide is the most popular flame retardant in the world, accounted for 38% of the total worldwide market. Aluminum trihydroxide is particularly favored by the European and the American as almost half of the market in these regions was dominated by this flame retardant type. Organophosphorus compounds, brominated compounds, and antimony oxides are the flame retardants that come behind aluminum trihydroxide, accounted for 18%, 17%, and 8%, respectively [25]. According to the forecast, phosphorus-based flame retardants, magnesium hydroxide, and melamine are expected to witness a gratifying market gains in the next few years (The Freedonia Group). However, halogenated retardants such as bromine- and chlorine-based flame retardants will face a dwindling market owing to their toxic nature that are detrimental to environment and human health. Especially in Western Europe and North America, bromate-based flame retardants have only 6.4% and 12% market shares, respectively, due to the implementation of stringent regulation in eliminating halogenated retardants [3].

Usage of hexabromocyclododecane (HBCD), a type of bromate-based flame retardant, was prohibited in several regions such as Japan in 2014, European Union in 2015, and Canada in 2016. In China, production and consumption of HBCD have been prohibited since December 26, 2016, with exception for its application of polystyrene in construction, where a buffering period until December 25, 2021 was given [25]. Therefore, consumption of flame retardants in these countries as well as the Middle East and Africa are very much driven by the regulations in their nation's respective industries. In comparison to the decreasing consumption in Japan, a growing market was observed in other Asian countries as a result of the expanding manufacturing industries in these countries. Increasing consumption on resins owing to the expanding manufacturing industries has increased the consumption of flame retardant in other Asian countries. As per sector, almost one third of the application of flame retardant is in building and construction sector. Flame retardants are widely used in the production of building and construction materials, such as insulating foams and rubber as well as in adhesives, paints, and varnishes. Electrical and electronics is also a very important sector in consuming flame retardants with demand shares of more than 40% [2].

3 Understanding the Behavior of Natural Fibers and Its Composite in Fire

As a nonthermoplastic material, natural fibers have lower thermal degradation temperature than its glass transition (T_g) and/or melting (T_m) temperatures. When subjected to an intense heat source, natural fibers tend to be combusted and decomposed. The thermal degradation of natural fibers involves a series of processes. Figure 2 illus-

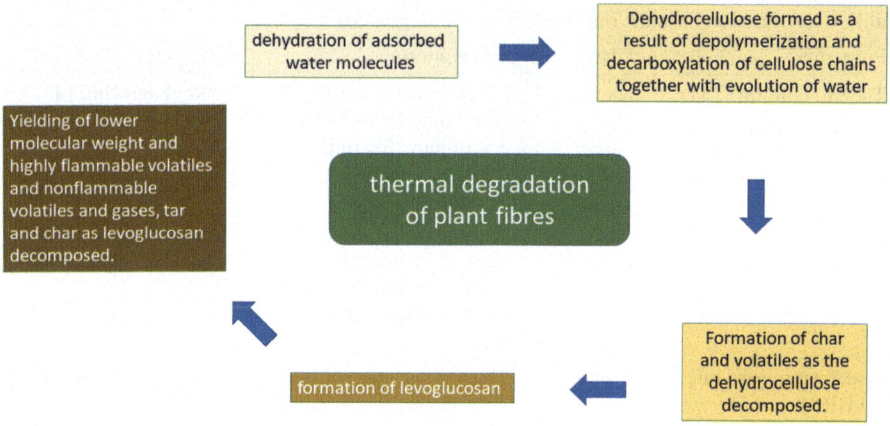

Fig. 2 Process of thermal degradation of plant fibers. Adapted from [14]

trates the process of the thermal degradation of plant fibers. During the first stage of thermal degradation, the water in the natural fibers was dehydrated. Next, dehydrocellulose was formed as a result of depolymerization and decarboxylation of cellulose chains with evolution of water. Further on, char and volatiles were generated as the dehydrocellulose were decomposed. As the thermal degradation process proceeds, levoglucosan was produced and later decomposed to generate both flammable and nonflammable volatiles and gases, tar, and char [14, 26].

Many factors have been known to affect the behavior of natural fibers and its composite in fire. Generally, natural fibers and its composites undergo a five-stage fire growth according to Mouritz and Gibson [40]. In the first stage, ignition occurs when intensity of the heat source and concentration of oxygen are sufficient to ignite the source and fuels and resulting in a sustainable combustion. In the second stage, the fire grows while consuming the fuel source. When the temperature reaches above 350 °C, ignition of the composite materials takes place. As the temperature exceeding 600 °C, a flashover stage succeeded, where the fire is fueled by all the combustible materials. As the fire is progressing to a fully developed fire, the temperature reaches over 900 °C in this stage and slowly proceeding to a post-flashover and decay stage where the fuels to sustain the combustion is diminishing continuously. The last stage of decay stage takes place after all the fuel is fully consumed, leaving behind burning residues and char.

Parameters such as time to ignition (TTI), heat release rate (HRR), peak heat release rate (pkHRR), total heat release/total heat evolved (THR/THE), etc., are important to quantify the combustion properties of a material. The important parameters used to quantify combustion properties of a material are summarized in Table 1.

Material decomposes during combustion. Volatiles gases, both flammable and nonflammable, are released during decomposition. Flammable gases released during decomposition of a material including carbon oxide, methane, and low molecular

Table 1 Description on some important combustion properties

Properties	Unit	Description	References
Time to ignition (TTI)	s	Describe the ease of ignition of a material by determining the time required for flaming combustion to starts	Schartel and Hull [47]
Heat release rate (HRR)	kW/m^2	The amount of thermal energy released (in kW) per a given unit area by a material when exposed to fire	Mouritz and Gibson [40]
Peak heat release rate (pkHRR)	kW/m^2	The highest point of the HRR curve	Mouritz and Gibson [40]
Mass loss rate	g/s	Mass loss of a material per unit time	Schartel and Hull [47]
Total heat release/Total heat evolved (THR/THE)	MJ/m^2	Total heat output to a specific point of combustion	Schartel and Hull [47]
Specific extinction area (SEA)	m^2/kg	To quantify smoke density	Mouritz and Gibson [40]
Petrella plot	–	A plot of pkHRR/TTI (x-axis) over THR or THE (y-axis), where x-axis represents fire growth rate and y-axis represents fire duration	Petrella [44]
Fire performance index (FPI)	sm^2/kW	Time to ignition (TTI)/Peak heat release rate (pkHRR)	Schartel et al. [46]
Fire growth rate (FIGRA)	kW/s	Heat release rate (HRR)/Time to ignition (TTI)	Schartel et al. [46]
Limiting oxygen index	vol.%	Minimum concentration of oxygen in a mixture of oxygen and nitrogen that is needed to support the flaming combustion of a material for 3 min or consumes a length of 5 cm of a sample. LOI $<21\%$ is combustible and $>21\%$ is self-extinguishing	Laoutid et al. [29]

organics. Meanwhile, carbon dioxide and water are among the nonflammable gases released. Other substances such as vapors and gases, solid carbonaceous char, and smoke are also the by-products of the decomposition [40].

4 Fire Properties Testing Methods

There are several techniques that can be used to characterize the flammability of a material including (i) cone calorimetry; (ii) pyrolysis combustion flow calorimetry (PCFC); (iii) limiting oxygen index (LOI); (iv) underwriters' laboratories (UL 94); (v) Ohio state university (OSU) heat release rate; and (vi) tunable diode laser absorption spectroscopy (TDLAS) oxygen analyzer.

4.1 Cone Calorimeter Test

Cone calorimeter is a fire testing method used to measure the flammability of materials by determining various flammability parameters. It is based on bench-scale heat release rate (HRR) and measures the gas flow and oxygen concentration. The data measured are then used for the determination of quantity of heat released per unit of time and surface area (HRR expressed in kW m^{-2}). Furthermore, other parameter such as of time to ignition (TTI), time to flame out (TFO), mass loss rate, levels of oxygen, carbon monoxide and carbon dioxide, and total smoke released (TSR) can also be measured. The principal underlaying the cone calorimeter test is the oxygen consumption calorimetry observed by Huggett which stated that the heat produced by most organic matters per kilogram of oxygen consumed is approximately 13.1 MJ [24]. To carry out basic cone calorimeter fire test, anoxygen is the only apparatus needed. However, to better demonstrate the burning process and understand the results which might be caused by other parameters, additional analyzers such as CO and CO_2 analyzers are normally fitted as well. A Fourier transform infrared (FTIR) spectrometer can also be fitted to further analyze combustion products and gaseous species. Despite its weakness reported by Carvel et al. [12], it remains the most practical medium-sized fire testing method to date.

4.2 Pyrolysis Combustion Flow Calorimetry (PCFC)

Pyrolysis combustion flow calorimetry (PCFC) also known as microscale combustion calorimetry (MCC), is a fire testing technique to determine the combustibility of macro-sized materials (materials with the dimension of milligram). PCFC uses the principal of the burning of a polymer solid. During the test, the sample undergoes controlled pyrolysis in an inert gas stream. The volatile pyrolysis products produced is then subjected to oxidation in high temperature. By using oxygen consumption calorimetry, the heat released during the combustion process is measured. This fire testing technique uses the combination of constant heating rate and flow characteristics of thermal analysis. The parameter PCFC is capable of measuring the heat release and heat release rate [36, 48].

4.3 Limiting Oxygen Index (LOI)

This method uses the combination of oxygen and nitrogen gas mixture during combustion test of a material. The minimum amount of oxygen in the mixture of oxygen and nitrogen which is required for the testing material to enable the combustion process is measured. This minimum concentration of oxygen is defined as the limiting oxygen index (LOI) and is expressed in volume percent (vol.%). It is used to illustrate the relative flammability of the materials tested in the present of oxygen–nitrogen gas mixture. Fenimore and Martin are the pioneers who invented this method in the year of 1966 [37, 57]. To conduct the test, a sample is placed vertically within a controlled atmosphere and its top inflamed with a burner. Then, the LOI, the minimum amount of oxygen in oxygen–nitrogen gas mixture that either maintains flame combustion of a material for 3 min or consumes a length of 5 cm of a sample is calculated. Materials are categorized as combustible if the LOI values are 21% or lower. Meanwhile, a material is considered as self-extinguishing if its LOI values exceed 21% where an external energy contribution is required to sustain the combustion at ambient temperature. Generally, materials with higher LOI value show better flame retardant property. The method is still one of the most vital, convenient, reproducible, and inexpensive technique of determining a various measure of flammability of macro-sized materials [41].

4.4 Underwriters Laboratories 94 (UL94)

UL 94 is a set of fire combustion tests approved by Underwriters Laboratories Inc. Some tests such as small and large flame vertical (*V*) tests, horizontal (*H*) tests for bulk and foamed materials, as well as radiant panel flame spread test can be conducted using UL 94. UL 94 V is among the most conventional and practical test, used to evaluate the plastics materials for parts in devices and appliances. This test is more or less the same with international standard IEC 60695-11-10 (Test method B) for small flames (50 W) and ASTM D3801-10 [29]. They are intended to serve as a preliminary indication of their acceptability with respect to flammability for a particular application. Firstly, a blue flame with 20 mm high central cone and a power of 50 W were produced using a burner for implementation of the test. The bottom surface of a sample was exposed to the flame (10 mm from the burner). The time needed for the samples to stop burning is recorded after the flame was applied to the surface for 10 s. The sample is subjected to the flame again for another 10 s after the initial burning has ceased. The after-flame time and the after-glow time are recorded. The standard specification is that at least five specimens must be tested. Based on the different recorded times (in second), the material tested can be classified with different fire rating such as V-0 (most flammable), V-1 (intermediate), and V-2 (least flammable) [42, 56].

4.5 *Ohio State University Heat Release Apparatus (OSU)*

Ohio State University heat release apparatus is a quantitative and forced flame combustion technique that describes how a material burn by measuring the rate of heat release (RHR) of specific materials. The sample that is subjected to the burning releases heat which causes the fluctuation of temperature of the constant air mass flowing through the apparatus. The fluctuation in enthalpy of the surrounding air is measured and the RHR is then calculated [51]. This apparatus setup can be divided two configurations (Configuration A and B) based on the orientation of the sample in the test. Configuration A (vertical sample orientation) is a flaming method used by Federal Aviation Administration (FAA) as a regulatory tool to determine the fire properties of aircraft interior cabin materials which are exclusively based on heat release. On the other hand, Configuration B (horizontal sample orientation) is commonly used for research and development purposes. The fundamental of the test is that it measure the heat release (RHR) and heat release rate (HRR) of a sample from the sensible enthalpy (temperature) rise of the air. During the test, with heat influx set at 35 kW m^{-2}, the specimen with an average maximum heat release (during the five minutes of the test) and average heat release (calculated within the first two minutes of the test) of <65 kW m^{-2} is considered meeting the requirement stipulated [11].

4.6 *Tunable Diode Laser Absorption Spectroscopy (TDLAS) Oxygen Analyzer*

Figure 3 displays a schematic of tunable diode laser absorption spectroscopy (TDLAS) oxygen analyzer, a new type of heat release rate testing system emerged in the recent years. This new testing system is based on the infrared absorption in accordance to the Beer-Lambert's law to record the gas concentration [21]. In comparison to the conventional and widely accepted cone calorimeter test method that have very complex and high maintenance gas handling systems and slow response time, TDLAS oxygen analyzer offers several advantages that are able to compensate the aforementioned matters of cone calorimeter [33]. TDLAS oxygen analyzer has higher sensitivity toward the change of heat release rate, having a very short response time of 3 s while cone calorimeter has a response time of 9 s.

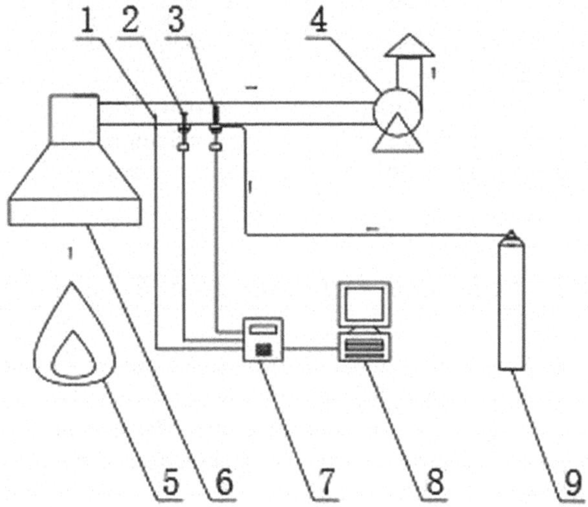

Fig. 3 Schematic of combustion heat release rate testing system based on TDLAS [33], with permission)

5 Types of Fire Retardants

Generally, fire retardants can be classified into reactive retardants, active fillers, and inert fillers. Understanding the mode of action of flame retardants is important to have a better picture on how different flame retardants act during fire attack as it is higher dependent to the chemical compounds of the fire retardant. The mode of action of a flame retardant can be conveniently classified into physical action and chemical action. Browne [10] has identified four major mode actions of flame retardants, namely chemical theories, coating theories, thermal theories, and gas theories. Flame retardants such as ammonium polyphosphates, guanidine phosphate, and guanylurea phosphate that exhibiting chemical action are capable to increase the production of water and char as well as reduce the generation of volatiles during combustion. Flame retardants that have coating ability or ceramification are able to form a protection layer on the fibers to inhibit the escape of flammable vapors and access of oxygen. Examples of this flame retardants including aluminum oxide, silica, iron oxide, sodium silicate, and potassium silicate. As for thermal action, flame retardants such as acetate, magnesium hydroxide, and cyanurates are able to absorb the surrounding heat or increase the thermal conductivity of the materials. On the other hand, gas released by magnesium chloride, potassium bicarbonate, ammonium bicarbonate, and ammonium sulfamate can dilute the flammable gases which in turn inhibit the formation of flammable mixture. Meanwhile, the function of bromine- and chlorine-based flame retardants is to inhibit the chain reaction in the flame, resulting in incomplete combustion and nonsustaining burning. However, it should be noted that such action did not occur independently but simultaneously. For example, phosphoric acid and dicyandiamide are involved in char formation (chemical) and gas dilution (gas), while aluminum hydroxide displaying gas and thermal action [35].

It is interesting to note that borax is a very special flame retardant, exhibiting three modes of action, which is chemical, gas, and coating action.

Fire retardants can be generally classified as reactive and additive [23]. Reactive compounds are added into the molecular structure of the polymer during processing to yield interaction with either matrix chemistry or fiber constituents or both [40]. Contrarily, additives are added into the polymer during or at the final stage of processing but did not react with polymer. Additives consist of active filler and inert filler in which both function as a heat sink and dilute mass fraction of organic material in hindering fire attack. The only difference is that the decomposition reaction of active filler further assists with reducing a fire [40]. Among these three types of retardants, reactive retardant is the most preferred type, followed by active filler and then inert filler.

One of the examples of reactive retardant is ammonium polyphosphate (APP) with a formula of $(NH_4PO_3)_n$. APP is an inorganic phosphorus compound which acts as a swelling material and char formers that inhibit the spread of fire. APP produces ammonia and phosphoric acid in solid state during decomposition and prevents the accessibility of oxygen to the material as well as insulates the emittance of flammable gases from the material [32]. Meanwhile, metal hydroxides and oxides such as aluminum trihydroxide (ATH, $Al(OH)_3$), magnesium hydroxide (MDH, $Mg(OH)_2$), and zinc borates (ZnB, $x ZnO.y B_2O_3.z H_2O$ are some examples of the active fillers. Metal hydroxides and oxides act as diluting agent by generating water during burning and smother the fuel and flammable gas that sustain a fire [59]. Another widely used active filler is melamine-based retardant. Melamine-based retardant is rich in nitrogen and release nonflammable gases such carbon dioxide and ammonia as well as water that are able to stop the propagation of a fire. One of the examples of melamine-based retardant is melamine cyanurate [32]. On the other hand, silicon dioxide (SiO_2) and calcium carbonate ($CaCO_3$) are the examples of inert filler. The presence of these inert filler aids the shielding of volatilization of thermally degrading products during burning by increases the melt viscosity [38]. Inert fillers are often being used together with other retardants such as ATH, MDH or APP.

Apart from the classification of reactive and additive fillers, flame retardants can also be categorized based on their chemical nature. In a review by Mngomezulu et al. [38], the authors have divided the types of flame retardants into five main categories, namely phosphorus-based, halogen-based, silicon-based, and mineral-based flame retardants as well as nanometric particles. The types and functions of the flame retardants are summarized in Table 2. Different types of flame retardants have different modes of action and, therefore, also functioned differently.

Table 2 Different types of flame retardants and its function

Categories	Type	Description	References
Phosphorus	Organic phosphorus – Resorcinol bis (diphenyl phosphate) (RDP) – Bisphenol-A bis (diphenyl phosphate) (BDP)	– Avoiding volatilization during thermal decomposition and migration toward the surface of a polymer	Faruk et al. [18]
	Inorganic phosphorus – Ammonium polyphosphate (APP)	– Produce ammonia and phosphoric acid in solid state during decomposition and prevent the accessibility of oxygen to the material – Insulate the emittance of flammable gases from the material	Lee et al. [32]
	Red phosphorus	– Used in small amount of <10% due to its highly concentrated form – Oxygen-containing radical scavenger for oxygen-containing polymers – Turns into phosphoric acid or phosphoric anhydride in oxygen- and nitrogen-containing polymers which later generates polyphosphoric acid during burning.	Laoutid et al. [29]
	Intumescent flame retardant system – Thick- or thin-film intumescent coatings	– Solvent or water-based system films – Forming an expanded carbonized layer during thermal degradation that function as an insulation layer in mitigating the transfer of heat to the material surface – Restricting diffusion of oxygen and accessibility of fuels that feel the fire.	Laoutid et al. [29]
Mineral	Hydroxycarbonates – Magnesite – Hydromagnesite	– Release water and carbon dioxide at temperature range of 200–550 °C	Laoutid et al. [29]
	Metal hydroxide – Aluminum trihydroxide – Magnesium dihydroxide	– The largest group of mineral-based fire retardant – Endothermic decomposition liberates water vapor and leave behind thermally stable inorganic residues – Smoke suppressant	Laoutid et al. [29]

(continued)

Table 2 (continued)

Categories	Type	Description	References
	Borates – Zinc borate	– Liberates water (H_2O), boric acid (H_3BO_3) and boron oxide (B_2O_3).	Pawlowski et al. [43]
Halogen	Halogenated flame retardant additives – Organochlorine – Organobromine	– Release bromine under fire conditions to inhibit free radical reactions	Morgan and Gilman [39]
Silicon	Silicone – Polydimethylsiloxane (PDMS)	– Excellent thermal stability and high heat release rate – Relatively environmentally friendly as no liberation of toxic gases	Chen and Wang [15]
	Silica – Silicon dioxide	– Increase the melt viscosity of polymer during pyrolysis and slow down volatilization and evolution of degradation products	Chen and Wang [15]
Nanometric particles	Nanoclay	– Nanoclay migrates to the surface of the material during combustion and forms a char residue that creates protective layer by limiting the volatilization of combustible products and the oxygen from diffuse into the material	Kiliaris and Papaspyrides [27]
	Carbon nanotube – Single-walled nanotubes (SWNTs) – Multi-walled nanotubes (MWNTs)	– Increase TTI and PHRR by improving the thermal conductivity of a polymer – Acts as a protective and insulation layer	Ye et al. [58]
	Graphene	– Layered graphene forms a barrier that retorts the emission of volatile flammable gases – Shield the material from the attack of fire	Wang et al. [55]

6 Improving Flame Retardancy of the Pineapple Leaf Fibers and Its Composites

High thermal stability is a prerequisite requirement for natural fibers to be applied as mechanical reinforcement in polymer composite. The thermal stability of pineapple leaf fibers (PALF) is differed between varieties. Sena Neto et al. [49] evaluated the thermal stability of leaves from 12 pineapple varieties by comparing their maximum working temperature, or called onset oxidation temperature (OOT). The results revealed that the OOT values of the 12 tested PALFs ranged between 240 and 272 °C. The reported values were higher than the other natural fibers and very close to the commercial curaua fibers [16], indicated that PALF is a potential mechanical reinforcement in polymer composite. This could be attributed to PALF that consists of a relatively high content of thermally more stable cellulose.

As polymer composite reinforced with natural fibers consists of a matrix and the reinforcement, three main methods are specifically catered to improve its flame retardancy. The first one includes the incorporation of flame retardant into the matrix or polymer. The second method is to conduct treatment on the fibers before reinforcing into the matrix. Last but not least is to apply surface treatment to the polymer composite [26]. In order to improve the flame retardancy of a PALF reinforced polymer composite, hybridization of PALF with other natural fibers with higher thermal stability is a feasible way. Siakeng et al. [50] investigated the thermal stability of raw coir fiber (CF) and pineapple leaf fiber (PALF) reinforced poly lactic acid (PLA) composites by means of thermogravimetric analysis (TGA). The results revealed that the hybridization of CF and PALF has increased the thermal stability of the PLA composite. PLA composite reinforced with 30% PALF has maximum degradation temperature (T_m) of 287 °C. However, the PLA composite reinforced with 15% CF and 15% PALF exhibited higher T_m of 290 °C, indicated better thermal stability was achieved. Similar findings were reported by Asim et al. [6] where kenaf (KF) was incorporated together with PALF into phenolic composite. The authors assessed the flame retardancy phenolic composites using horizontal and vertical under writer laboratory (UL-94) tests. Lower burning rate of 15.56 mm/min was recorded in the phenolic composites reinforced with 35% KF and 15% PALF as compared to the phenolic composites reinforced with 50% PALF in which the burning rate was recorded as 15.71 mm/min. Asim et al. [5] compared the fire resistance of pineapple leaf fiber (PALF) and kenaf (KF) phenolic composites and observed that the KF composites had better fire resistance compared to that of the PALF composites. The author attributed the findings to the high cellulose content of PALF than KF fibers. Silane treatment was found that it did not helps in improving the fire retardancy of the phenolic composites. Instead, lower fire resistance was recorded in the composites as the treatment exposed the cellulose of the fibers.

Threepopnatkul et al. [54, 53] used two phosphate derivative flame retardants, namely bisphenol-A bis (diphenyl phosphate) (BDP) and 9, 10-Dihydro-9-oxa-10-phosphaphenanthrene-10-oxide (DOPO) to improve the fire retardancy of the pineapple leaf fiber (PALF) reinforced acrylonitrile butadiene styrene (ABS). These two flame retardants were compounded with ABS using 1% wt diisononyl phthalate as plasticizer and the resultant composites were tested for limiting oxygen index (LOI) and thermo gravimetric analysis (TGA). The results revealed that the addition of BDP and DOPO has increased the LOI values of the composites. The neat ABS composites recorded a LOI value of 19.2%. When added with 20% DBP, the LOI value increased to 21.5% and in the case of 20% DOPO, the LOI value was 22.0%. As for burning rate, the burning rate of neat ABS was 0.0013 mm/min. After the reinforcement of PALF, the burning rate increased to 0.0028 mm/min as the PALF has lower thermal stability. However, after addition of 20% BDP, the burning rate reduced to 0.0025 mm/min and 0 mm/min burning rate was achieved when added with 20% DOPO, indicated that DOPO is a more efficient flame retardants compared to that of the BDP. Phosphorus existed in DOPO could be decomposed to phosphoric acid and form a molten viscous surface layer on the composites to protect the polymer within. 20% DOPO resulted in 9.61% wt residue of protective char layer, while the residual of 20% BDP was 3.59%. On the other hand, based on the TGA analysis, the first stage degradation temperature of the DOPO- and BDP-added composites were lower than PALF/ABS composites.

7 Future Prospects

A review done by Asim et al. [7] pointed out that PALF has been mainly utilized in the fabric and textile industries for the past couple of years. Nevertheless, it shows promising prospects in various areas such as furniture, aerospace, automobile, building, and sport industries. In the near future, composite materials with low flammability are expected to gain favor in various industries such as public transport and construction sectors. In term of quantifying the fire performance, Fan et al. [17] stated that the researchers around the world are still unable to reach a consensus on the actual fire resistance of composite materials. The authors suggested that the mechanism of thermal decomposition during a fire is highly dependent on the types of fiber and matrix as well as the interaction between these two elements. Unfortunately, understanding on the subject is still remained ambiguous. Most of the studies are based on cone calorimetry experiments which create some disparities between thermal and fire performance of composite materials among researchers. Therefore, it is recommended that a thermomechanical model should be developed in the future in order to describe the fiber–matrix interaction and understand the decomposition kinetics of the materials.

Apart from that, owing to the constantly changing and stringent environmental regulations and policies, application of environmentally safe compounds as fire retardants will be the key for a period of time that is to come [28]. Due to the implementation of stringent regulation in eliminating halogenated flame retardants, halogen-free flame retardants will be a sought-after product in the global flame retardants market. Halogenated retardants such as bromine- and chlorine-based flame retardants not only are toxic in nature, but also generate dense and heavy smoke as well as corrosive combustion by-products during burning [34]. In view of this, the market of phosphorus-based flame retardants is expected to continue growing in the next few years. New vapor phase flame retardants consist of phosphorus and a few metals such as tin, iron, and manganese are predicted to be entered the market in the coming years [39]. For now, lower environmentally impact seems to be the major driving force in flame retardants technology. Therefore, the removal or replacement of the currently in use flame retardants are revolving around the environmental issue.

8 Conclusions

Pineapple leaf fibers (PALF) are a promising reinforcing agent for polymer composite owing to its good mechanical properties. Nevertheless, due to the inheritance nature of natural fibers, PALF has poor resistance to fire and, therefore, susceptible to fire attack. Flammability is a very important parameter for a material was it to install in an area that poses possible fire threat. Material with poor fire resistance inevitably have restricted applications in some area. Regarding to this, numerous methods can be adopted to protect the material from fire hazard. One of the existing methods is to apply flame retardants. The global demands and consumption of fire retardant have grown steadily over the recent years, reflecting the importance of flame retardants in a variety of industries. As a nonthermoplastic material, natural fibers combust and decompose when exposed to an intense source of heat. In order to improve their flame retardancy, flame retardants with different chemical composition can be added to the matrix. Understanding the mode of action of flame retardants is important to have a better picture on how different flame retardants act during fire attack as it is higher dependent to the chemical compounds of the fire retardant. Four major modes of actions of flame retardants, namely chemical theories, coating theories, thermal theories, and gas theories have been identified. The types of flame retardants and its corresponding modes of action have been highlighted and discussed in this chapter. In order to enhance the flame retardancy of PALF and its composites, hybridization of PALF with other natural fibers with higher thermal stability such as coir fibers can be done. Of course, addition of flame retardants such as bisphenol-A bis (diphenyl phosphate) (BDP) and 9, 10-Dihydro-9-oxa-10-phosphaphenanthrene-10-oxide (DOPO) is a more direct and easy way to improve the flame retardancy of the PALF-based composites. However, studies on the enhancement of flame retardancy of PALF and its composites are relatively scarce. With the development of the times and the increasing interests in application of PALF in polymer composites, many future researches in improving the flame retardancy of PALF-based composites are anticipated.

References

1. Abdul Motaleb KZM, Islam MS, Hoque MB (2018) Improvement of physicomechanical properties of pineapple leaf fiber reinforced composite. Int J Biomater 2018:7384360
2. Anonymous (2017) Global flame retardant market projected to reach US$11.96 billion by 2025. Addit Polym 1:10–11
3. Anonymous (2018) Ceresana updates flame retardants market study. Addit Polym 2018(3):8–9
4. Arib RMN, Sapuan SM, Hamdan MAMM, Paridah MT, Zaman HMDK (2004) A literature review of pineapple fibre reinforced polymer composites. Polym Polym Compos 12:341–348
5. Asim M, Jawaid M, Nasir M, Saba N (2018) Effect of fiber loadings and treatment on dynamic mechanical, thermal and flammability properties of pineapple leaf fiber and kenaf phenolic composites. J Renew Mater 6(4):383–393
6. Asim M, Paridah MT, Jawaid M, Nasir M, Saba N (2018) Physical and flammability properties of kenaf and pineapple leaf fibre hybrid composites. IOP Conf Ser Mater Sci Eng 368:012018

7. Asim M, Abdan K, Jawaid M, Nasir M, Dashtizadeh Z, Ishak MR, Hoque ME (2015) A review on pineapple leaves fibre and its composites. Int J Polym Sci 2015:950567
8. Babrauskas V, Peacock RD (1992) Heat release rate: the single most important variable in fire hazard. Fire Saf J 18:255–272
9. Bourbigot S, Duquesne S (2007) Fire retardant polymers: recent developments and opportunities. J Mater Chem 17:2283–2300
10. Browne FL (1958) Theories of the combustion of wood and its control—a survey of the literature. FPL report no. 2136. Forest Products Laboratory, Madison, WI
11. Carpenter K, Janssens M (2005) Using heat release rate to assess combustibility of building products in the cone calorimeter. Fire Technol 41:79–92
12. Carvel R, Steinhaus T, Rein G, Torero JL (2011) Determination of the flammability properties of polymeric materials: a novel method. Polym Degrad Stab 96:314–319
13. Chai MW, Bickerton S, Bhattacharyya D, Das R (2012) Influence of natural fibre reinforcements on the flammability of bio-derived composite materials. Compos Part B-Eng 43:2867–2874
14. Chapple S, Anandjiwala R (2010) Flammability of natural fibre-reinforced composites and strategies for fire retardancy: a review. J Thermoplast Compos Mater 23:871–893
15. Chen L, Wang YZ (2010) A review on flame retardant technology in China. Part I: development of flame retardants. Polym Adv Technol 21:1–26
16. Correa AC, de Morais Teixeira E, Pessan LA, Mattoso LHC (2010) Cellulosenanofibers from curaua fibers. Cell 17:1183–1192
17. Fan M, Naughton A, Bregulla J (2017) Fire performance of natural fibre composites in construction. In: Fan M, Fu F (eds) Advanced high strength natural fibre composites in construction. Woodhead Publishing, Cambridge, pp 375–404
18. Faruk O, Bledzki AK, Fink HP, Sain M (2012) Biocomposites reinforced with natural fibers: 2000–2010. Prog Polym Sci 37:1552–1596
19. Fu F, Lin X, Xu E (2017) Functional pretreatments of natural raw materials. In: Fan M, Fu F (eds) Advanced high strength natural fibre composites in construction. Woodhead Publishing, Cambridge, pp 87–114
20. Grexa O, Poutch F, Manikova D, Martvonova H, Bartekova A (2003) Intumescence in fire retardancy of lignocellulosic panels. Polym Degrad Stab 82:373–377
21. Hangauer A, Spitznas A, Chen J, Strzoda R, Hans L, Fleischer M (2009) Laser spectroscopic oxygen sensor for real time combustion optimization. Proc Chem 1(1):955–958
22. Hazarika D, Gogoi N, Jose S, Das R, Basu G (2017) Exploration of future prospects of Indian pineapple leaf, an agro waste for textile application. J Clean Prod 141:580–586
23. Horrocks AR, Price D (eds) (2001) Fire retardant materials. CRC Press, Boca Raton
24. Huggett C (1980) Estimation of rate of heat release by means of oxygen consumption measurements. Fire Mater 4:61–65
25. IHS Markit (2017) Flame retardants: specialty chemicals update program. Accessed on 16 Jan 2019 from https://ihsmarkit.com/products/chemical-flame-retardants-scup.html
26. Kandola BK (2012) Flame retardant characteristics of natural fibre composites. RSC Green Chem 1:86–117
27. Kiliaris P, Papaspyrides CD (2010) Polymer/layered silicate (clay) nanocomposites: an overview of flame retardancy. Prog Polym Sci 35:902–958
28. Kozlowski RM, Muzyczek M, Walentowska J (2014) Flame retardancy and protection against biodeterioration of natural fibers: state-of-art and future prospects. In: Papaspyrides CD, Kiliaris P (eds) Polymer green flame retardants. Elsevier B.V., Amsterdam, pp 801–836
29. Laoutid F, Bonnaud L, Alexandre M, Lopez-Cuesta JM, Dubois P (2009) New prospects in flame retardant polymer materials: from fundamentals to nanocomposites. Mater Sci Eng R Rep 63:100–125
30. Leao AL, Cherian BM, Narine S, Souza SF, Sain M, Thomas S (2015) The use of pineapple leaf fibers (PALFs) as reinforcements in composites. In: Faruk O, Sain M (eds) Biofiber reinforcements in composite materials. Woodhead Publishing, Cambridge, pp 211–235
31. Leao AL, Souza SF, Cherian BM, Frollini E, Thomas S, Pothan LA, Kottaisamy M (2010) Agro-based biocomposites for industrial applications. Mol Cryst Liq Cryst 522:18–27

32. Lee CH, Salit MS, Hassan MR (2014) A review of the flammability factors of kenaf and allied fibre reinforced polymer composites. Adv Mater Sci Eng 2014:514036
33. Liu J, Zhao X, Gao S, Ma X (2016) Research on combustion heat release rate testing technology based on TDLAS. Proc Eng 135:107–111
34. Loredo NU, Bermejo JS (2016) Enhanced flame retardancy of flax bio-composites for the construction market. J Facade Des Eng 4:67–76
35. Lowden LA, Hull TR (2013) Flammability behaviour of wood and a review of the methods for its reduction. Fire Sci Rev 2:4
36. Lyon RE, Walters RN (2004) Pyrolysis combustion flow calorimetry. J Anal Appl Pyrol 71:27–46
37. Lyon RE, Walters RN, Stoliarov SI (2007) Screening flame retardants for plastics using microscale combustion calorimetry. Polym Eng Sci 47:1501–1510
38. Mngomezulu ME, John MJ, Jacobs V, Luyt AS (2014) Review on flammability of biofibres and biocomposites. Carbohydr Polym 111:149–182
39. Morgan AB, Gilman JW (2013) An overview of flame retardancy of polymeric materials: application, technology, and future directions. Fire Mater 37:259–279
40. Mouritz AP, Gibson AG (eds) (2007) Fire properties of polymer composite materials. Springer Netherlands, Dordrecht
41. Nelson MI (2001) A dynamical systems model of the limiting oxygen index test: II. Retardancy due to char formation and addition of inert fillers. Combust Theor Model 5:59–83
42. Patel P, Hull TR, Moffatt C (2012) PEEK polymer flammability and the inadequacy of the UL-94 classification. Fire Mater 36:185–201
43. Pawlowski KH, Schartel B, Fichera MA, Jäger C (2010) Flame retardancy mechanisms of bisphenol A bis(diphenyl phosphate) in combination with zinc borate in bisphenol A polycarbonate/acrylonitrile-butadiene-styrene blends. Thermochim Acta 498:92–99
44. Petrella RV (1994) The assessment of full-scale fire hazards from cone calorimeter data. J Fire Sci 12:14–43
45. Ramamoorthy SK, Skrifvars M, Persson A (2015) A review of natural fibers used in biocomposites: plant, animal and regenerated cellulose fibers. Polym Rev 55:107–162
46. Schartel B, Bartholmai M, Knoll U (2006) Some comments on the main fire retardancy mechanisms in polymer nanocomposites. Polym Adv Technol 17:772–777
47. Schartel B, Hull TR (2007) Development of fire-retarded materials—interpretation of cone calorimeter data. Fire Mater 31:327–354
48. Schartel B, Pawlowski KH, Lyon RE (2007) Pyrolysis combustion flow calorimeter: a tool to assess flame retarded PC/ABS materials? Thermochim Acta 462:1–14
49. Sena Neto AR, Araujo MAM, Barboza RMP, Fonseca AS, Tonoli GHD, Souza FVD, Mattoso LHC, Marconcini JM (2015) Comparative study of 12 pineapple leaf fiber varieties for use as mechanical reinforcement in polymer composites. Ind Crop Prod 64:68–78
50. Siakeng R, Jawaid M, Arrifin H, Sapuan SM (2018) Thermal properties of coir and pineapple leaf fibre reinforced polylactic acid hybrid composites. IOP Conf Ser Mater Sci Eng 368:012019
51. Smith EE (1996) Heat release rate calorimetry. Fire Technol 32:333–347
52. The Freedonia Group (2018) World flame retardants. Accessed on 16 Jan 2019 from https://www.freedoniagroup.com/industry-study/world-flame-retardants-3258.htm
53. Threepopnatkul P, Krachang T, Kulsetthanchalee C (2014) Phosphate derivative flame retardants on properties of pineapple leaf fiber/ABS composites. Polym Polym Compos 22(7):591–597
54. Threepopnatkul P, Krachang T, Teerawattananon W, Suriyaphaparkorn K, Kulsetthanchalee C (2013) Effect of flame retardants on performance of PALF/ABS composites. Adv Mater Res 747:351–354
55. Wang Z, Wei P, Qian Y, Liu J (2014) The synthesis of a novel graphene-based inorganic–organic hybrid flame retardant and its application in epoxy resin. Compos Part B-Eng 60:341–349
56. Wang Y, Zhang F, Chen X, Jin Y, Zhang J (2010) Burning and dripping behaviors of polymers under the UL94 vertical burning test conditions. Fire Mater 34:203–215

57. White RH (1979) Oxygen index evaluation of fire-retardant-treated wood. Wood Sci 12:113–121
58. Ye L, Wu Q, Qu B (2009) Synergistic effects and mechanism of multiwalled carbon nanotubes with magnesium hydroxide in halogen-free flame retardant EVA/MH/MWNT nanocomposites. Polym Degrad Stab 94:751–756
59. Zhang S, Horrocks AR (2003) A review of flame retardant polypropylene fibres. Prog Polym Sci 28:1517–1538
60. Zheng C, Li D, Ek M (2019) Improving fire retardancy of cellulosic thermal insulating materials by coating with bio-based fire retardants. Nord Pulp Pap Res J 34:96–106

Green Acoustic Absorber from Pineapple Leaf Fibers

Azma Putra, Iwan Prasetiyo and Zulkefli Selamat

Abstract The environmental issue becomes the central topic in the discussion for the last decade. Scientific works to overcome the problems are thus progressing including finding the alternative natural materials to replace the conventional synthetic ones. This chapter discusses the use of natural fibers extracted from pineapple leaf as natural acoustic absorber. The preparation for absorber samples is discussed, and the effects of fiber density, sample thickness, and introduction of backed air gap were measured using normal field incidence method in an impedance tube. The measured results reveal that the pineapple fibers can have good sound absorption above 500 Hz for thickness of 30 mm and density of 117 kg/m^3. Almost the same performance can be achieved for thickness of 20 mm by introducing backed air gap of 20 mm. The effect of the quarter-wavelength with the presence of backed air gap can be clearly observed where this can be used as the design guide to determine the required thickness of the absorber.

Keywords Sound absorber · Absorption coefficient · Acoustic material · Pineapple fiber

1 Introduction

It has been known that noise is harmful to human health and well-being [29] where some negative symptoms and cardiovascular are pronounced [6, 30]. Meanwhile, the number of noise sources and their intensity level are growing as consequences of

A. Putra (✉)
Centre for Advanced Research on Energy, Universiti Teknikal Malaysia Melaka, Hang Tuah Jaya, 76100 Durian Tunggal, Melaka, Malaysia
e-mail: azma.putra@utem.edu.my

I. Prasetiyo
Acoustic Laboratory, Department of Engineering Physics, Institut Teknologi Bandung, Ganesha 10, 40132 Bandung, Indonesia

Z. Selamat
Fakulti Kejuruteraan Mekanikal, Universiti Teknikal Malaysia Melaka, Hang Tuah Jaya, 76100 Durian Tunggal, Melaka, Malaysia

M. Jawaid et al. (eds.), *Pineapple Leaf Fibers*, Green Energy and Technology,
https://doi.org/10.1007/978-981-15-1416-6_8

population increase, traffic noise, and massive industrial activities. Effective noise controls are thus necessary to suppress the noise level to the acceptable limit. From the noise control perspective, sound absorbers are a crucial element to be developed.

Absorption of an incident sound can exist through viscous, thermal, and structural mechanisms with the basic phenomena as illustrated by Fig. 1. Those mechanisms can be realized by a resonant system, a fibrous or porous material, or combination of both [19]. Each system has its own advantage and drawbacks for example the resonant system, which is superior in absorbing sound at low frequency, but it comes with

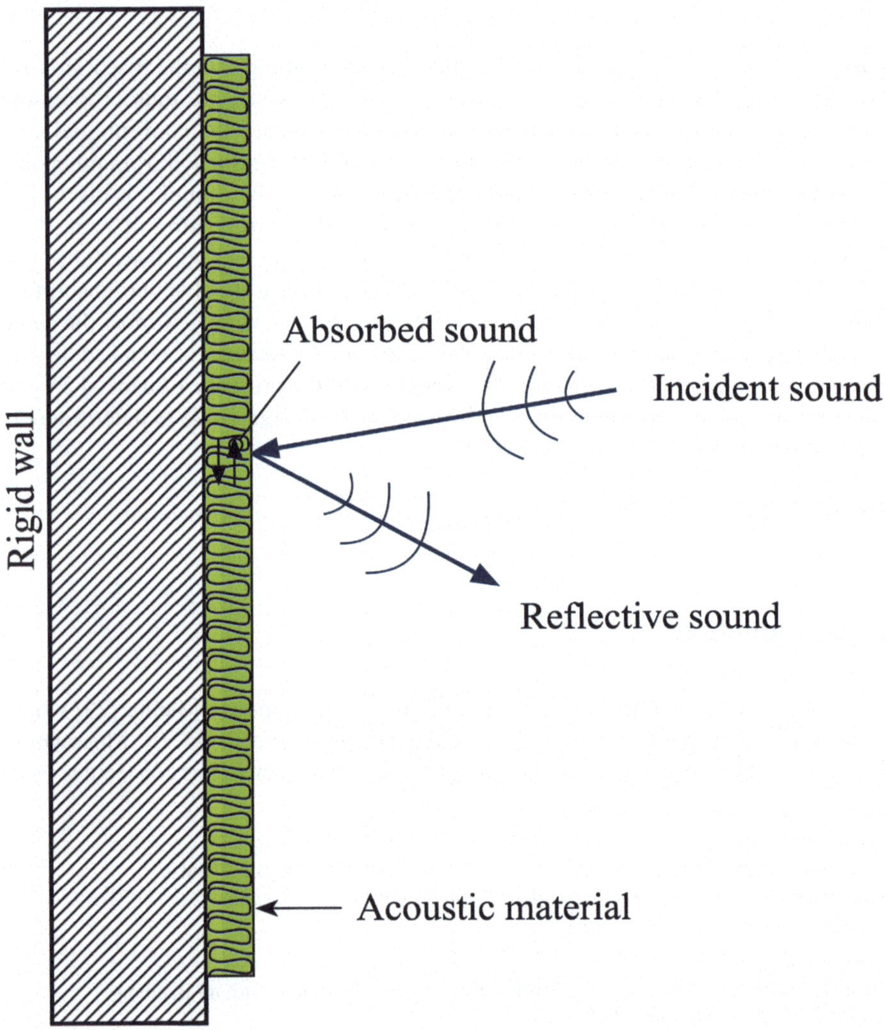

Fig. 1 Sound absorption mechanism to incident sound by an acoustic material

narrow absorption bandwidth. Conversely, the fibrous material system forming interconnected small diameter pores is typically poor in absorbing low-frequency sound, but it has wide absorption bandwidth covering mid and high frequency. However apart from this, the fibrous porous material is still preferable to be used for practical purposes due to its easy constructions and attractive cost/performance ratio.

Man-made mineral fibrous porous materials like fiber glass and rock wool are typical commercial sound absorber materials. Such absorbers are typically applied on the interior side of building walls to absorb incoming waves so that excessive reflected waves can be reduced. See Fig. 2.

However, these materials are non-degradable where the disposal of the materials can cause an environmental burden. These facts drive researchers to seek alternative fibers that meet criteria of biodegradable, eco-friendly, renewable, health-safe, and easily available in nature. Considering the requirements, natural fibers become strong candidate to meet such criteria. Moreover, typical natural fibers have hollow strand structures so that their porosity is high with pore–to-pore distance greater than mean free path of air molecules [8]. This is an ideal property for absorbing sound as substantial portion of the pressure wave is able to penetrate the material before encountering a solid section.

The sources of natural fibers are mainly from animal (e.g., wool and fur felt), vegetables (e.g., kenaf and hemp), and wood. In particular, the vegetable fibers can be obtained from various parts of plants and can be categorized into seed fiber (e.g., cotton and kapok); bast fiber or skin fiber (e.g., flax, jute, kenaf, hemp, ramie); leaf (or hard) fiber (e.g., sisal, palm, agave, cocos, bromelia); stalk or wood fiber (e.g., straw of wheat, rice, softwood, or hardwood); and fruit fiber (e.g., coconut) [1, 8].

Since the past ten years, researchers have focused their studies on natural materials. Numerous works on acoustic materials from natural materials have been published. Fouladi et al. [18] measured natural organic multi-layer coir fibers which showed good absorption coefficient of 0.85 in average from 1 kHz. The performance at lower frequency was improved by adding granular materials from rice husk grain in the coir fibers sample [26]. Oil palm fibers from the empty fruit bunch have similar characteristics with coir fibers, where they are type of hard fibers with relatively large diameter. The sound absorption of the former has been studied, and measurement showed similar sound absorption performance with the coir fibers [31]. Other natural materials have been shown to also have good sound absorption performance. These include tea-leaf fibers [38, 15], paddy fibers [27, 33], jute fibers [17], ijuk (Arenga Pinnata) [20], kapok fibers [39], kenaf fibers [25], sisal-kenaf composite [14], date-palm fibers [36], and numerous other natural materials in the form of composite.

In this chapter, we report our work on pineapple leaf fibers (PALFs) as sound absorber. The PALFs have been shown to have high Young's modulus and tensile strength [5], including its thermal conductivity [4] among other natural fibers. The works concerned on the PALF as the reinforced fibers in composites have also been published [3, 4, 13]. Moreover, the fibers are also well known to have high content of cellulose and thus can be used as the source of cellulose nanofibrils for biomedical applications and biotechnological applications [10].

(a)

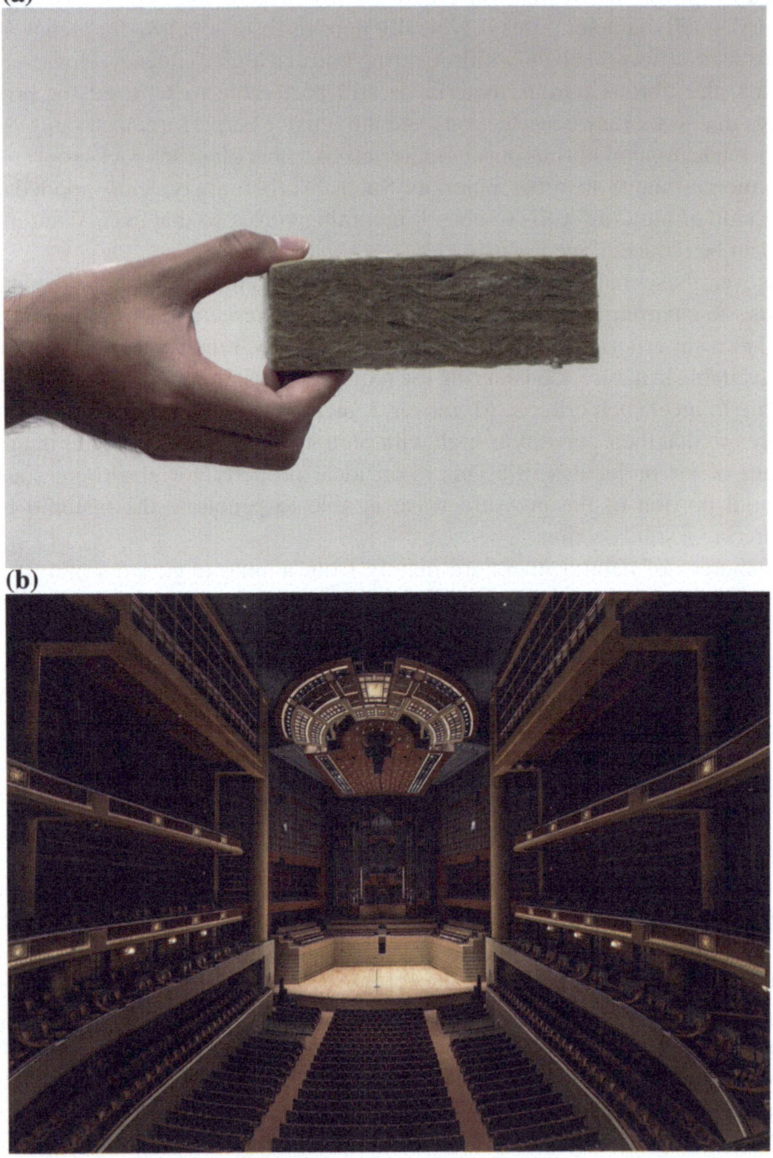

(b)

Fig. 2 **a** Rock wool acoustic absorber specimen and **b** a symphony hall auditorium where the use of acoustic materials is important to produce excellent room acoustic quality [35]

2 Fundamental of Sound Absorption

Fundamental theory of sound absorption mechanism can be traced back to the work of Zwikker and Konten [41]. Porous mediums are considered consisting of solid phase (frame) and fluid phase (air). Viscous and thermal interactions between the sound field and the fibers introduce energy dissipation in rigid and limp frame. For this case, the primary energy dissipation is in the fluid phase. Extra dissipation can also be present due to the flexibility of the solid frame resulting in inter-fiber friction as well as fiber vibration, as this induces the increase of temperature. In this case, the solid structure also contributes to the energy transfer [7].

To significantly absorb the incoming wave, an absorber needs to have thickness about a minimum of a tenth-wavelength and maximum of a quarter-wavelength, equivalent to the incident dominant frequencies [11]. From macroscopic point of view, some physical parameters of the absorbing material are relevant to determine the absorption behavior [16]. They are

(1) porosity ϕ: the ratio of void area to the total area.
(2) static flow resistivity σ: the ratio between the pressure difference across the thickness and the induced normal velocity through it under the condition of steady low-speed flow.
(3) tortuosity (or geometric tortuosity) α_∞: the degree of irregularity for pore direction.

The presence of visco-thermal phenomenon in absorbing mechanism alters the effective density ρ_e and the effective bulk modulus K_e (or compressibility) of a gas undergoing oscillatory motion within rigid frame. This leads to the difference of sound speed compared to that in the free space. In other words, the consequences of the visco-thermal effect are in the form of the effective density and the bulk modulus to contain the attenuation factor so that both parameters are now complex quantities; the sound speed in the absorbing material is slower than in the free space.

The concept of impedance is found to be useful in modeling sound absorption as wave propagates through different media. The behavior of pressure, particle velocity, and sound speed is attributed to characterize the impedance Z_c and the wavenumber k_c. For an absorptive layer, both parameters are related with the complex ρ_e and K_e. This yields [2]

$$Z_c = \sqrt{\rho_e K_e} \quad (1)$$

$$k_c = \omega\sqrt{\frac{\rho_e}{K_e}} \quad (2)$$

where ω is the frequency (in radian/second). It should be noted that Z_c and k_c are also complex quantities indicating that particle velocity and pressure are no longer in phase.

The complex ρ_e and K_e can be further determined depending on the pore geometry of the acoustic material where the complex mass density and bulk density are present as the consequences of viscous and thermal layer interaction in the pores. General expressions of ρ_e as the function of frequency are expressed as [9, 23]

$$\rho_e(\omega) = \rho_0\alpha_\infty\left(1 + \frac{\sigma\phi}{j\omega\rho_0\alpha_\infty}G_j(\omega)\right) \tag{3}$$

where ρ_0 is the density of air in free space (1.2 kg/m^3), α_∞ is the tortuosity, which represents the complexity of propagation path in the material, ϕ is the porosity, σ is the flow resistivity, and Λ is the viscous characteristic length and

$$G_j(\omega) = \left(1 + j\frac{4\omega\rho_0\alpha_\infty^2\eta}{\sigma^2\phi^2\Lambda^2}\right)^{1/2} \tag{4}$$

where η is the viscosity of air. The effective bulk modulus is expressed as

$$K_e(\omega) = \frac{\gamma P_0/\phi}{\gamma - (\gamma - 1)G'(\omega)} \tag{5}$$

where $P_0 = 1.0132 \times 10^5$ Pa, $\gamma = 1.4$ for air at 18 °C, and with $G'(\omega) = \left[1 - j\frac{8\kappa}{\Lambda' C_p \rho_0 \omega}\sqrt{1 + j\frac{\Lambda'^2 C_p \rho_0 \omega}{16\kappa}}\right]$ where Λ' is the thermal characteristic length.

The preceding models are also well known as the Johnson–Champoux–Allard model. Although the model provides good prediction of sound absorption coefficient, it is also known that the parameters of viscous characteristic length Λ and the thermal characteristic length Λ' are the most challenging quantities to be measured.

Alternatively, Z_c and k_c can be estimated based on empirical model proposed by Delney and Bazley [12] where

$$Z_c = \rho_0 c\left[1 + 0.078\left(\frac{\rho_0 f}{\sigma}\right)^{-0.623} - j0.074\left(\frac{\rho_0 f}{\sigma}\right)^{-0.66}\right] \tag{6}$$

$$k_c = \frac{\omega}{c}\left[0.0987\left(\frac{\rho_0 f}{\sigma}\right)^{-0.7} - j0.189\left(\frac{\rho_0 f}{\sigma}\right)^{-0.595}\right] \tag{7}$$

The model was developed based on experimental data from acoustic materials with small diameter of fibers such as rock wool and glass wool, and thus, the validity of Eqs. (6) and (7) only extends over the range $0.01 < \rho_0 f/\sigma < 0.1$.

The surface impedance Z_s of a sound absorber with thickness D backed up by a rigid wall can be defined by making use of Z_c and k_c in Eqs. (1) and (2) and also in Eqs. (6) and (7) written as [2]

$$Z_s = -jZ_c\cot(k_c D) \tag{8}$$

The normal sound absorption coefficient is thus calculated by

$$\alpha_n = \frac{4\mathrm{Re}(Z_s)}{(1 + \mathrm{Re}(Z_s))^2 + (\mathrm{Im}(Z_s))^2} \tag{9}$$

Making use of Eqs. (6)–(9), we introduce the effect of flow resistivity and absorber thickness on sound absorption coefficient and the results are shown in Fig. 3a. It is clear that the flow resistivity σ affects the behavior of sound absorption in which greater σ increases sound absorption coefficient. However, the increment is not linear to the flow resistivity value. Hence, doubling the flow resistivity does not mean doubling gain in absorption. Moreover, excessive flow resistivity value can reduce wave penetration into porous media that leads to excessive reflection.

The absorber thickness is also a key factor in sound absorption as shown in Fig. 3b. For the same flow resistivity, doubling the thickness can increase sound absorption coefficient by 0.2. Thicker absorber enables the incoming wave to propagate into porous medium through longer path than in the thinner one. Moreover, the thicker the absorber layer, the lower the frequency of sound can be absorbed. Although it is not always the case, the rule of thumb of a quarter-wavelength is often used in practice to design the thickness of the absorber depending on the dominant frequency of sound to be absorbed [24].

3 Methodology

3.1 Preparation of Sound Absorber Samples

3.1.1 Fiber Extraction

The pineapple leaf contains large fiber bundles (100–460 μm) in the middle and smaller technical fiber (30–80 μm) arranged longitudinally at the bottom of the leaf. The fiber surfaced is covered by waxy layer, so to extract the fiber the waxy layer must first be removed [28].

There are two methods to extract the pineapple leaf fibers (PALF). The first one is by using a conventional method by scrapping the leaf on a bench. The tool is known as "ketam." We applied the advanced method by using a machine called "Mesin Pemacah Daun Nenas" or the Pineapple Leaf Scrapper Machine as shown in Fig. 4. To obtain the extracted fibers using this machine, only two steps are required instead of six steps using the "ketam" tool.

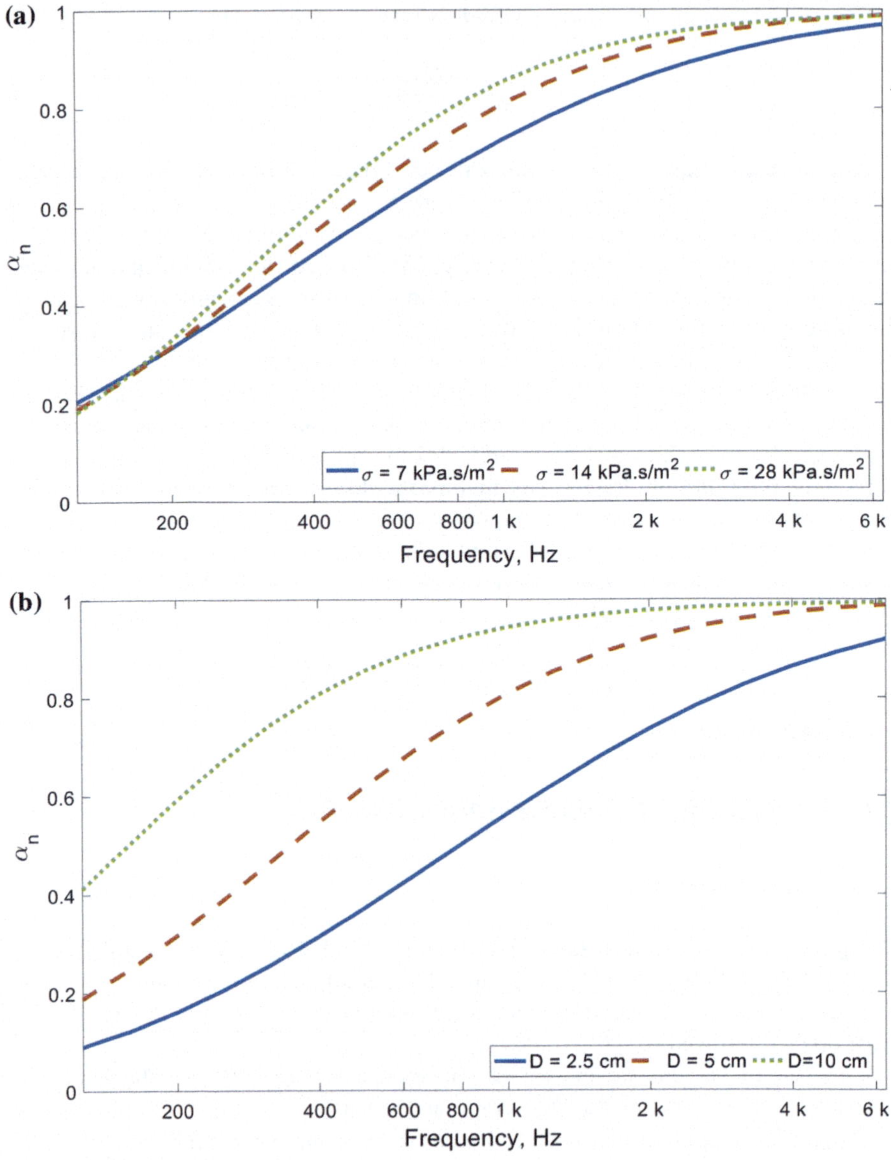

Fig. 3 Effect of physical property variation on absorption characteristics: **a** flow resistivity, σ, and **b** panel thickness, D

Fig. 4 Mesin Pemecah Daun Nenas or Pineapple Leaf Scrapper Machine

In the machine, by means of "grinding" process, the pineapple leaves went through a feed roller and a scratching roller. In this step, parts of the cellulose, waxy layers, and water from the pineapple leaves were scrapped and removed. The extracted raw PALF is shown in Fig. 5a.

According to the study in Ref. [37], PALF contains chemical entities like holocellulose (87.56%), alpha-cellulose (78.11%), hemicellulose (9.45%), and lignin (4.78%). All the chemical constituents and also dirt and particle must be detached by using alkaline treatment. The PALF was then immersed in distilled water with 1% natrium hydroxide (NaOH) for 1 hour. The fibers were washed by distilled water in order to neutralize the remaining alkali. Lastly, the fibers were dried directly under the sun to remove the moisture content. Figure 5b shows the PALF after the alkaline treatment process and to be used for the sound absorber.

3.1.2 Sample Fabrication

To observe the effect of density of the PALF and the thickness of the absorber on the sound absorption performance, the test samples were designed with several densities and thickness. For the density, the bulk density was used and is defined as the mass of the PALF, m over the total volume of the cylindrical casing, V containing the fibers given by $\rho = m/V$. The fibers were first weighted and then inserted into an

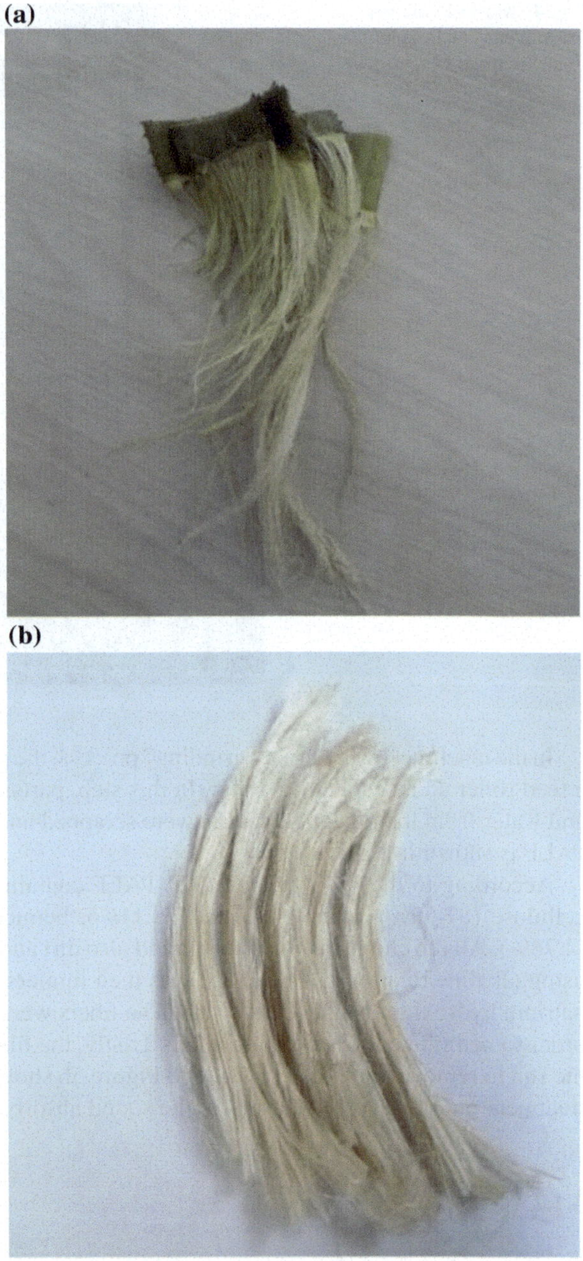

Fig. 5 **a** Raw PALF extracted from the machine and **b** PALF after the alkaline treatment

Fig. 6 Test samples of PALF sound absorber

aluminum web casing having diameter of 33 mm to fit inside the impedance tube in the sound absorption measurement. The thickness of the casing defines the thickness of the absorber sample. See Fig. 6.

3.2 Measurement of Sound Absorption Coefficient

3.2.1 Methods in Measurement

The objective of sound absorption measurement is to measure the sound absorption coefficient [as in Eq. (9)] of the acoustic material. Figure 1 can be used as a schematic diagram to illustrate the sound power propagation in an acoustic material. By assuming, the sound power of the incident sound is W_{i} and the reflected power is W_{r}, and the sound power absorbed by the material is thus

$$W_{\mathrm{a}} = W_{\mathrm{i}} - W_{\mathrm{r}} \tag{10}$$

The absorption coefficient of the material is therefore the fraction of absorbed sound power to the incident sound power expressed as

$$\alpha = \frac{W_{\mathrm{a}}}{W_{\mathrm{i}}} = 1 - r \tag{11}$$

where $r = W_{\mathrm{r}}/W_{\mathrm{i}}$ is denoted as the reflection coefficient.

It is almost impossible to measure directly the absorbed sound power inside the material, but it is convenient to measure the incident and the reflected sound on the surface of the material. Based on Eq. (11), the reflection coefficient, *r*, is therefore the most important quantity to be measured to indirectly determine the absorption coefficient, α. The reflection coefficient, r; is varied with frequency of the incident sound and so does the absorption coefficient, α.

Measurement of sound absorption coefficient can be divided into two methods based on the incident sound field, namely

a. Diffuse field method: the material under test is impinged by the sound coming from all possible angles of incidence.
b. Normal incident field method: the material under test is impinged by the sound where the propagation is perpendicular to the surface of the material.

The illustration is shown in Fig. 7.

For the diffuse field method, a special room with highly reflective surfaces is required to maximize the reflective sound energy (minimum reflection loss) inside the room as illustrated in Fig. 7a. The room is usually constructed with non-parallel walls to further optimally build up the diffuse field energy. One can hear the sound from a clapped hand reverberated across the walls in the room before the sound finally dies off. This room is usually called reverberant chamber. A good chamber is where the reverberation time is sufficiently long (can achieve up to 10 s or more), and thus, a large room is preferable. In addition, the greater the volume of the chamber, the lower the starting frequency of the valid diffuse field, which means the broader the valid frequency of the measurement [34].

This method is regulated in the ISO 354:2003 [22] where the volume of the room must be 150 m^3 at minimum and not greater than 500 m^3. At least 10 m^2 of test material is required for the measurement. The absorption coefficient is determined from the measured reverberation times, with and without the test material in the chamber.

For the normal incidence method, a circular tube (called impedance tube) is used where the sound energy can be conditioned to excite the test material at 0° angle of incidence. The circular tube diameter usually ranges from 30 to 100 mm as illustrated in Fig. 7b. For this purpose, the diameter of the tube must be smaller than a quarter-wavelength of the incoming sound incidence, which determines the highest valid frequency of the measurement. The 30 mm diameter is used for mid- to high-frequency measurement, while the 100 mm diameter is used for low to mid frequencies. Two acoustic microphones are placed in front of the material to measure the sound pressure generated inside the tube. From signal processing, the transfer function between the two sensors is obtained as a function of the reflection coefficient to calculate the absorption coefficient of the test material. The procedure of this method is documented in ISO 10534-2:2001 [21].

The diffuse field method can be seen to closely represent the performance of the acoustic materials as in the practical installation where the sound can excite the absorber panel from different angles of incidence. The method, however, requires a special room and a large amount of test material. The normal incidence using the impedance tube offers a simple and cheaper method with convenient preparation of the test samples. Although the normal incidence does not represent the sound field in practice, the test results are valid to be used to represent the performance of sound

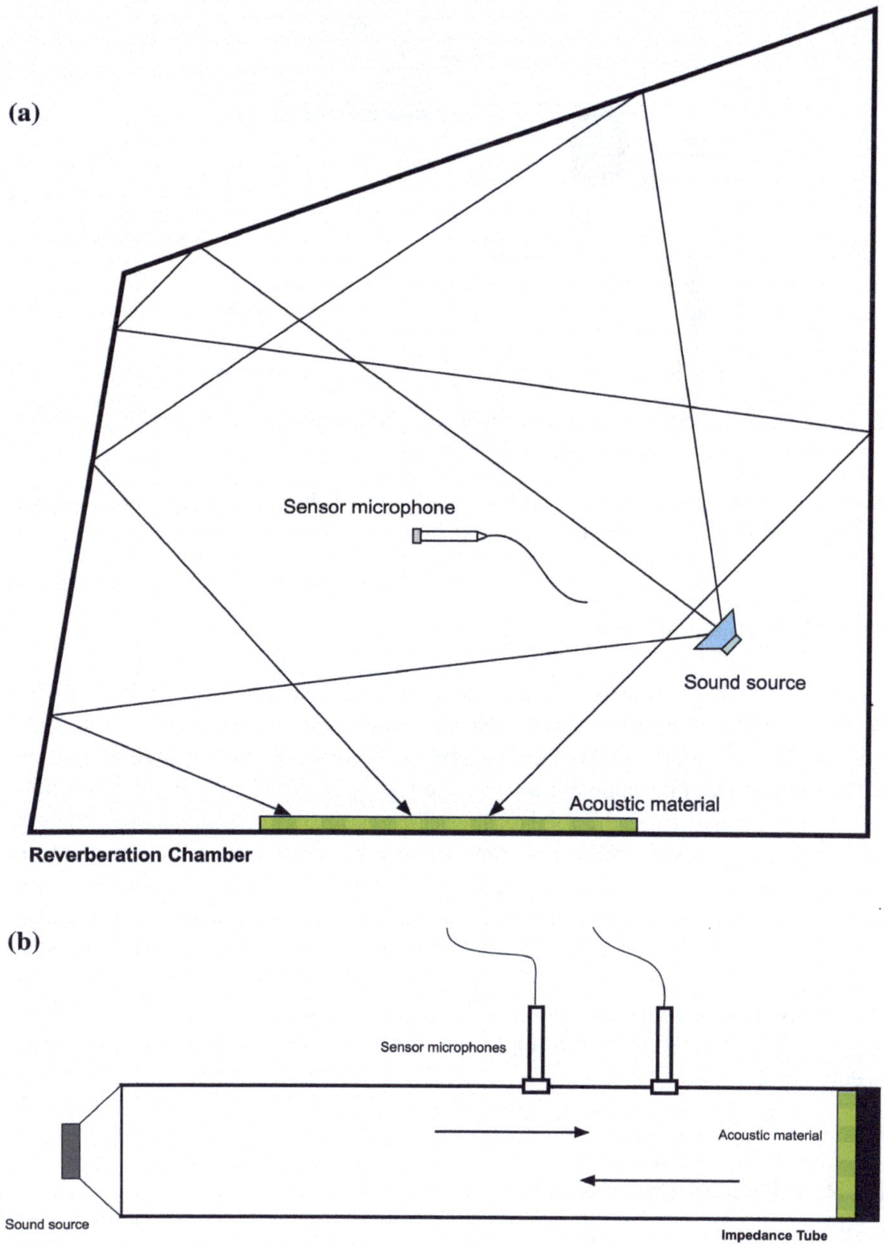

Fig. 7 Schematic diagram of two methods to measure the absorption coefficient of sound absorber: **a** diffuse field method and **b** normal incident field method

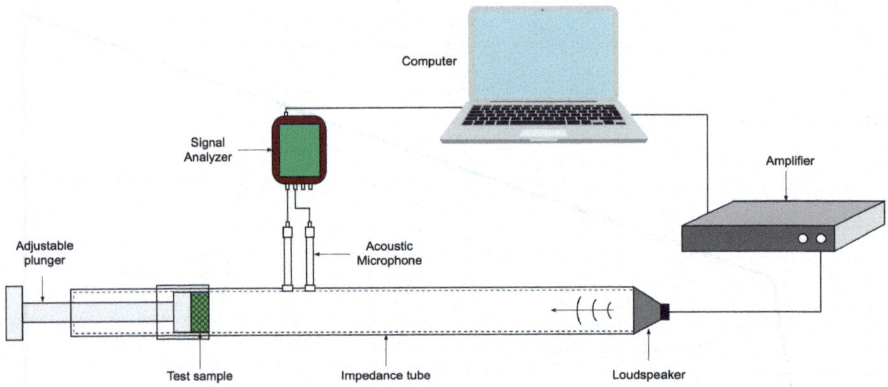

Fig. 8 Experimental setup for sound absorption coefficient measurement using an impedance tube

absorption of the test materials, and it usually noted that the absorption coefficient is usually greater if measured under the diffuse field method.

3.2.2 Experimental Setup

Figure 8 shows diagram of the experimental setup using the impedance tube following the ISO 10534-2. The impedance tube has inner diameter of 33 mm. The PALF test sample was fitted tightly into the sample holder and was located at one end of the impedance tube opposite to the location of the loudspeaker. A white noise was then generated from the loudspeaker. Two acoustic microphones recorded the built-up sound pressure inside the impedance tube. Prior to the measurement, the microphones were calibrated by using a sound calibrator to ensure that they measured the correct level of sound pressure. A signal analyzer was used as the data acquisition system to process the recorded signals and to obtain the transfer function between the two microphones. These transfer functions were later used to calculate the sound absorption coefficient as a function of frequency. Some of the results of measured absorption coefficient of the pineapple leaf fibers presented here are those presented in Putra et al. [32].

4 Results and Discussion

4.1 Effect of Density

Density of fibers has been shown to play an important role to significantly improve the absorption coefficient [31]. The greater the amount of fibers in the sample, the more effective the sound trapped inside the absorber. The flow resistivity and tortuosity

Table 1 Parameters of test samples for various fiber densities (constant thickness)

No.	Mass of fibers (g)	Thickness (mm)	Bulk density, ρ (kg/m^3)
1	1	20	58
2	2	20	117
3	3	20	175

increase as the inner porosity increases together with the complexity of the sound path. Measured absorption coefficients of PALF with constant absorber thickness, t, and various bulk densities, ρ_{bulk} (as listed in Table 1) is plotted in one-third octave bands in Fig. 9. It can be observed that by increasing the density, it increases the absorption coefficient toward low frequency.

However, it has also been shown that if the density is too high, the porosity reduces with the formation of close pores, and thus, it becomes difficult for the sound wave to penetrate the absorber [31].

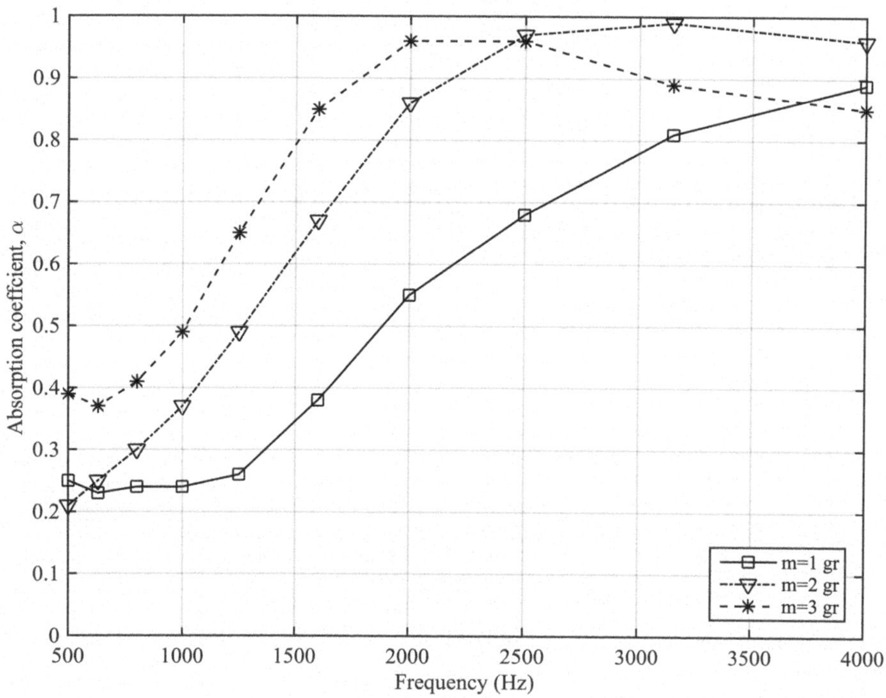

Fig. 9 Measured absorption coefficient of PALF absorber with various densities (□: $\rho_{\text{bulk}} = 58$ kg/m^3, ∇: $\rho_{\text{bulk}} = 117$ kg/m^3, ∗: $\rho_{\text{bulk}} = 175$ kg/m^3, $t = 20$ mm)

4.2 *Effect of Thickness*

Increasing the thickness of a sound absorber panel has been shown as an effective way to increase the sound absorption performance. See again Fig. 3b. The thicker the absorber means the longer it takes for the sound (with the wavelength shorter than the thickness) to travel through the depth of the absorber.

Figure 10 shows the measured absorption coefficient of the PALF fibers with different thicknesses of sample. To have consistent results on the effect of thickness, the density is kept constant as listed in Table 2. The improvement of absorption coefficient towards low frequency can be obviously observed as the thickness is increased. With 30 mm-thick sample, the PALF shows good sound absorption performance ($\alpha > 0.5$) above 500 Hz.

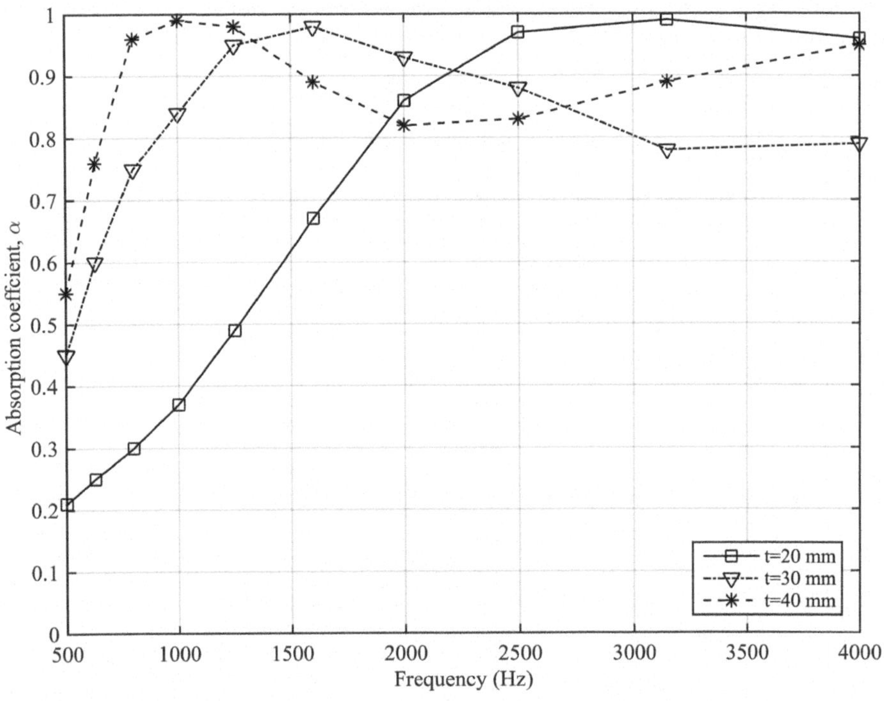

Fig. 10 Measured absorption coefficient of PALF absorber with various thicknesses ($\rho_{\text{bulk}} = 117$ kg/m^3)

Table 2 Parameters of test samples for various absorber thicknesses (constant density)

No.	Mass of fibers (g)	Thickness (mm)	Bulk density, ρ (kg/m^3)
1	2	20	117
2	3	30	117
3	4	40	117

4.3 Effect of Air Gap

Besides increasing the thickness of the absorber panel, another popular method to improve low-frequency absorption is by introducing air gap between the absorber and the rigid wall. This is useful for a thin absorber panel, and thus, it saves the amount of fibers and also cost.

The "quarter-wavelength" rule of thumb is usually used in practice to determine the required air gap (and the thickness) of the absorber to control the target frequency of absorption. It determines the expected peak frequency of absorption which is equivalent to the quarter-wavelength, λ of the absorber thickness and the air gap, $d + D$. At the quarter-wavelength, the sound has maximum velocity at the surface of the absorber to penetrate the absorber with maximum energy [24]. The illustration is shown in Fig. 11.

In the experiment, the plunger in the impedance tube (see Fig. 8) can be adjusted to provide an air gap at the back of the test sample.

Figure 12 shows the measured absorption coefficient for sample with thickness $t = 10\,\text{mm}$ and 20 mm with different depths of air gap. It demonstrates how the introduction of air gap can effectively improve the absorption coefficient effectively at low frequency.

The peaks of absorption can be observed to obey the quarter-wavelength rule. For example in Fig. 12a, for thickness of $t = 10\,\text{mm}$ and air gap $D = 30\,\text{mm}$, the peak is at around 2 kHz. The quarter-wavelength, in this case, is thus $(1/4)\lambda = t + D = 40\,\text{mm}$, which gives $\lambda = 0.16\,\text{m}$. The frequency at this wavelength is therefore $f = c/\lambda = 343/0.16 \approx 2144\,\text{Hz}$, where $c = 343\,\text{m/s}$ is the speed of sound. Similarly, in Fig. 12b for sample thickness of 20 mm with air gap of 30 mm, by applying the quarter-wavelength rule, this gives $\lambda = 0.2\,\text{m}$. The frequency is $f = 343/0.2 \approx 1700\,\text{Hz}$. The measured peak can be seen to be around this frequency.

However, care must be taken at the frequency equivalent to the half-wavelength where the sound velocity is minimum at the front surface of the absorber, and thus, it has low sound absorption. In the sound absorption coefficient curve, this is indicated by a dip at twice the peak frequency of absorption. Although the dip is not so obvious, this can be seen in Fig. 12b for air gap of 30 mm, where for peak frequency at around 1.5–1.7 kHz, the absorption coefficient reduces to form a dip at around 3–3.5 kHz.

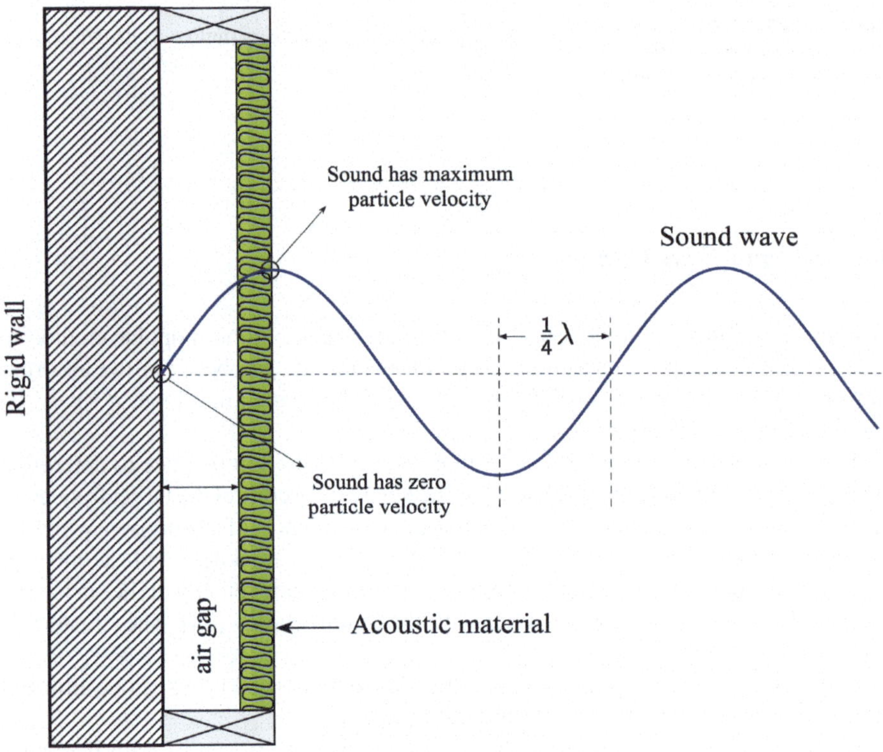

Fig. 11 Illustration of peak sound absorption at the quarter-wavelength

4.4 Effect of Fabric Cover

In practice, most of the fibrous acoustic absorbers are not directly exposed visually, but they are covered by a fabric layer for protection or artistic purpose. Here, a cotton fabric was used to cover the front surface of the sample as seen in Fig. 13. The measured absorption coefficient can be seen in Fig. 14 to increase due to the fabric layer. However, as the thickness of the sample increases, the effect becomes smaller. This is also found for sound absorption of paddy fibers [33].

5 Conclusion

Sound absorption performance of natural pineapple leaf fibers (PALF) has been discussed. The measurement results show that the PALF can have good absorption coefficient of $\alpha = 0.7$–1 above 800 Hz for thickness of 30 mm and density of 117 kg/m^3. By introducing the air gap backing the absorber panel, 10–mm-thick

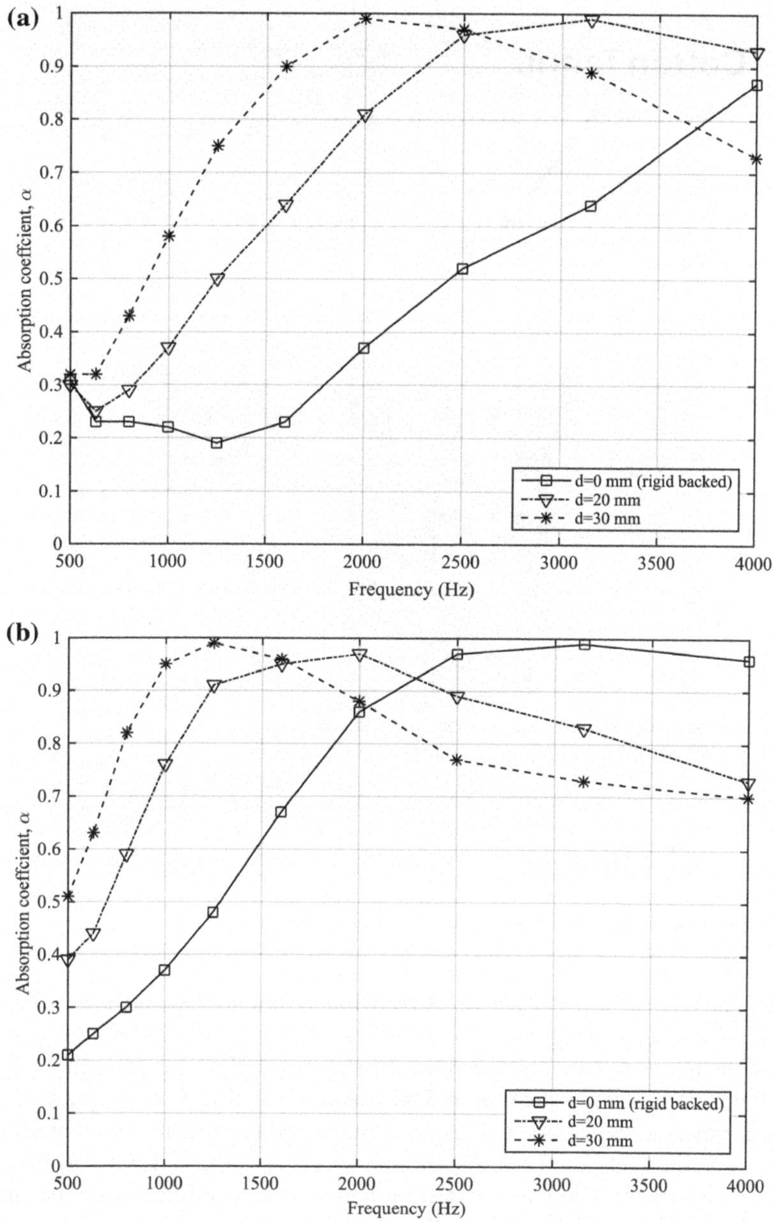

Fig. 12 Measured absorption coefficient of PALF absorber with various air gaps ((a) t = 10 mm, (b) t = 20 mm, $\rho_{\text{bulk}} = 117$ kg/m^3)

Fig. 13 Test sample of PALF absorber with the front cotton fabric

absorber can have $\alpha = 0.7$–1 above 1 kHz with 30 mm air gap and almost the same absorption coefficient range for 20-mm-thick absorber with 20 mm air gap. The effect of the quarter-wavelength can be clearly observed where this can be used as the guideline to determine the required thickness of the absorber. The pineapple leaf fibers can therefore be a potential green sound absorber to replace the conventional synthetic and mineral absorbers. As the characteristic of the PALF is soft fibers, the fibers can also be the fillers and as a front layer for the hard fibers, such as natural coir and palm oil fibers to enhance the sound absorption performance as already discussed by Ying et al. [40], as the latter are found to be more conveniently shaped (compression process) to become a more compact panel. The applications can range from room acoustic material to industrial sound absorber for noise control.

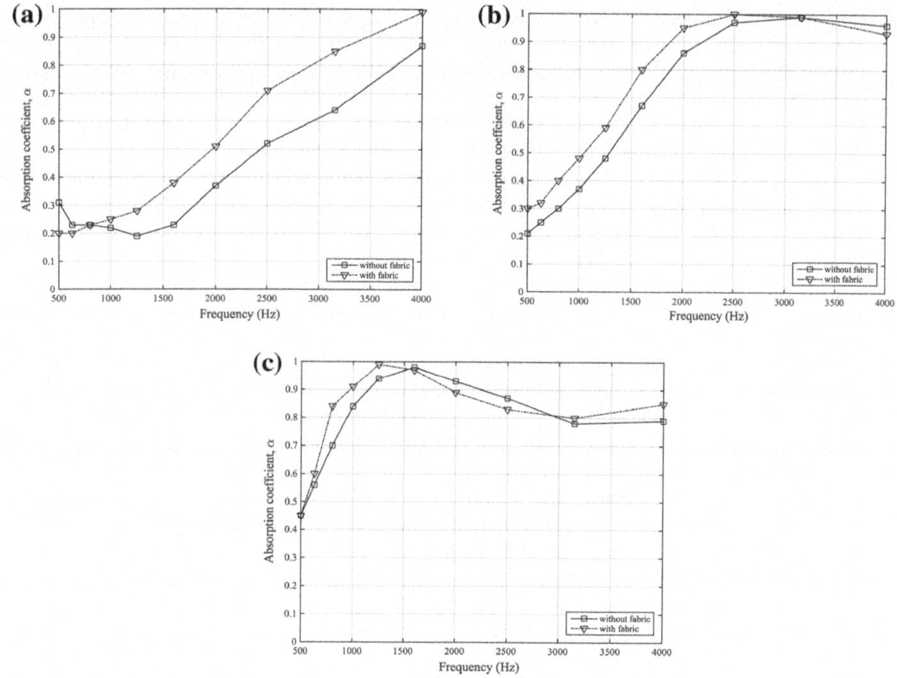

Fig. 14 Measured absorption coefficient of PALF absorber with fabric cover ((a) t = 10 mm, (b) t = 20 mm, (c) t = 30 mm, ρ_{bulk} = 117 kg/m^3)

References

1. Ahmad F, Choi HS, Park MK (2015) A review: natural fiber composites selection in view of mechanical, light weight, and economic properties. Macromol Mater Eng 300(1):10–24
2. Allard J, Atalla N (2009) Propagation of sound in porous media: modelling sound absorbing materials, 2nd edn. Wiley
3. Arib R, Sapuan S, Ahmad M, Paridah M, Zaman HK (2006) Mechanical properties of pineapple leaf fibre reinforced polypropylene composites. Mater Des 27(5):391–396
4. Asdrubali F, D'Alessandro F, Schiavoni S (2015) A review of unconventional sustainable building insulation materials. Sustain Mater Technol 4:1–17
5. Asim M, Abdan K, Jawaid M, Nasir M, Dashtizadeh Z, Ishak M, Hoque ME (2015) A review on pineapple leaves fibre and its composites. Int J Polym Sci 2015:950567 (16 pages). https://doi.org/10.1155/2015/950567
6. Basner M, Babisch W, Davis A, Brink M, Clark C, Janssen S, Stansfeld S (2014) Auditory and non-auditory effects of noise on health. Lancet 383(9925):1325–1332
7. Bell LH, Bell DH (1994) Industrial noise control: fundamentals and applications. Marcel Dekker, New York
8. Berardi U, Iannace G (2015) Acoustic characterization of natural fibers for sound absorption applications. Build Environ 94:840–852
9. Champoux Y, Allard JF (1991) Dynamic tortuosity and bulk modulus in air-saturated porous media. J Appl Phys 70(4):1975–1979
10. Cherian BM, Leão AL, De Souza SF, Thomas S, Pothan LA, Kottaisamy M (2010) Isolation of nanocellulose from pineapple leaf fibres by steam explosion. Carbohyd Polym 81(3):720–725

11. Cox T, D'Antonio P (2016) Acoustic absorbers and diffusers: theory, design and application, 3rd edn. CRC Press, Boca Raton
12. Delany M, Bazley E (1970) Acoustical properties of fibrous absorbent materials. Appl Acoust 3(2):105–116
13. Devi LU, Bhagawan S, Thomas S (1997) Mechanical properties of pineapple leaf fiber-reinforced polyester composites. J Appl Polym Sci 64(9):1739–1748
14. Dunne R, Desai D, Sadiku R (2017) Material characterization of blended sisal-kenaf composites with an abs matrix. Appl Acoust 125:184–193
15. Ersoy S, Kucuk H (2009) Investigation of industrial tea-leaf-fibre waste material for its sound absorption properties. Appl Acoust 70(1):215–220
16. Fahy F (2013) Foundations of engineering acoustics. Elsevier Academic Press, London
17. Fatima S, Mohanty A (2011) Acoustical and fire-retardant properties of jute composite materials. Appl Acoust 72(2–3):108–114
18. Fouladi MH, Nor MJM, Ayub M, Leman ZA (2010) Utilization of coir fiber in multilayer acoustic absorption panel. Appl Acoust 71(3):241–249
19. Ingard U (1994) Notes on sound absorption technology. Noise Control Foundation, New York
20. Ismail L, Ghazali MI, Mahzan S, Zaidi AA (2010) Sound absorption of arenga pinnata natural fiber. World Academy of Science, Engineering and Technology 67:804–806
21. ISO-10534-2 (2001) Acoustics-determination of sound absorption coefficient and impedance in impedances tubes-part 2: transfer-function method. ISO, Brussels, Belgium
22. ISO-354 (2003) Acoustics-measurement of sound absorption in a reverberation room. British Standards Institution
23. Johnson DL, Koplik J, Dashen R (1987) Theory of dynamic permeability and tortuosity in fluid-saturated porous media. J Fluid Mech 176:379–402
24. Kuttruff H (2016) Room acoustics, 6th edn. CRC Press
25. Lim ZY, Putra A, Nor MJM, Yaakob MY (2018) Sound absorption performance of natural kenaf fibres. Appl Acoust 130:107–114
26. Mamtaz H, Fouladi MH, Nuawi MZ, Namasivayam SN, Ghassem M, Al-Atabi M (2017) Acoustic absorption of fibro-granular composite with cylindrical grains. Appl Acoust 126:58–67
27. Mediastika CE (2009) Jerami sebagai bahan baku panel akustik pelapis dinding. DIMENSI J Archit Built Environ 36(1):20–27
28. Mohamed A, Sapuan S, Shahjahan M, Khalina A et al (2009) Characterization of pineapple leaf fibers from selected malaysian cultivars. J Food Agric Environ 7(1):235–240
29. Moszynski P (2011) WHO warns noise pollution is a growing hazard to health in Europe. British Medical Journal Publishing Group
30. Münzel T, Gori T, Babisch W, Basner M (2014) Cardiovascular effects of environmental noise exposure. Eur Heart J 35(13):829–836
31. Or KH, Putra A, Selamat MZ (2017) Oil palm empty fruit bunch fibres as sustainable acoustic absorber. Appl Acoust 119:9–16
32. Putra A, Or KH, Selamat MZ, Nor MJM, Hassan MH, Prasetiyo I (2018) Sound absorption of extracted pineapple-leaf fibres. Appl Acoust 136:9–15
33. Putra A, Abdullah Y, Efendy H, Mohamad W, Salleh N (2013) Biomass from paddy waste fibers as sustainable acoustic material. In: Advances in acoustics and vibration 2013
34. Schroeder MR (1996) The "schroeder frequency" revisited. J Acoust Soc Am 99(5):3240–3241
35. Skeeze (2019) Symphoni hall. https://pixabay.com/photos/symphony-hall-auditorium-concert-893342/. Access 14 Aug 2019
36. Taban E, Khavanin A, Ohadi A, Putra A, Jafari AJ, Faridan M, Soleimanian A (2019) Study on the acoustic characteristics of natural date palm fibres: experimental and theoretical approaches. Build Environ. https://doi.org/10.1016/j.buildenv.2019.106274
37. Trevor J, D'Antonio P (2016) Acoustic absorbers and diffusers, 3rd edn. Taylor & Francis
38. Wong K, Ahsan Q, Putra A, Subramonian S, Nor MJM (2017) Preliminary study on the sound absorption behavior of spent tea leaves filled with natural rubber latex binder. J Teknol 79(5–2)

39. Xiang H, Wang D, Liua H, Zhao N, Xu J (2013) Investigation on sound absorption properties of kapok fibers. Chinese J Polym Sci 31(3):521–529
40. Ying LZ, Putra A, Nor MJM, Muhammad N, Yaakob MY (2016) Sound absorption of multilayer natural coir and kenaf fibers. In: Proceedings of 23rd international congress on sound and vibration. International Institute of Acoustics and Vibrations, Greece
41. Zwikker C, Kosten CW (1949) Sound absorbing materials. Elsevier

Physicochemical Properties of Nanocellulose Extracted from Pineapple Leaf Fibres and Its Composites

Ismail Muhamad Fareez, Nazmul Haque, Der Juin Ooi, Ainil Hawa Jasni and Fauziah Abd Aziz

Abstract Significant advancement on cellulose-based biomaterial research has also led to the development of nano-sized pineapple leaf cellulose fibres with wide application potentials. The present chapter presents the comprehensive review of cellulose fibre structure extracted from different pineapple varieties, covering some aspects related to the structure of this natural cellulose in terms of its morphology, chemical, physical and mechanical properties. This chapter also briefly introduces the fundamentals of nanocellulose and discussed the isolation and properties of pineapple leaf cellulose nanofibrils, nanofibrillated cellulose and cellulose nanocrystals in view to open further areas of composite study on the ideal selection of these nanomaterials for industrial use.

Keywords Cellulose · Pineapple leaf fibre (PALF) · Nanocellulose · Crystallinity · Mechanical properties

1 Introduction

The use of natural fibres for composites, as alternatives to traditional reinforcement materials such as glass fibres, carbon and talc, offers numerous potential advantages such as low cost, high specific properties, low energy consumption, low density, biodegradability, flexibility and availability of a wide variety of fibres around the globe [1]. These materials not only promote agriculture-based economy and rural

I. M. Fareez (✉) · N. Haque · D. Juin Ooi
Department of Oral Biology and Biomedical Sciences, Faculty of Dentistry, MAHSA University, Saujana Putra Campus, 42610 Jenjarom, Selangor Darul Ehsan, Malaysia
e-mail: muhamadfareez@mahsa.edu.my

A. H. Jasni
Chemical Defense Research Centre (CHEMDEF), National Defence University of Malaysia, Kem Sg. Besi, 57000 Kuala Lumpur, Malaysia

F. A. Aziz
Department of Physics, Centre of Defense Foundation Studies, National Defence University of Malaysia, Kem Sg. Besi, 57000 Kuala Lumpur, Malaysia

M. Jawaid et al. (eds.), *Pineapple Leaf Fibers*, Green Energy and Technology,
https://doi.org/10.1007/978-981-15-1416-6_9

income, but also minimize the use of processing machineries and hazardous chemicals, thus reducing health risks. Indeed, the use of natural fibres as the reinforcement for polymeric matrix has demonstrated the importance in a broad range of applications. They are used across varied industries including automotive, aerospace, construction and food processing [2].

Natural fibres, also referred to as cellulosic fibres, mainly comprise cellulose, lignin, hemicellulose and trace amount of pectin, pigments and extractives. These cellulose fibres can be derived from different plant parts including bast (jute, hemp, flax, mesta and kenaf), leaf (pineapple, henequen, screw pine and sisal) and fruit or seed fibres (cotton, coir and oil palm) [3, 4]. The properties of the cellulose in these natural fibres vary depending on the source materials. In particular, its properties are strongly affected by the chemical composition, microfibrillar angle, internal fibre structure and cell dimensions [5, 6]. The cellulose molecules have a wide range of different diameters (0.1–3.0 nm) that arranged into microfibrils (diameter around 10–30 nm) and subsequently forming the cell wall of the plant fibre. The plant fibre has distinctive lengths (1–50 mm) and diameters (10–50 μm). Each type of cellulose present in the fibre structure possesses own crystalline organization and affects the mechanical properties of the structure.

In tropical countries, fibrous plants are available in abundance as some of them are a waste product of the plant cultivation that are cheaply available in inexhaustible amount. Pineapple (*Ananas comosus*) is one of them [7]. Interestingly, PALF exhibits excellent mechanical properties among other plant fibres. Figure 1 shows the scheme of crystalline cellulose prepared from raw PALF and its microfibril organization.

In general, pineapples are classified into Smooth Cayenne, Abacaxi, Queen, Singapore Spanish, Cabezona, Sugar Loaf and Red Spanish varieties. The different varieties are further crossbred, and dozens of pineapple cultivars with own distinctive flavours and growth characteristics have now been produced [9]. For example, the Josapine variety is a hybrid between the 'Sarawak' variety ('Smooth Cayenne') and the 'Johor' variety ('Singapore Spanish' × 'Smooth Cayenne') produced by the Malaysian Agriculture Research and Development Institute (MARDI) [10]. It is known that these varieties are clearly illustrated in terms of crystallite size, cellulose content amount of non-cellulosic residues, microfibril angle and quality [11, 12]. In fact, it is worth pointing out that the characterization of the PALF from different geological sources is crucial in order to efficiently compare and exploit these resources [11].

The present chapter is geared towards highlighting the groundbreaking findings on the study of cellulose fibre and its characterization in micro and nanometric dimension, also called cellulose micro/nanofibre (CMNF), derived from various pineapple cultivars and varieties. The review of PALF at various properties is discussed in view to open further areas of composite study on ideal selection of these CMNF for both academic and industrial use.

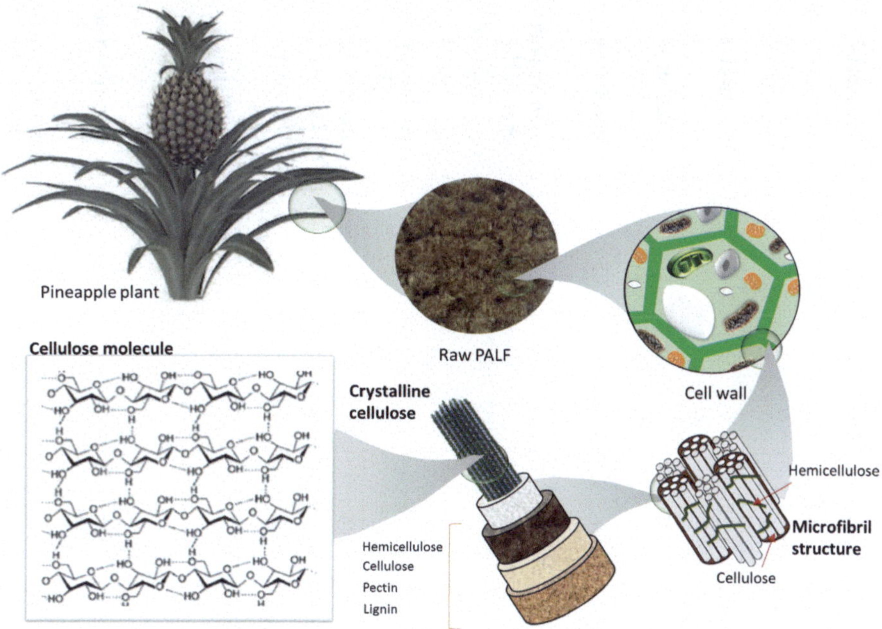

Fig. 1 Scheme of the cellulose cell wall and microfibril organization. Adapted from Fareez et al. [8]

2 Structure of Cellulose from Different Pineapple Leaf Fibre Varieties

Typically, this PALF consists of a vascular bundle system that forms a ribbon-like structure made up of bunches of fibrous cells which produced after mechanical removal of all the epidermal tissues [13]. Their high specific strength and stiffness are associated with its relatively high cellulose content (70–82%) and comparatively low microfibrillar angle (14°). Due to these properties, PALF is difficult to degrade in nature, and these can have an adverse effect to the environmental. About 70–85% of PALF is glucose polymer cellulose which is surrounded by a matrix of lignin (4–15%) and hemicelluloses (6–19%) in an amorphous state [7]. Both of these van der Walls and intermolecular hydrogen network create a relatively stable complex structural hierarchy of cellulose with high thermal degradations and superior mechanical properties [8, 14]. Table 1 listed the chemical composition of extracted PALF of different geographical regions and varieties. This is measured according to a standard method of the Technical Association of Pulp and Paper Industry (TAPPI). Regardless of the extraction methods, a comparison revealed that the PALF extractive content of cellulose, hemicellulose and lignin are ranging between 66–86, 11–23 and 2–15%, respectively. Neto et al. [14] on the hand, who studied 12 PALF varieties from Brazil which prepared solely through mechanical extraction without any chemical

Table 1 Chemical composition of pineapple leaf fibre (PALF) from different geographical regions

Origins	Variety name	Cellulose (%)	Hemicellulose (%)	Lignin (%)	Pectin (%)	Ash (%)	Fat and wax	Moisture content (wt%)	References
India	Unknown	85	–	12	–	–	–	–	Verma et al. [15]
Central Java, Indonesia	Unknown	67.1–69.3	–	14.5–15.4	–	1.21	–	–	Siregar et al. [16]
Penang, Malaysia	Unknown	73.4	16.6	10.5	–	2.0	–	–	Khalil et al. [17]
Tamil Nadu, India	Unknown	81.0	–	12.7	–	–	–	13.5	Idicula et al. [18]
Penang, Malaysia	Unknown	74.33	–	10.41	–	4.73	–	–	Wan Nadirah et al. [19]
Sao Paulo, Brazil	Unknown	81.27	12.31	3.46	–	–	–	10.52	Cherian et al. [7]
Minas Gerais, Brazil	Unknown	74.5	20.4	8.72	–	2.28	–	–	dos Santos et al. [20]
India	Unknown	69.5	17.8	4.4	1.1	0.9	3.3	–	Banik et al. [21]
Johor, Malaysia	Unknown	66.2	19.5	4.28	–	4.5		–	Daud et al. [22]
Uthai Thani, Thailand	Unknown	71.0	20.0	2.0	–	–	–	–	Kengkhetkit and Amornsakchai [23]
Assam, India	Queen	67.65	11.35	4.2	1.2	0.63	–	–	Hazarika et al. [24]
Johor, Malaysia	Josapine	85.53	22.5	11.8	–	–	–	–	Fareez et al. [8]

treatment, had recorded cellulose content of only 49–63%; whereas, hemicellulose and lignin had 6–13 and 20–36%, respectively. The CMNF structure may be made of the same plant species but the properties and the specific applications of CMNF are unique to the source of the cellulose, and it can be possibly altered through a different physical and chemical treatment.

The promising reinforcing capabilities of CMNF are strongly attributed to their intrinsic chemical natures, degree of crystallinity and hydrogen bonding. Generally, cellulose is a linear natural homopolymer composed of highly ordered stereo-regular chain of cellobiose, the D-anhydroglucose units, aggregated through oxygen covalently by 1, 4, β-glycosidic bonds at C1 and C4 carbon atoms. This hydrophilic polymer is linked together by strong intermolecular hydrogen bonds presented on three hydroxyl groups [primary hydroxymethyl (CH_2OH) at the C-6 position and secondary OH at the C-2 and C-3] of different reactivities [25, 26]. During biosynthesis, these hydroxyl groups are essential in placing the highly ordered (crystalline) structure and disordered region (amorphous-like), aligning the chains parallel to each other forming cellulose microfibrils (5–50 nm in diameter and several microns in length). Both of these van der Walls and intermolecular hydrogen network generate a relatively stable complex structural hierarchy of cellulose with a high axial stiffness [27]. In PALF, these cellulose structures provide an important reinforcement phase for the plant particularly on the leaves.

Furthermore, cellulose exists in several polymorphs of crystalline forms, namely cellulose II, III, IV, V and VI. They are varied in their chain packing schemes, unit cell dimensions, and hydrogen-bonding relationships. The naturally occurring cellulose, cellulose I, has the highest axial elastic modulus and can be further divided into two distinct crystalline phases or allomorphs, namely cellulose Iα (triclinic structure), a dominate allomorph for most bacteria and Iβ (monoclinic structure), dominant allomorph for hardwood and tunicates [28]. These allomorphs coexisted in various proportions depending on the source of the cellulose, and it can be possibly altered through different physical and chemical treatments. Due to intensive hydrogen-bonding system, the crystallographic structure of these allomorphs can be accurately measured [29]. There is also a correlation between crystallinity index of cellulose allomorphs (cellulose I and II) and conversion of cellulose. Crystallinity index of cellulose I negatively affects cellulose conversion, and crystallinity index of cellulose II positively affects cellulose conversion [30].

Indeed, the crystal and molecular structure, together with the hydrogen-bonding system [31] in cellulose, that will be discussed here is fundamental in order to understand the superior structure, thermal and mechanical properties of the cellulose.

2.1 *Crystallinity Analysis by X-Ray Diffraction (XRD)*

As discussed, cellulose has six polymorphs (I–VI), and their properties of crystallinity depend strongly on the molecular structures of the chains. Therefore, the crystallinity study of isolated CMNF is relevant to explore the precise effect of the source and the extraction methods on the crystal structure of the cellulose particle. Varied chain configurations and packing will create different crystallinity or polymorphs structures with varied unit lattice parameters. As cellulose possesses an intricate intra/intermolecular hydrogen-bonding network, it provides high mechanical strength of cellulose. The crystallinity of cellulose is often studied in its nanometric dimension, which is also called nanocellulose (NC) by means of X-ray diffraction (XRD) technique and rarely through solid-state carbon-13 nuclear magnetic resonance (13C NMR), Fourier-transform infrared (FTIR) spectroscopy, Raman spectroscopy and single angle X-ray scattering (SAXS) [32].

As cellulose is in nature a mix of highly crystalline and disordered amorphous region, the relative amount, namely crystallinity index of these two different domains and theoretical crystallite size can be determined. The variation is all due to variable ratio of crystalline and amorphous domain found in native cellulose or cellulose I. To date, isolation of (i) cellulose nanofibrils (CNFs), also known as nanofibrillated cellulose (NFC) and (ii) cellulose nanocrystals (CNC) or cellulose whiskers has been the primary form of NC since early reports in the late 1940s [33]. CNCs, which isolated by acid hydrolysis or enzymatic methods, have relatively higher crystallinity nanoparticles ranging about 65–95% when compared to mechanically prepared CNF, with or without treatment [34]. Removal of the disordered amorphous region by acid treatment forms a dense crystalline packing of cellulose chains into highly ordered crystal structures hold by an extensive intra and intermolecular hydrogen network of glucose domain in the molecule that enhance the crystallinity ratio of CNC [32]. In addition, lower crystallite ratio of CNF is attributable to its XRD diffraction pattern [34, 35]. It presented a main peak at $2\theta = 22.5°$ (200) shoulder in the region $2\theta = 14.9°$ (110) that is attributable to typical cellulose I (native cellulose) diffractogram. The excess of amorphous portion of the NC that bound to the crystallite domain in CNF has less ordered structure, thus had a very limited crystalline packing process [36]. Further removal of these residues by means of alkali treatment and bleaching technique can, however, improve the crystallinity of the CNF sample [34].

Mercerization or alkali treatment is commonly adopted in natural fibres treatment. The function of NaOH is to disrupt hydrogen bonding within the cellulosic fibres structure and enriching the fibre surface coarseness as it eliminated lignin, wax and oils, initially shielding the fibre exteriors thus depolymerises the cellulose structures, disclosing the short length crystallites [37]. Fareez et al. [8], for example, had studied the effect of bleaching time on the cellulose extracted from Josapine PALF in which cellulose I in the form of monoclinic was widely prominent. As compared to non-treated cellulose (62.8%), bleaching with 0.2% sodium hydroxide (NaOH) and 4.0% hypochlorite from 1 to 4 h demonstrated considerably a higher percentage of crystallinity (82.90–85.41%) in all samples while retaining the original structure of cellulose in the PALF. The bleached PALF samples' crystal size was found to be in the range of 3.564–3.807 nm.

2.2 Hydrogen Bonding by Fourier-Transform Infrared (FTIR) Analysis

As discussed earlier, the CNCs are produced through intermolecular hydrogen bonding among the cellulose in an ordered manner while the interconnected CNFs formed by alternating crystalline cellulose bonded together by amorphous domain forming nanofibrils strings in the plant cell wall structure [38]. In order to further understand this, Fourier-transform infrared spectroscopy (FTIR) is applied to investigate the varying structure of cellulose in their various forms using attenuated total reflection to produce infrared spectra [8]. Generally, cellulose peak is identified by C–H deformation vibration located at 900–895 cm^{-1} representing α-glucosidic bonding among glucose monomer unit in cellulose. The FTIR spectra showing a functional group of CNC are similar to CNF. However, CNCs spectra have been reported to have stronger absorption intensity on cellulose peak in the spectra showing higher cellulose content in CNC in comparison with CNF. This is due to a lack amount of hydroxyl groups present on amorphous region of CNF.

Meanwhile, cellulose has many hydroxyl groups available for interaction with water by hydrogen bonding causing the fibre to be highly hygroscopic [39]. They interact with water not only at the surface but also in the bulk [40]. The quantity of water absorbed depends on the relative humidity of the confined atmosphere with which the fibre is in equilibrium. The sorption isotherm of cellulosic material depends on the purity of cellulose and the degree of crystallinity [41]. All –OH groups in the amorphous phase are accessible to water; whereas, only a small amount of water interacts with the surface –OH groups of the crystalline phase.

The transformations in the chemical composition in the CNF of PALF samples before and after chemical treatments were conducted with FTIR instrument and Raman spectroscopy by Fareez et al. [8]. Within FTIR–Raman spectra, cellulose I was extensively being the principal form of the treated PALF. The peak at 3423 cm^{-1} was responsible to the presence of hydrogen bond of O–H groups at C-6 and sequentially stretched at C-2 due to O–H absorption. Another vibration stretch occurred at 3352 cm^{-1} of C-3 which indicated the formation of intramolecular hydrogen bonds with oxygen (O-5) followed by intermolecular hydrogen bond at C-6 and C-3 situated at the band 3301 cm^{-1}. These changes were caused by the reduction in hydrogen bonding as hydroxyl groups were vanished by alkali treatment. Treated PALF has much heightened –OH saturation which can be seen in the range 3300–3500 cm^{-1}. Lastly, the asymmetrical peak of methyl group –CH was identified at 2851 cm^{-1} in raw untreated PALF only.

A study conducted by Aziz et al. [42] is an interaction between polyethylene terephthalate PET/PALF electrospun nanocomposite fibres based on FTIR spectrum generated. The interaction involved the lone pair oxygen with the hydrogen group in PALF which was reflected in the FTIR spectra at 3400 cm^{-1} due to the presence of hydrogen bonding and hydrophilic properties. The presence of C=O bond is visible at 1715 cm^{-1} for PET/PALF which resembled hemicellulose in PALF and conjugated ester group in PET. The benzene and aromatic lignin presence within the skeletal vibration were observed through the stretching of C=C at 1050 cm^{-1}. The vibration at 725 cm^{-1} was due to C–H bond originated from cellulose glucose ring of PALF. The intensity decline of the peak after the integration of PALF indicated the occurrence of interactions between both PET and PALF.

3 Characteristics of Cellulose Extracted from Pineapple Leaf Fibre (PALF)

The varying form of CMNF shows distinct properties and functionality for specific applications. The superior characteristics of PALF CMNF include high thermal stability, high Young's modulus/tensile strength and high surface compatibility with other materials, such as polymer, protein and living cells opening up a wide range of possibilities of this PALF cellulose to be the main component of a composite polymer. This chapter encompasses the main characteristic of CMNF extracted from PALF, with a special focus on morphology of various forms of CMNF, their rheology properties, thermal stability, optical properties, surface chemistry and superior mechanical strength with some important recent advances in cellulose nanocomposite research.

3.1 Morphology

Morphology assessment is crucial in determining the structural identity such as dimension, shape, rigidity and defect of a particular material. Microscopic observations using various high technological approaches such as SEM to observe finer scale structure and surface are common to analyse the topography, crystallinity and composition of a material.

3.1.1 Extracted Raw PALF

Vodounon et al. [43] performed a comparative study of different treatment on the effects of PALF treated (T-PALF) with sodium hydroxide (NaOH) and untreated PALF (N-PALF). Apparently, the T-PALF possesses much smoother morphological surface in comparison with N-PALF as NaOH impurities had been eliminated from the cellulose fibres, thus creating a more flexible fibre with increased adhesion of the soil matrixes. The morphology comparison of untreated and treated PALF has been reported by Jaafar et al. [44]. The untreated PALF has impurities, wax and fatty acids. The alkali treatment implementation had changed the morphology of the fibre bundles into smaller fibrillation fibres which enhanced the surface areas to volume ratio. The diameter of the fibrillations fibre had decreased after alkali treatment thus increasing the interfacial adhesion of the matrix with the fibres. Due to the dissolution of and the leaching of other component in the plant fibres such as hemicellulose, fatty acids and soluble lignin, the surface of the fibres became coarse.

Teles et al. [45] validated that PALF's mechanical strengths were dependent on the diameter size of the fibre using Weibull statistic method. Based on the SEM morphological features of PALF, thin fibre with uniform structure of 0.11 mm and the simultaneous breakage of the microfibrils were observed, some displaying thicker fibre diameter of 0.25 mm after rupture effect with prominent defects and breakage of the structure. Thus, thinner PALF microfibrils are likely to be more compacted which attributed to the packed cross-sectional areas, lesser porosity resulting in higher mechanical strength. Other than that, statistical evidences proved that lower stress could easily rupture PALF with thicker microfibrils compared to a thinner PALF.

In addition, Fareez et al. [8] investigated the characteristic of cellulose extracted untreated PALF fibre from Josapine cultivar. Based on the SEM observation, the PALF microfibrils are 36 μm in diameter; meanwhile, the defibrillated PALF after alkaline and bleaching process has the average diameter between 4.28 and 3.12 μm.

3.1.2 PALF Composite

Biopolymer reinforcement can be done by using cellulose from PALF which acts as a filler or support material in a composite. It is expected to elevate mechanical, biodegradability, thermal and barrier performance of composite materials in comparison with synthetic polymers due to nanometre scale which has a large surface area to volume ratio [46]. The interfacial adhesion between untreated PALF and the reinforced tapioca biopolymer (TBP) possesses gap between fibre matrices and has pull-out fibre which is an unfavourable outcome. It was attributed to the fact that adhesion between the fibre surfaces with the TBP resin was low due to the presence of impurities and wax on the fibre surfaces lowering the adhesion capability of this composite in the untreated PALF. The treated PALF sample had improved in terms of morphological features as impurities and wax had been vanished from the fibre surfaces, improving the fibre matrix adhesion. It was demonstrated that there is a reduced amount of fibre gaps matrix between treated PALF-TBP. Even after fracture test, the fibre cracks and breakages were attributed to the even distribution of load imposed throughout the surface [47].

3.1.3 Electrospun PALF Composite

Electrospun PET/PALF nanocomposites were studied showing that PALF has smaller diameter 40–80 nm in comparison with PET fibres networks of 160–180 nm. It was calculated that the average diameter of PALF fibres was 52.1 ± 10 nm intersecting upon PET fibre networks and coalesces each other. With the addition of PALF into PET electrospinning solution, a reaction had occurred where the PALF had thickened the solution concentration, thus creating a larger fibre diameter of PET fibres [42].

3.2 Rheological Properties and Degree of Polymerization

There are many stresses which effect on coating during the production, thermal process, oxidation storage and application [48]. Thus, it is important to know, how a product will behave, when it is affected with different deformation and that is what we can find out from rheology test using rhemometer. Rheology is a science of deformation and flow. It is a branch of physics and physical chemistry where it describes the deformation of a body under the influence of stresses and the flow behaviour of liquids and also deformation behaviour of solids. Meanwhile, the degree of polymerization, or DP, is usually defined as the number of monomeric units in polymer molecule. Specifically in cellulose, it is a measure of the average length of the cellulose macromolecules. Normally, cellulose ageing status can be determined using DP. In virgin cellulose, the DP value is at the highest as insulation is still intact. Gradually, overexposure, service and treatments, the macromolecules become short due to chain scission, hence lowering insulation and DP [49].

Analysis done by Fareez et al. [8] to determine the molecular weight (MW) of PALF from the viscosity tests indicated the MW at 3.8671×10^{-4}–1.0560×10^{-4} with polymerization degree of 447–107 after 4 h of bleaching. Lower polymerization degree was attributed to the fact that longer bleaching period in comparison with other samples undergoes lesser bleaching time.

3.3 Thermal Stability

Many industrial applications had integrated cellulosic components in their materials, thus its thermal stability is the key performance. The type of atmosphere in the cellulosic thermal process plays a very important role in determining the reactions and products involved as the presence of nitrogen will determine how many steps it will require. Generally, heating temperature, time, impurity presence and degree of crystallinity are the variables responsible for the cellulose pyrolysis [50]. The degradation onset temperature of PALF usually was measured using thermogravimetric analysis (TGA), and the thermal transition phase is measured using differential scanning calorimetry (DSC) to assess the overall thermal stability.

Santosha et al. [51] investigated the thermal stability of pineapple leaf fibres and banana fibres at the ratio of 1:1 reinforced with polyester. In the curing process, there were 1.5% volume of accelerator and catalyst supplementation of polyester resin at room temperature. They used hand lay-up technique with 0.05 MPa compressive pressure to be left cured for 24 h. After the fibre had been removed from the mould, the composites were again cured for 2 h at 70 °C. The resulting specimens of various fiver volume fractions of 0.112, 0.166, 0.213, 0.274 and 0.346 were tested for thermal conductivity. They concluded that the thermal conductivity decreases as the fibre content increased. This composite property was deduced from the banana and pineapple leaf fibre which have low thermal conductivity. Opposite trend was observed in the specific heat capacity from differential scanning calorimeter. However, the thermal conductivity of the reinforced polyester composites had significantly improved as temperature increased, with the variations of thermal conductivity rate from 0.171 to 0.213 Wm^{-1} K^{-1} within 30–120 °C.

In this case, banana/PALF/polyester composites were proven to possess improved thermal insulation, lighter and cheaper suitable to be utilized in applications of electronic packages, building material, sport equipment, automobile interior part, etc. [51]. The thermal properties of PALF reinforced with polylactic acid (PLA) laminated composites were developed by Huda et al. [52]. TGA results in Fig. 2 of pre-treated PALF of two different chemical treatments showed that the composite possesses improved thermal stability, thus improving the mechanical properties due to the enhancement in the fibre matrix entanglement between them. This is because the silane and alkali-treated PALF-reinforced composites had showcased elevated mechanical performance compared to untreated PALF from mechanical testing. The impact of the chemical treatments, namely alkaline treatment using 5 wt% NaOH or silane treatment using 5 wt% 3-amino propyl triethanoxy silane (3-APS) had been studied to investigate the interfacial entanglement and mechanical properties of the composites. The compression moulding technique was adopted in the fabrication process followed by air drying of the PLA pellets for 10 h under vacuum at 80 °C. PLA films with 1 mm thickness were produced with 250 mm × 200 mm compression frame mould at 190 °C and 260 psi within 10 min then to 100 °C and 624 psi. Later, the varied fibres loading of PALF was stacked and compressed using moulding machine resulting to laminated composites with 140 mm × 180 mm dimensions and 2 mm thickness.

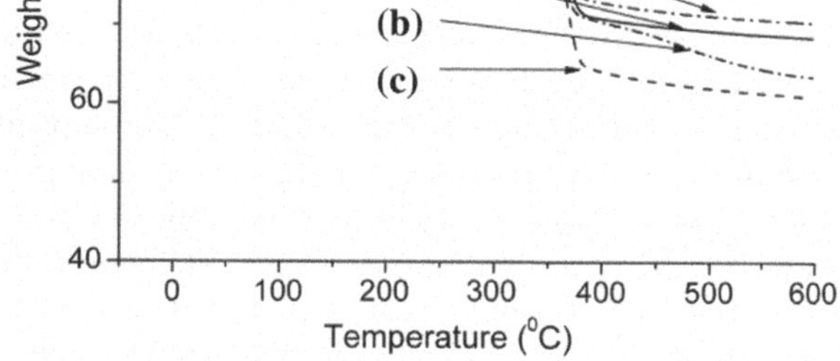

Fig. 2 Thermogram of (a) non-treated PALF (PALF), (b) PALF NaOH (PALFNA), (c) PALF (PALFSI) and (d) NaOH and PALF silane treatment (PALFNASI) fibres (reprinted by permission of Taylor and Francis Ltd. http://www.tandf.co.uk)

3.4 Strength and Stiffness

High specific strength and stiffness of PALF have been acknowledged by several researchers [53, 54]. Though the strength and stiffness of PALF differ from variety to variety and extraction procedures [55, 56], they have the potential to be used as a sound absorber and thermal insulator as a replacement of artificial fibres [53]. The PALF also has the potential to be used as a composite material. However, it is needed to be chemically treated to change its hydrophilic nature. Besides, chemical treatments, lengths and loads of PALF play important role in increasing the strength and stiffness of the composite [53, 55, 56].

In recent years, CNF and CNC from PALF have been received considerable attention to be used as reinforcing bio-composite materials, because of their small sizes that increase their surface to volume and aspect (length to diameter) ratio. The increased aspect ratios of the CNF and CNC from PALF could help to improve the mechanical properties like strength and stiffness in thermoplastic starch (TPS), polyvinyl alcohol (PVA), polylactic acid (PLA), polycarbonate (PC), polyurethane (PU) and polymethyl methacrylate (PMMA), eventually could help to produce highly durable and greener nanocomposite materials [57–59, 20].

Cherian et al. [57] have shown that addition of 5% w/w of CNF from PLAF with having average aspect ratio of 67 to PU approximately increased 300 and 2600% of the strength and stiffness of the composite, respectively [57]. While, CNC produced by acid hydrolysis of PLAF for 30 min found to have higher average aspect ratio ($\approx$60) compared to the CNC produced by 60 min ($\approx$50) of acid hydrolysis [20], which emphasize the importance of extraction procedure on the potential of CNC and CNF. Though CNF and CNC have the reinforcing potential, further research has shown that higher aspect ratio along with network structure of CNF could have better reinforcing efficiency over CNC [60].

3.5 Optical Properties

CNF or CNC from different sources has shown more transparency compared to the microfibres [59]. CNF, at very low concentration (0.5–1%) have shown to exhibit an excellent transparent appearance in the PMMA composite [61]. However, the effect of CNF from PALF on transparency is inversely proportional to its w/w concentration in the PVA or PMMA composite films [46, 61]. It is noteworthy to mention that better transparency of PLAF-derived CNC compared to PLAF-derived CNF has been reported by several researchers [60]. This optical property of CNC as well as CNF has shown their potential in manufacturing flexible electronic film, coating agent and receptacles for drug delivery and optical digital storage media [59].

3.6 Surface Chemistry

Usually, CNF contains both amorphous and crystalline region, while CNC contains only the crystalline region due to successful removal of the amorphous component from the cellulose structure. Hence, higher crystallinity is found in CNC [60]. CNF and CNC have a high surface to volume ratio along with a lot of hydroxyl group. Therefore, surface modification of this nanocellulose is needed to make them compatible to both hydrophilic and hydrophobic components [62].

For instance, starch and cellulose are polysaccharide in nature. Compatibility between them does not necessitate nanocellulose surface modification to use them as TPS reinforcing component [60]. Surface modification of nanocellulose is not required to use them as PVA reinforcing material because of compatibility between them as well [46]. However, highly polar nature of nanocellulose due to high hydroxyl group presence on the surface makes it difficult to disperse them in the PMMA matrix. This necessitates 2,2,6,6-tetramethylpiperidin-1-oxyl radical (TEMPO)-mediated oxidation followed by the surface modification with polymerized methyl methacrylate (MMA) to improve the hydrophobicity of nanocellulose and compatibility to PMMA matrix [61]. Hence, besides extraction, surface modification of PLAF-derived nanocellulose plays an important role in the efficient use of them as reinforcing materials.

4 Conclusion and Future Perspective

With increasing environmental consciousness, innovation that adopts eco-friendly approach stands to be rewarded in the marketplace. Commercial innovative solution to meet the awareness for the use of renewable natural resources, environmental friendly materials and sustainable ways of operating has led to the development and use of natural cellulose fibres. PALF productions represent one such effort. Utilization of pineapple leaf as a source of cellulose-based biomaterials could significantly improve pineapple leaf waste management. At present, substantial research on PALF and its nanoscale forms focuses on the characterization, mechanistic analysis and improving the properties of the biomaterials. While the present applications of natural cellulose fibre and nanocellulose are limited due to cost and availability, the outlook for the field is promising. In particular, producing the organic cellulose materials with tunable and biomimetic characteristics may be of particular interest for the future, especially in the coating and biomedical industries. Other than the challenging technological breakthrough, standardization and regulatory safety review are also necessary for further commercialization. The development of new advanced surface modification, functionalization, bioconjugation and crosslinking procedures will further boost the application of natural cellulose fibres. It is envisioned that the cellulose and nanocellulose structures can be further assembled for use in tissue engineering, tissue repair and controlled drug delivery system. Other than the challenging technological breakthrough, standardization and regulatory safety review are, however, necessary for further commercialization purposes.

References

1. Madhu P, Sanjay M, Senthamaraikannan P, Pradeep S, Saravanakumar S, Yogesha B (2018) A review on synthesis and characterization of commercially available natural fibers: part-I. J Nat Fiber 1–13
2. Mohammed L, Ansari MN, Pua G, Jawaid M, Islam MS (2015) A review on natural fiber reinforced polymer composite and its applications. Int J Polym Sci
3. Sanyang M, Sapuan S, Jawaid M, Ishak M, Sahari J (2016) Recent developments in sugar palm (*Arenga pinnata*) based biocomposites and their potential industrial applications: a review. Renew Sustain Energy Rev 54:533–549
4. Trache D, Hussin MH, Haafiz MM, Thakur VK (2017) Recent progress in cellulose nanocrystals: sources and production. Nanoscale 9(5):1763–1786
5. Khakalo A, Vishtal A, Retulainen E, Filpponen I, Rojas OJ (2017) Mechanically-induced dimensional extensibility of fibers towards tough fiber networks. Cellulose 24(1):191–205
6. Satyanarayana K, Pillai C, Sukumaran K, Pillai S, Rohatgi P, Vijayan K (1982) Structure property studies of fibres from various parts of the coconut tree. J Mater Sci 17(8):2453–2462
7. Cherian BM, Leão AL, de Souza SF, Thomas S, Pothan LA, Kottaisamy M (2010) Isolation of nanocellulose from pineapple leaf fibres by steam explosion. Carbohydr Polym 81(3):720–725
8. Fareez IM, Ibrahim NA, Yaacob WMHW, Razali NAM, Jasni AH, Aziz FA (2018) Characteristics of cellulose extracted from Josapine pineapple leaf fibre after alkali treatment followed by extensive bleaching. Cellulose 25(8):4407–4421
9. Fernández G, Pomilio AB (2003) Optimized growth conditions and determination of the catalytic type of the peptidase complex from a novel callus culture of pineapple (*Ananas comosus*). Mol Med Chem 1:39–49
10. Chan Y, Coppens DEG, Sanewski GM (2002) Breeding and variety improvement. In: Bartholomew DP, Paull RE, Rohrbach KG (eds) The pineapple, botany, production and uses. CABI Publishing, New York, pp 33–35
11. Deepa B, Abraham E, Cordeiro N, Mozetic M, Mathew AP, Oksman K, Faria M, Thomas S, Pothan LA (2015) Utilization of various lignocellulosic biomass for the production of nanocellulose: a comparative study. Cellulose 22(2):1075–1090
12. Klemm D, Kramer F, Moritz S, Lindström T, Ankerfors M, Gray D, Dorris A (2011) Nanocelluloses: a new family of nature-based materials. Angew Chem Int Ed 50(24):5438–5466
13. Jose S, Salim R, Ammayappan L (2016) An overview on production, properties, and value addition of pineapple leaf fibers (PALF). J Nat Fiber 13(3):362–373
14. Neto ARS, Araujo MA, Barboza RM, Fonseca AS, Tonoli GH, Souza FV, Mattoso LH, Marconcini JM (2015) Comparative study of 12 pineapple leaf fiber varieties for use as mechanical reinforcement in polymer composites. Ind Crop Prod 64:68–78
15. Verma D, Gope P, Shandilya A, Gupta A, Maheshwari M (2013) Coir fibre reinforcement and application in polymer composites. A Environ Sci 4(2):263–276
16. Siregar J, Sapuan S, Rahman M, Zaman H (2008) Characterization and chemical composition of short pineapple leaf fibres (PALF). In: Proceeding of postgraduate seminar on natural fibre composites. Faculty of Engineering, Universiti Putra Malaysia, Serdang, Selangor
17. Khalil HSA, Alwani MS, Omar AKM (2007) Chemical composition, anatomy, lignin distribution, and cell wall structure of Malaysian plant waste fibers. BioResources 1(2):220–232
18. Idicula M, Boudenne A, Umadevi L, Ibos L, Candau Y, Thomas A (2006) Thermophysical properties of natural fibre reinforced polyester composites. Compos Sci Technol 66(15):2719–2725

19. Wan Nadirah WO, Jawaid M, Al Masri AA, Abdul Khalil HPS, Suhaily SS, Mohamed AR (2012) Cell Wall Morphology, Chemical and Thermal Analysis of Cultivated Pineapple Leaf Fibres for Industrial Applications. J Polym Env 20 (2):404–411
20. Santos RMD, Flauzino Neto WP, Silvério HA, Martins DF, Dantas NO, Pasquini D (2013) Cellulose nanocrystals from pineapple leaf, a new approach for the reuse of this agro-waste. Ind Crop Prod 50:707–714
21. Banik S, Nag D, Debnath S (2011) Utilization of pineapple leaf agro-waste for extraction of fibre and the residual biomass for vermicomposting. In: Proceeding Paper, Semantic Scholars
22. Daud Z, Mohd Hatta MZ, Mohd Kassim AS, Awang H, Mohd Aripin A (2013) Exploring of agro waste (pineapple leaf, corn stalk, and napier grass) by chemical composition and morphological study. BioResources 9(1):872–880
23. Kengkhetkit N, Amornsakchai T (2014) A new approach to "Greening" plastic composites using pineapple leaf waste for performance and cost-effectiveness. Mater Design 55:292–299
24. Hazarika D, Gogoi N, Jose S, Das R, Basu G (2017) Exploration of future prospects of Indian pineapple leaf, an agro waste for textile application. J Clean Prod 141:580–586
25. Fan LT, Gharpuray MM, Lee Y-H (1987) Nature of cellulosic material. Cellulose hydrolysis. Springer, Berlin Heidelberg, Germany
26. Khalil HA, Bhat A, Yusra AI (2012) Green composites from sustainable cellulose nanofibrils: a review. Carbohydr Polym 87(2):963–979
27. Moon RJ, Martini A, Nairn J, Simonsen J, Youngblood J (2011) Cellulose nanomaterials review: structure, properties and nanocomposites. Chem Soc Rev 40(7):3941–3994
28. Atalla RH, Vanderhart DL (1984) Native cellulose: a composite of two distinct crystalline forms. Science 223(4633):283–285
29. Nishiyama Y, Langan P, Chanzy H (2002) Crystal structure and hydrogen-bonding system in cellulose Iβ from synchrotron X-ray and neutron fiber diffraction. J Am Chem Soc 124(31):9074–9082
30. Song Y, Zhang J, Zhang X, Tan T (2015) The correlation between cellulose allomorphs (I and II) and conversion after removal of hemicellulose and lignin of lignocellulose. Bioresour Technol 193:164–170
31. Razali M, Amira N, Azraaie N, Abidin NAMZ, Ibrahim NA, Abdul Aziz F, Abdul Rahman S (2015) Effect of chemical treatment on crystalline cellulose: changes in crystallinity and functional groups of cellulose. Adv Mater Res 35–39
32. Lee K-Y, Santmartí A (2018) Crystallinity and thermal stability of nanocellulose. In: Nanocellulose and sustainability. CRC Press, pp 67–86
33. Nickerson R, Habrle J (1947) Cellulose intercrystalline structure. Ind Eng Chem 39(11):1507–1512
34. Hammel E, Tang X, Trampert M, Schmitt T, Mauthner K, Eder A, Pötschke P (2004) Carbon nanofibers for composite applications. Carbon 42(5–6):1153–1158
35. French AD, Cintrón MS (2013) Cellulose polymorphy, crystallite size, and the Segal crystallinity index. Cellulose 20(1):583–588
36. Nam S, French AD, Condon BD, Concha M (2016) Segal crystallinity index revisited by the simulation of X-ray diffraction patterns of cotton cellulose Iβ and cellulose II. Carbohydr Polym 135:1–9
37. Mtibe A, Linganiso LZ, Mathew AP, Oksman K, John MJ, Anandjiwala RD (2015) A comparative study on properties of micro and nanopapers produced from cellulose and cellulose nanofibres. Carbohydr Polym 118:1–8
38. Yu H-Y, Qin Z-Y, Liu L, Yang X-G, Zhou Y, Yao J-M (2013) Comparison of the reinforcing effects for cellulose nanocrystals obtained by sulfuric and hydrochloric acid hydrolysis on the mechanical and thermal properties of bacterial polyester. Compos Sci Technol 87:22–28
39. Wei Z, Sinko R, Keten S, Luijten E (2018) Effect of surface modification on water adsorption and interfacial mechanics of cellulose nanocrystals. ACS Appl Mater Interface 10(9):8349–8358
40. Makarem M, Lee CM, Sawada D, O'Neill HM, Kim SH (2017) Distinguishing surface versus bulk hydroxyl groups of cellulose nanocrystals using vibrational sum frequency generation spectroscopy. J Phys Chem Lett 9(1):70–75

41. Gauthier R, Joly C, Coupas A, Gauthier H, Escoubes M (1998) Interfaces in polyolefin/cellulosic fiber composites: chemical coupling, morphology, correlation with adhesion and aging in moisture. Polym Compos 19(3):287–300
42. Aziz FA, Surip S, Bonnia N, Sekak K (2018) The effect of pineapple leaf fiber (PALF) incorporation into polyethylene terephthalate (PET) on FTIR, morphology and wetting properties. IOP Conf Ser Earth Environ Sci 1:012082
43. Vodounon NA, Kanali C, Mwero J (2018) Compressive and flexural strengths of cement stabilized earth bricks reinforced with treated and untreated pineapple leaves fibres. Open J Compos Mater 8(4):145–160
44. Jaafar J, Siregar JP, Oumer AN, Hamdan MHM, Tezara C, Salit MS (2018a) Experimental investigation on performance of short pineapple leaf fiber reinforced tapioca biopolymer composites. BioResources 13(3):6341–6355
45. Teles MCA, Glória GO, Altoé GR, Amoy Netto P, Margem FM, Braga FO, Monteiro SN (2015) Evaluation of the diameter influence on the tensile strength of pineapple leaf fibers (PALF) by Weibull method. Mater Res 18:185–192
46. Wahyuningsih K, Iriani ES, Fahma F (2016) Utilization of cellulose from pineapple leaf fibers as nanofiller in polyvinyl alcohol-based film. Indones J Chem 16(2):181–189
47. Jaafar J, Siregar JP, Piah MBM, Cionita T, Adnan S, Rihayat T (2018b) Influence of selected treatment on tensile properties of short pineapple leaf fiber reinforced tapioca resin biopolymer composites. J Polym Env 26(11):4271–4281
48. Munajad A, Subroto C (2018) Fourier transform infrared (FTIR) spectroscopy analysis of transformer paper in mineral oil-paper composite insulation under accelerated thermal aging. Energies 11(2):364
49. Krause C, Dreier L, Fehlmann A, Cross J (2014) The degree of polymerization of cellulosic insulation: review of measuring technologies and its significance on equipment. In: 2014 IEEE electrical insulation conference (EIC). IEEE, pp 267–271
50. Santmartí A, Lee K-Y (2018) Chapter 5: crystallinity and thermal. In: Nanocellulose and sustainability: production, properties, applications, and case studies. CRC Press, Boca Raton, p 67
51. Santosha PCR, Gowda ASSS, Manikanth V (2018) Effect of fiber loading on thermal properties of banana and pineapple leaf fiber reinforced polyester composites. Mater Today Proc 5(2):5631–5635
52. Huda MS, Drzal LT, Mohanty AK, Misra M (2008) Effect of chemical modifications of the pineapple leaf fiber surfaces on the interfacial and mechanical properties of laminated biocomposites. Compos Interface 15(2–3):169–191
53. Asim M, Abdan K, Jawaid M, Nasir M, Dashtizadeh Z, Ishak MR, Hoque ME (2015) A review on pineapple leaves fibre and its composites. Int J Polym Sci 2015:1–16
54. George J, Sreekala MS, Thomas S (2001) A review on interface modification and characterization of natural fiber reinforced plastic composites. Polym Eng Sci 41(9):1471–1485
55. Chollakup R, Tantatherdtam R, Ujjin S, Sriroth K (2011) Pineapple leaf fiber reinforced thermoplastic composites: effects of fiber length and fiber content on their characteristics. J Appl Polym Sci 119(4):1952–1960
56. Hujuri U, Chattopadhay SK, Uppaluri R, Ghoshal AK (2008) Effect of maleic anhydride grafted polypropylene on the mechanical and morphological properties of chemically modified short-pineapple-leaf-fiber-reinforced polypropylene composites. J Appl Polym Sci 107(3):1507–1516
57. Cherian BM, Leão AL, de Souza SF, Costa LMM, de Olyveira GM, Kottaisamy M, Nagarajan ER, Thomas S (2011) Cellulose nanocomposites with nanofibres isolated from pineapple leaf fibers for medical applications. Carbohydr Polym 86(4):1790–1798
58. George J, Sabapathi SN (2015) Cellulose nanocrystals: synthesis, functional properties, and applications. Nanotechnol Sci Appl 8:45–54
59. Mahardika M, Abral H, Kasim A, Arief S, Asrofi M (2018) Production of nanocellulose from pineapple leaf fibers via high-shear homogenization and ultrasonication. Fibers 6(2)

60. Balakrishnan P, Gopi S, Geethamma VG, Kalarikkal N, Thomas S (2018) Cellulose nanofiber vs nanocrystals from pineapple leaf fiber: a comparative studies on reinforcing efficiency on starch nanocomposites. Macromol Symp 380(1)
61. Shih YF, Chou MY, Lian HY, Hsu LR, Chen-Wei SM (2018) Highly transparent and impact-resistant PMMA nanocomposites reinforced by cellulose nanofibers of pineapple leaves modified by eco-friendly methods. Express Polym Lett 12(9):844–854
62. Abdul Khalil HPS, Davoudpour Y, Saurabh CK, Hossain MS, Adnan AS, Dungani R, Paridah MT, Islam Sarker MZ, Fazita MRN, Syakir MI, Haafiz MKM (2016) A review on nanocellulosic fibres as new material for sustainable packaging: process and applications. Renew Sustain Energy Rev 64:823–836

Cellulose Nanostructures Extracted from Pineapple Fibres

Karen S. Prado, Asaph A. Jacinto and Márcia A. S. Spinacé

Abstract The fibres from pineapple plant leaves and pineapple crown leaves are unprocessed wastes with high cellulose content (74–83 wt%) that are environmentally and economically interesting as source to extract cellulose nanostructures (CNs). CNs are materials with unique and remarkable properties that can be used in several high value-added applications. This chapter covers the main topics related to the isolation of CN from pineapple fibres. The main types of CN (cellulose nanocrystals, cellulose nanofibres, amorphous nanocellulose, hairy nanocrystalline cellulose and cellulose nanoyarn) are presented. The advantages of using pineapple fibre wastes to produce CN are addressed. The hierarchical structure and chemical composition of the lignocellulosic fibres are described, as well as the existing pretreatment and treatment methods (chemical, enzymatic and mechanical) for extracting CN from pineapple fibres. This chapter also covers the characterization, modification and potential applications of CN isolated from pineapple fibres, such as in the production of biomedical devices and biodegradable bio-nanocomposites.

Keywords Pineapple fibres · Lignocellulosic waste · Nanocellulose · Cellulose nanostructure · Extraction methods · Structure · Characterization · Properties

1 Introduction

Cellulose has plenty of industrial applications, including the traditional ones in the paper, pulp and textile industries. However, in the last twenty years a new type of cellulosic material called "nanocellulose" has attained the attention of numerous researchers. This generic term describes a range of nanostructured cellulosic materials that have at least one dimension from 1 to 100 nm. These cellulose nanostructures (CNs) show unique features that combine important properties of cellulose, such as hydrophilicity and broad chemical modification capacity, with the specific features

---

K. S. Prado · A. A. Jacinto · M. A. S. Spinacé (✉)
Centre of Natural and Human Sciences, Federal University of ABC, Av. dos Estados, 5001, Santa Terezinha, 09.210-170, Santo André, São Paulo, Brazil
e-mail: marcia.spinace@ufabc.edu.br

M. Jawaid et al. (eds.), *Pineapple Leaf Fibers*, Green Energy and Technology,
https://doi.org/10.1007/978-981-15-1416-6_10

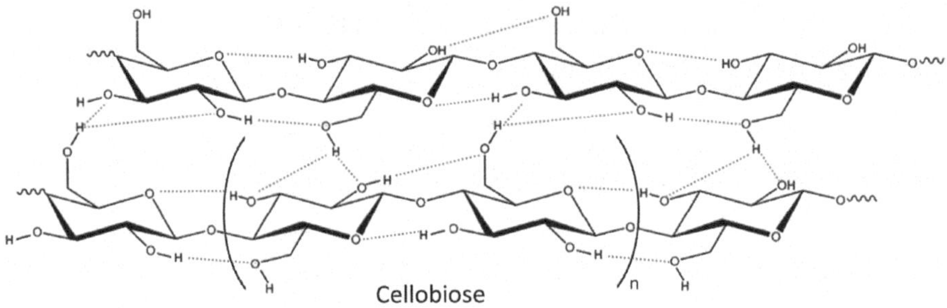

Fig. 1 Chemical structure of cellulose, showing the cellobiose repeating unit ("*n*" is the degree of polymerization). Intra- and intermolecular hydrogen bonding are drawn with dotted lines

of the nanoscale materials, mainly related to their large surface area [102], resulting in a material with low weight, and high strength and stiffness [102, 164].

In order to understand the origin and properties of nanocellulose, it is worth to recall the hierarchical structure of cellulose. Cellulose, the most abundant renewable polymer produced by nature [74], is composed by successive D-anhydroglucose units rotated 180° around the β-1,4-glycosidic linkages [49, 98]. Cellulose chains may be composed by up to 20,000 units of glucose, and each dimer of glucose forms the repeating structural unit of cellulose known as cellobiose (Fig. 1) [35, 74].

During the biosynthesis, about 36 individual cellulose chains assemble forming elementary fibrils with diameters from 2 to 20 nm (Fig. 2). The elementary fibrils have organized regions where the glucose units are packed together in a parallel fashion forming crystallites of about 5 nm wide, which are stabilized by intra- and intermolecular hydrogen bonds [2, 74, 98]. According to the model of paracrystalline structure, the elementary nanofibrils are formed by a highly ordered crystalline core that can contain some defects such as vacancies caused by ends of chains, and a lower-ordered paracrystalline surface around the core formed by a thin monomolecular layer with thickness of around 0.4 nm [84, 85].

The arrangements of cellulose chains can give rise to six interconvertible cellulose polymorphs (namely I, II, III_I, III_{II}, IV_I and IV_{II}) that depend on source, method of extraction or treatment [74]. In its natural form, the crystalline structure of native cellulose is classified as cellulose I, which presents a parallel alignment of the chains. Native cellulose occurs in two suballomorphs of cellulose I, cellulose Iα and cellulose Iβ that differ in the longitudinal arrangement of the hydrogen bonds. In general, cellulose Iα is predominant in primitive organisms such as algae and bacteria, while cellulose Iβ is predominant in higher organisms such as plants [74]. Cellulose I easily converts into other polymorphs when subjected to different chemical treatments.

In cellulose nanofibrils, the crystalline domains interpose among amorphous regions, which are characterized by chain dislocations and weak hydrogen bonds [11, 74]. Cellulose nanofibrils hierarchically aggregate via Van der Waals forces and both intra- and intermolecular hydrogen bonds into bundles known as microfibrils, with diameters of 8–50 nm. The assembling of these cellulose microfibrils may result in the cellulosic fibres usually found in higher plants [11, 62, 74, 98, 163].

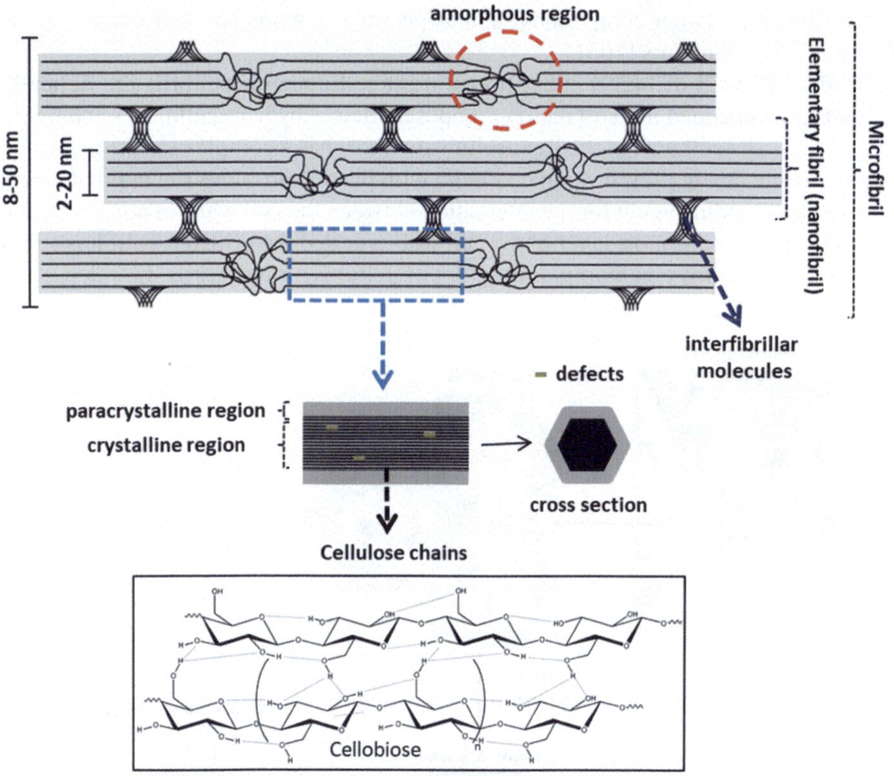

Fig. 2 Scheme of a cellulose microfibril formed by bundles of nanofibrils. Each nanofibril is formed by a crystalline core, amorphous domains and a paracrystalline surface [22, 84]

Higher plants are the main source of CN. Other sources include bacteria, algae, animals (such as tunicates), fungi and protozoa [102]. Examples of higher plant fibres include wood (softwood or hardwood), seed (e.g. cotton), fruit (e.g. coir), bast (e.g. jute), straw (e.g. rice), grass (e.g. bagasse) and leaf (e.g. pineapple leaves) [90, 92]. Higher plant fibres are also known as "vegetal fibres" and consist in lignocellulosic fibres whose main components are cellulose, hemicelluloses and lignin.

Cellulose corresponds to 35–50 wt% of dry lignocellulosic fibres [68, 141]. Hemicelluloses are non-crystalline heteropolymers which composition is dependent on the plant source and origin. They represent about 20–35 wt% of the lignocellulosic fibre [141]. The role of hemicelluloses is to fill the voids around cellulose fibrils, influencing the aggregation of cellulose nanofibrils during the biosynthesis and providing coupling to the lignin [10]. Lignin consists in a complex group of aliphatic and aromatic constituents that form a matrix around the cellulosic fibres and hold the natural structure of plant cell wall together [49]. It corresponds to 10–25 wt% of the lignocellulosic fibre and provides stiffness and compressive strain to the fibre [141]. Besides, lignin acts as a protection against biological attacks and keeps the water in the fibre [92]. Other compounds such as pectin, pigments and extractives can be found in lower amounts [163].

In the cell wall of higher plant fibres, rigid cellulose microfibrils are helically wound and embedded in a soft matrix composed mainly by hemicelluloses and lignin (Fig. 3) [2]. The cell wall of the lignocellulosic fibres has a complex structure formed by concentric layers [92]. A middle lamella with pectin surrounds the internal layers and acts as a binding agent for the interaction between the cell walls of adjacent plant cells [189]. The outermost layer of the cell wall is called the primary wall. It is a thin layer (100–200 nm) containing a loose net of cellulose microfibrils, first deposited

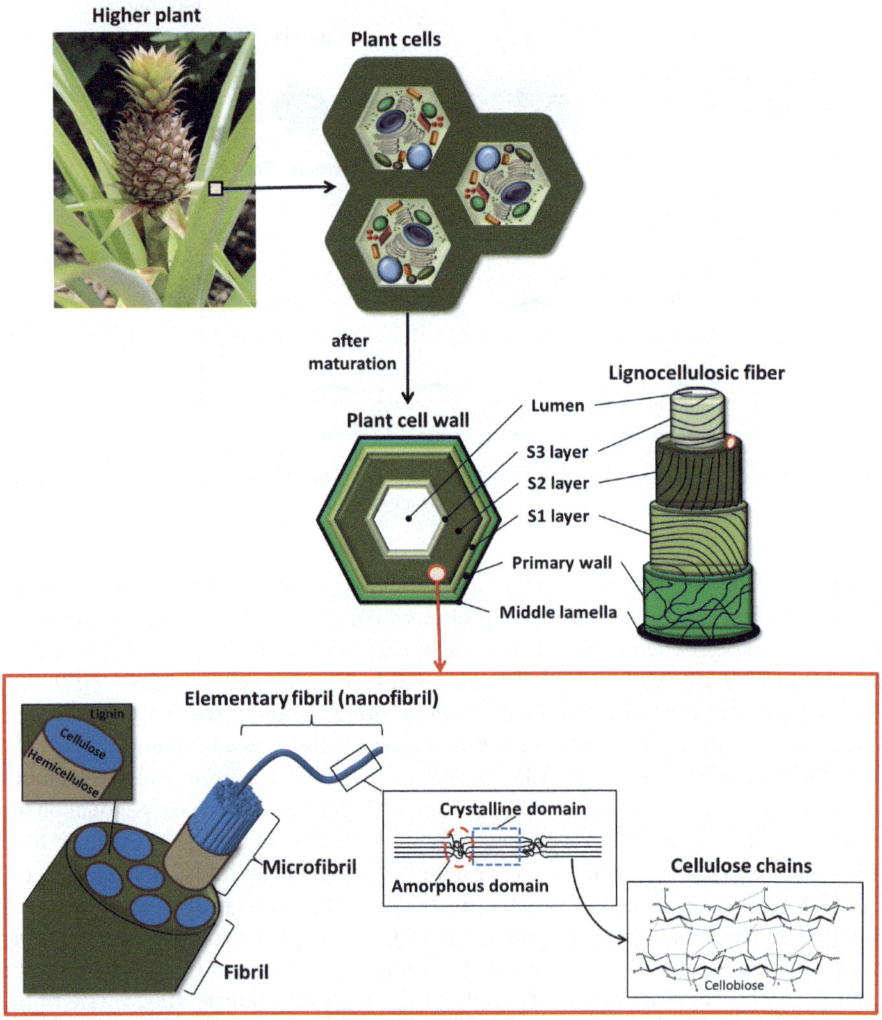

Fig. 3 Structural organization of cellulose in higher plants

during the plant cell growth around the secondary wall [84]. The secondary cell wall (thickness of 2–4 μm) consists of three layers that differ in terms of their cellulose content: an outer layer (S1), a thicker middle layer with more cellulose content (S2) and an inner layer (S3) surrounding the lumen. Within each layer, cellulose microfibrils are packed in a series of helical thread that are oriented at characteristic angles in relation to the fibre axis. These angles are called microfibril angles and vary depending on the cell wall layer and the plant type [92, 102, 185]. The lumen is the central vacuole of the plant cell during the growing of the fibre [165]. After the fibre matures, the organelles start to degenerate, the protoplast dries up and the lumen naturally collapses, leaving an empty central cavity in the fibre. The area of the lumen ranges from 20 to 70% of the cross-sectional area for wood fibres and from 0 to 5% from annual plants such as hemp [117]. Consequently, the structure of the lignocellulosic fibres can be considered as elongated and thickened single plant cells [75].

CN can be isolated from lignocellulosic fibres after removal of non-cellulosic compounds. The relative abundance of cellulose, hemicellulose and lignin is the key factor in determining how easily CNs are isolated [107]. The content and properties of cellulose are strongly influenced by many factors affecting the original fibres such as variety, microfibril angle, climate, harvest, maturity, fibre modification and technical processes (such as spinning and carding) [90, 92]. Other factors that influence the isolation of CN from lignocellulosic fibres are the crystallinity and degree of polymerization of cellulose, accessibility of the surface area of cellulose and fibre strength [107].

Among the many possible lignocellulosic sources, pineapple fibres arise as an outstanding source due to a combination of factors that favours the isolation of CN, such as high cellulose content, low microfibril angle and high crystallinity [6, 148].

Besides, significant amounts of pineapple fibre wastes are disposed each year, which are mostly burned or used for low value-added purposes, such as in farming activities (e.g. animal feeding) and composting [64, 90]. When disposed, these lignocellulosic residues pose serious environmental damage due to their high biological and chemical oxygen demand [144]. The use of these pineapple fibre wastes as alternative sources for the production of CN arises as a promising solution for the management and valorization of these agro-industrial residues [64]. Besides, the use of agro-industrial biomass can improve the energy security and reduce the greenhouse emissions [107]. Additionally, large profits can be achieved when producing CN, thus turning the negative cost of plant wastes into positive-earning materials [107].

As shown in Fig. 4, the first studies on the production of CN from pineapple fibres were published in 2010 [33, 108], and since 2015 the number of publications has been experiencing a significant rise.

Considering the relatively novelty of the use of pineapple fibres to produce CN, this chapter addresses the main processes currently employed to obtain CN from pineapple fibres, considering the possible pretreatments and the extraction methods. In addition, other possible methods are briefly explained. Next, we show the

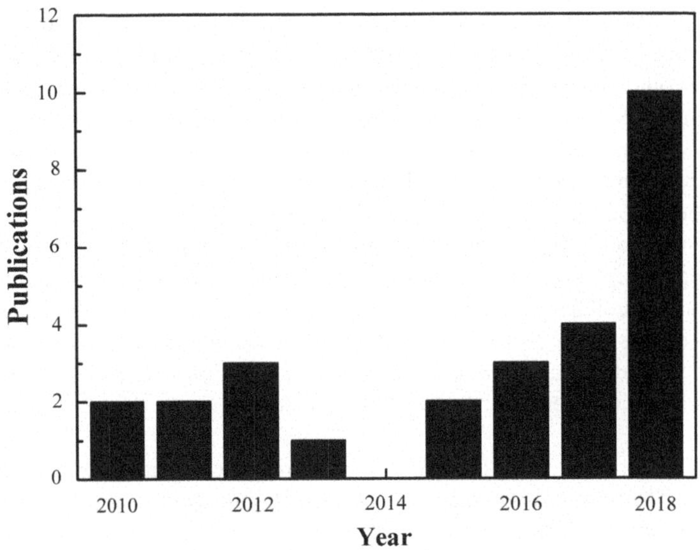

Fig. 4 Publications from 2010 to 2018 about CN from pineapple fibres in Web of Science

main properties of these nanostructures and how to characterize them. The last part concerns the applications of these nanostructures and future trends in the area.

2 Pineapple Fibres as Promising Source to Extract Cellulose Nanostructures

Pineapple (*Ananas comosus*) is a member of the *Bromeliaceae* family and is an herbaceous perennial plant typical from tropical regions. In 2017, the global production of pineapples amounted to 27.4 × 10^6 tons, and the five leading producing countries were Costa Rica (~3.06 × 10^6 tons), Philippines (~2.67 × 10^6 tons), Brazil (~2.25 × 10^6 tons), Thailand (~2.12 × 10^6 tons) and India (~1.86 × 10^6 tons) [52].

Pineapple plants are short, with heights ranging from 0.75 to 1.5 m. The main parts of a pineapple plant are the crown, the fruit, a short and stout stem, the slips, the sucker and a rosette of long (0.50–1.8 m), narrow, fibrous and spiny leaves [173, 180]. Fresh leaves yield about 2–3% of fibres [8]. Figure 5 shows a representation of the pineapple plant and its different parts.

The fruit is the only part of the pineapple plant that currently has commercial value. It is industrialized as food articles (juice, pulp, jelly, etc.), for edible purposes, or as source of bioactive compounds [8, 108]. The other parts (crown, stem and leaves) consist in lignocellulosic residues that are still discarded. This occurs because the pineapple plant is not cultivated for the commercialization of lignocellulosic residues of crown, stem and leaf due to the lack of knowledge of their economic potential.

Fig. 5 Scheme of the pineapple plant and its different parts

There is still little research on the fibres of the pineapple plant leaves, and their industrial use is recent compared to other lignocellulosic fibres such as jute, flax and sisal [156].

The residues generated from the cultivation of pineapple plant could be used as an alternative raw material for new applications. Besides, the reduction of agricultural residues results in a decrease of the environmental degradation [180]. In the literature was reported the use of the lignocellulosic fibre wastes of pineapple crown leaves [173] and mainly pineapple plant leaves [77, 79, 86, 108] as raw materials.

The pineapple leaf fibres (PLFs) have a ribbon-like structure and consist of a vascular bundle system present in the form of bunches of fibrous cells. The fibre bundles are usually mechanically extracted from the leaves using a decorticator machine to remove all the epidermal tissues [33, 34, 157]. Sena Neto et al. [156] characterized six botanical varieties of PLF extracted manually and observed a regular and smooth surface morphology of the fibre cells and low crystallinity index ranging from 49 to 64%. Sena Neto et al. [157] characterized 12 varieties of PLF extracted mechanically (decortification method) and observed similar morphological aspects of fibre bundle in the presence of some mucilage from the leaves, although in lower quantities than previously found [156]. X-ray diffraction analysis indicated the presence of cellulose I polymorphic form for all samples of PLF and values of cellulose crystallinity index between 61 and 78.7% [157].

PLF is an interesting source to extract CN because it has a higher α-cellulose content (74–83 wt%) compared to other fibres (Table 1) and low microfibrillar angle (14°). This higher quantity of cellulose in the PLF is to support the higher weight of the fruit and its lower perishability [8].

Abraham et al. [1] studied the extraction of cellulose nanofibrils from PLF, banana and jute fibres, and concluded that PLF resulted in CN with higher quality and yield. Deepa et al. [42] carry out similar studies using sisal, kapok, banana rachis, PLF and coir fibres. The results were analogous, where the pineapple fibres presented the highest degree of crystallinity and extraction yield. Additionally, leaf and crown

Table 1 Chemical composition of different lignocellulosic fibres

Source	α-Cellulose, wt%	Hemicellulose, wt%	Lignin, wt%	Extractive, wt%
Cotton[a]	95	2	1	0.4
PLF[b,c,d]	74–83	6.35	10.41	6.68
Ramie[a]	76	17	1	6
Flax (retted)[a]	71	21	2	6
Jute[a]	71	14	13	2
Hemp[a]	70	22	6	2
Sisal[a]	73	14	11	2
Bagasse[a]	40	30	20	10
Hardwood[c]	31–64	25–40	14–34	0.1–7.7
Softwood[c]	30–60	20–30	21–37	0.2–8.5

a: Khalil et al. [96]; b: Cherian et al. [33]; c: Wan Nadirah et al. [180]; d: Jonoobi et al. [90]

fibres of the pineapple residues stand out among the different agricultural residues due to the large amount generated during the cultivation of the pineapple plant [8].

In this context, a question arises about which countries have more publications and citations about CN extracted from pineapple fibres. Thus, we applied the bibliometric procedure and the Boolean search expression suggested by Milanez et al. [121, 122] to analyse the scientific papers about CN in the Web of Science database on 26 November 2018, searching since 1945.

This search resulted in 12,607 papers. Then, we added the term "pineapple" to the expression, resulting in 51 publications. Among them, only 27 were actually about CN from pineapple fibres, instead of cellulose in the micro-scale or of bacterial cellulose produced from pineapple juice or pineapple peels. The authors were from 42 different institutions, in 17 distinct countries. Figure 6a shows the countries with more publications, while Fig. 6b presents those with more citations.

Milanez et al. [120] and García et al. [64] analysed the countries with more publications about the CN extracted from different sources. Comparing their results with those shown in Fig. 6a, we verified that China and Sweden are among the countries with more papers about CN from distinct sources, including pineapple fibres. Further, in Fig. 6a, the presence of Costa Rica, Brazil, India, Indonesia and China is possibly justified because they are among the ten countries with the highest pineapple production worldwide in 2017 [52].

Considering the world region, the Asian countries (India, China, Malaysia, Indonesia and Taiwan) were the majority among those shown in Fig. 6a while the European countries (Slovenia and Sweden) were the minority. However, the situation was different concerning the citations. In Fig. 6b, there were four Asian countries (Malaysia, India, Iran and Israel) and four European countries (Slovenia, Sweden, France and Portugal), and there is no longer dominance of Asian countries in this field as in Fig. 6a. Even China was no longer among the most cited ones. It is interesting

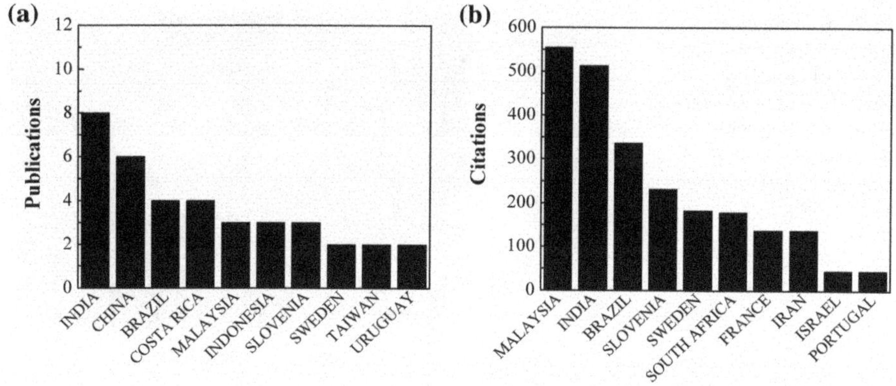

Fig. 6 Results from the search in Web of Science (1945–2018) about CN isolated from pineapple fibres: **a** countries which published more papers and **b** countries whose papers were most cited

to study it again in future aiming to confirm if the development is in an emerging state.

Besides Web of Science, other databases such as Scopus and the Google Scholar were included in our search and resulted in 36 distinct publications about CN extracted from pineapple fibres. We used those data to explain the following topics about pretreatment, extraction and characterization of CN.

3 Classification of Cellulose Nanostructures

Cellulose nanostructures with diverse morphologies and properties can be prepared from lignocellulosic sources such as pineapple fibres using different methods of isolation [50]. In this section, we describe the two main classes of CN: cellulose nanocrystals (CNCs) and cellulose nanofibres (CNFs). It is worth to mention that different nomenclatures could be found in works from different authors, since there was not a uniform manner to refer to these nanostructures over the years [102]. Other classes of CN also found in the literature are briefly described in the end of this section.

3.1 Cellulose Nanocrystals

CNCs, also known as nanocrystalline cellulose or *whiskers*, consist of rod-like shaped cellulose crystals with diameters (d) usually between 10 and 70 nm and lengths (L) from 100 nm to several micrometres (Fig. 7) [102]. The geometry and aspect ratio (L/d) of CNC depend on the source and could range from 1 to 100 [11]. In case

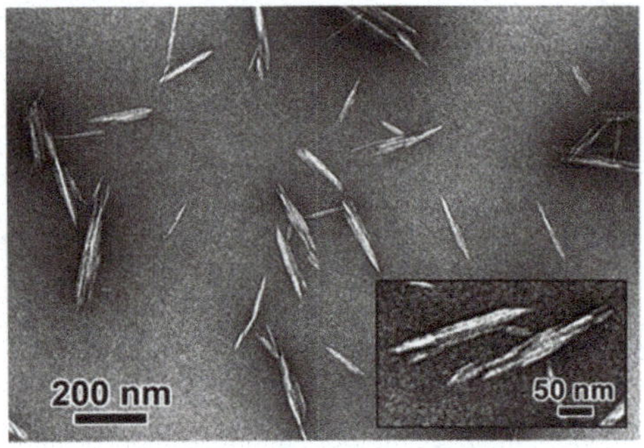

Fig. 7 Transmission electron microscopy image of CNC obtained by sulphuric acid hydrolysis of cotton. Adapted with permission from [50]. Copyright 2019 American Chemical Society

of CNC isolated from PLF, Dos Santos et al. [45] reported aspect ratio around 60, while Prado and Spinacé [144] reported aspect ratio of 6.3 for CNC extracted from pineapple crown fibres.

The extraction of CNC from cellulose fibres is usually performed by acid hydrolysis, but could also occur by oxidation or other processes (TAPPI WI 3021 [170]). The isolation of CNC by acid hydrolysis occurs through a selective degradation process where the amorphous regions of the cellulose nanofibril are more susceptible to the acid attack, leading to the cleavage of glycosidic bonds [16, 163]. As the amorphous domains are removed, the crystalline regions are left intact, as represented in Fig. 8 [163]. The acid molecules attack firstly the paracrystalline and then the crystalline core of cellulose [136]. This results in CNC with near perfect crystalline arrangement that present high modulus [164] and high mechanical strengths [11]. Consequently, CNCs have a significant potential as a reinforcing material in polymeric composites [19, 164].

The surface functionality of CNC may be modulated according to the mineral acid used in the hydrolysis. When sulphuric acid is used, the sulphate groups react with the surface hydroxyl groups of CNC resulting in negatively charged crystals. Consequently, CNCs are uniformly dispersed via electrostatic repulsion in aqueous suspension [74]. CNCs are randomly oriented in the dilute region and align

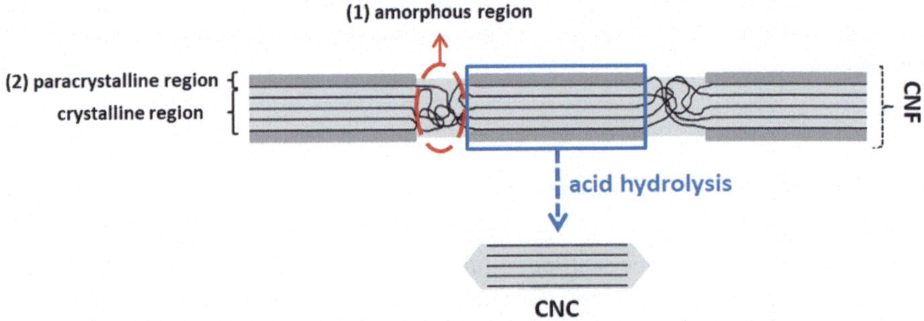

Fig. 8 Sequence of acid molecule attack during the acid hydrolysis reaction: (1) amorphous and (2) paracrystalline region, leaving the crystalline region in the CNC

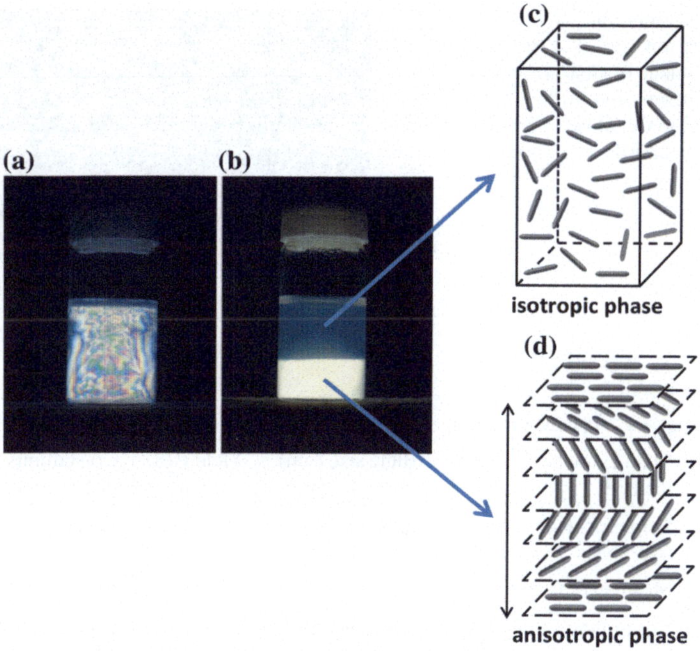

Fig. 9 CNC aqueous 0.63% (w/w) suspension observed between crossed polarizers: **a** just after shearing, showing iridescent birefringence; **b** after 1 week, showing the separation into the **c** upper isotropic and the **d** lower anisotropic phases. **a** and **b** Adapted with permission from [5]. Copyright 2019 American Chemical Society; **c** and **d** [74]

forming a chiral nematic liquid crystalline phase when a critical concentration is reached (Fig. 9). Beyond this critical concentration, aqueous CNC suspensions show iridescent birefringence after shearing [74, 102]. Because of the negative diamagnetic susceptibility of cellulose, CNC becomes orientated when exposed to magnetic field, allowing the production of highly regular single domains that have potential applications in optical devices [44, 149].

3.2 Cellulose Nanofibrils

CNFs (Fig. 10), also called by some authors as "nanofibres", are the elementary fibrils of cellulose with diameters of 2–20 nm and length of several tens of micrometres formed during cellulose biosynthesis, as represented in Fig. 2. CNF can be produced by some bacteria or isolated from cellulosic fibres, usually by TEMPO-mediated oxidation (2,2,6,6-tetramethylpiperidine-1-oxyl radical), enzymatic hydrolysis or mechanical processes involving delamination or fibrillation [102, 184]. However, these mechanical processes generally produce bundles of fibrils instead of individual

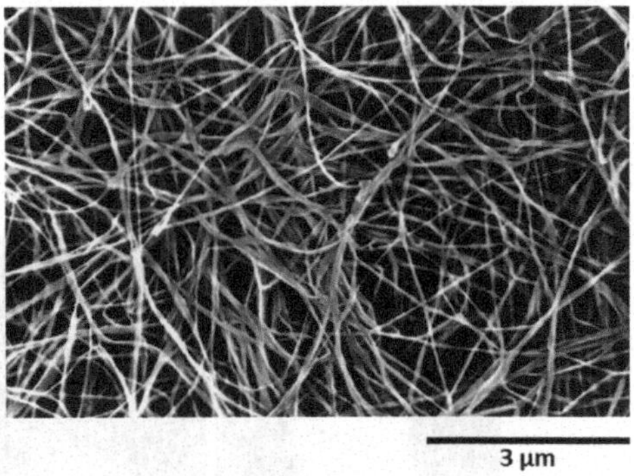

Fig. 10 Field emission scanning electron microscopy image of CNF. Adapted by permission from: Springer Nature Cellulose [100], Copyright 2019

Table 2 Comparison between the properties of CN extracted from PLF

CN type	Main extraction method	Particle size (nm)	Yield (%)	Crystallinity index (%)
CNC[a–d]	Sulphuric acid hydrolysis	4–250	55–77	72–84
CNF[c,e–k]	Steam explosion	5–285	40–76	55–92

a: Gao et al. [63]; b: Dos Santos et al. [45]; c: Balakrishnan et al. [14]; d: Moreno et al. [126]; e: Cherian et al. [33]; f: Abraham et al. [1]; g: Cherian et al. [34]; h: Costa et al. [37]; i: Deepa et al. [42]; j: Balakrishnan et al. [13]; k: Wahyuningsih et al. [179]

fibrils, since a high amount of energy is required to cleave the intra- and intermolecular hydrogen bonds between the individual cellulose fibrils [153]. Depending on the method of isolation, CNF isolated from PLF may have diameters in the range of 5–150 nm [42, 159, 179].

CNFs are flexible nanostructures with relatively high aspect ratio and usually lower crystallinity compared to CNC. CNF can form networks and have a potential to be used as a rheology modifier in foods, paints, cosmetics and pharmaceutical products [163, 164, 184]. Table 2 shows a comparison between the properties of CNC and CNF extracted from PLF.

Balakrishnan et al. [14] compared the stability of CNC and CNF isolated from PLF in aqueous suspensions and showed that CNCs have higher stability than CNF (Fig. 11). After 2 weeks of storage, CNF suspension tends to settle down at the bottom while the CNC suspension is found unchanged due to the higher stability [14]. It was attributed to the negatively charged groups inserted in the surface of CNC during the hydrolysis with mineral acids [14].

However, CNCs extracted with sulphuric acid are expected to have lower thermal stability than CNF, since the sulphate groups catalyse cellulose degradation. This is an advantage when using CNF as reinforcement in thermoplastic matrices. Differently from CNC transparent nanocomposites, CNF leads to the formation of translucent nanocomposites due to higher agglomeration and entanglement [184]. The type of CN also interferes in the processing methods used to produce polymeric nanocomposites [47].

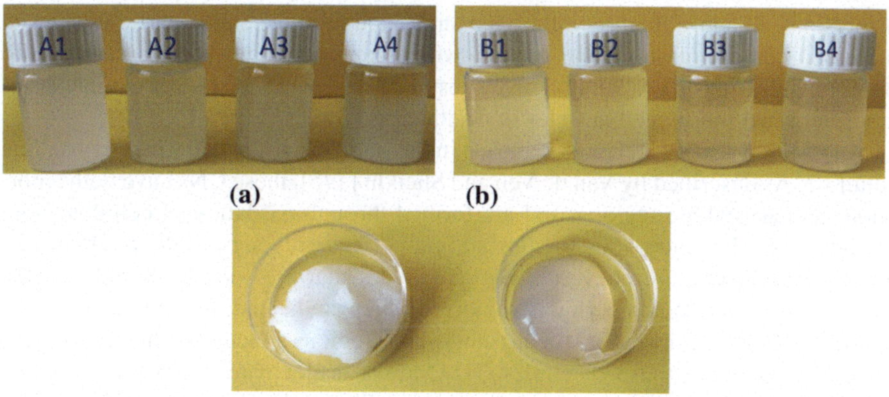

Fig. 11 Stability of **a** CNF and **b** CNC aqueous suspensions. (A1, A2, A3 and A4) and (B1, B2, B3 and B4) are related to 0, 1, 2 and 3 after weeks of preparation, respectively [14]. Copyright © 2019 by John Wiley Sons, Inc. Reprinted by permission of John Wiley & Sons, Inc.

3.3 Other Types of Cellulose Nanostructures

Electrospinning process produces CNF usually denominated as cellulose nanoyarns. They consist in nano-scaled filaments produced when an external electric field is applied on the surface of a solution of cellulose or cellulose derivatives, such as cellulose acetate [103, 146]. Several systems for dissolving cellulose have been studied, including *N,N*-dimethylacetamide (DMAc)/LiCl, dimethyl sulphoxide/triethylamine/SO_2, *N*-methylmorpholine-N-oxide (NMMO)/water, trifluoroacetic acid, liquid ionics such as 1-butyl-3-methylimidazolium and NaOH/urea aqueous solution [87, 99, 146]. Surip et al. [167] reported the use of trifluoroacetic acid to dilute cellulose from PLF and produce electrospun CNF. The size of the cellulose electrospun fibres depends on the concentration of the solution and electrospinning conditions [166]. In general, cellulose nanoyarn has low crystallinity and lower thermal stability than the initial source [95]. The advantage of electrospun CNF is that they are homogeneous and can be produced directly from the cellulose solution. Electrospun membranes have received attention in the last years due to their potential use in biomedical, electronic and advanced filtration industries [146]. However, cellulose nanoyarns have not been widely studied till date [103].

Other types of CN reported in the literature have not been produced from PLF yet. They include amorphous cellulose nanoparticles and hairy nanocrystalline cellulose. Amorphous cellulose nanoparticles consist in spherical cellulose nanoparticles with average sizes of less than 200 nm and with a proportion of crystalline regions less than 20 wt% [128]. They can be produced either by mechanical methods as grinding (e.g. in ball mill) or by dissolution and regeneration of amorphous cellulose in different solvents, followed by a hydrolysis step with subsequent ultrasound disintegration [83, 128]. Ioelovich [82] reported that concentrated paste of amorphous cellulose showed thickening properties that can be used to prevent phase separation of water dispersions. In addition, the synthesized amorphous cellulose nanoparticles showed increased content of sulphonic groups, which can enable amorphous cellulose to immobilize various therapeutic-active substances, allowing the use of complexes of amorphous cellulose nanoparticles in health care.

Another class of CN recently reported in the literature is the hairy nanocrystalline cellulose. As described by Van de Ven and Sheikhi [175], these CNs have both amorphous and crystalline regions, and are formed through oxidation of cellulose with periodate. As the amorphous regions are more kinetically accessible, the hydroxyl groups of cellulose are converted into aldehydes that are soluble in water in the room temperature. When periodate-oxidized fibres are heated, they fall apart in the nanocrystalline cellulose that still has amorphous regions attached to their ends in the form of dialdehyde modified cellulose (DAMC). The aldehyde chains provide steric stability to the CN, which are also called "sterically stabilized nanocrystalline cellulose". These nanostructures can act as intermediaries that can be functionalized with other chemical groups, producing electrosterically stabilized CNC. Hairy nanocrystalline cellulose has numerous potential applications including superhydrophobic films, security packaging, transparent films, metal ion and non-ionic scavengers, polymeric nanocomposites and rheology modifiers [175].

4 Extraction of Cellulose Nanostructures

Similar to other lignocellulosic fibres, pineapple fibres are constituted of cellulose and other non-cellulose compounds such as hemicelluloses, lignin, ashes and waxes. In order to extract pure CN, normally a pretreatment step prior to the CN isolation is required to remove these non-cellulosic compounds [95, 150].

4.1 Pretreatments

The hierarchical structure of cellulose in the lignocellulosic fibres is complex as described in Sect. 1 and shown in Fig. 3. Consequently, the process of fibrillation is difficult and takes high energy consumption to overcome the interfibrillar hydrogen bonds and release the CN [95]. Proper pretreatment promotes accessibility of the cellulose fibres, increases the internal surface, alters crystallinity, breaks hydrogen bonds and increases cellulose reactivity. Therefore, these pretreatments decrease the energy demand and facilitate the CN production [95, 150].

Different methodologies can be used in the pretreatment of lignocellulosic fibres aiming to remove hemicellulose, lignin and other non-cellulosic materials. The choice of the pretreatment method depends mainly on the knowledge of the composition of the cellulose source and the morphology desired for the cellulose after further treatments process [64, 150]. This section describes the most well-known pretreatments used for lignocellulosic fibres, including pulping process, alkaline, bleaching, alkaline-acid-alkaline treatments and ionic liquid. A combination of these methodologies and other methodologies can also be used. However, it is important to keep in mind that some methodologies can be used not only as pretreatment, but also for the extraction of CN such as ionic liquid treatment.

4.1.1 Pulping Process

Pulping process is generally used to isolate cellulose fibres from wood, but can also be used in other lignocellulosic sources such as pineapple fibres [118]. The pulping process can be done mechanically or chemically.

The mechanical pulping method is mainly applied to wood fibres due to their higher non-cellulosic content. In this process, the fibres are obtained by grinding wood logs using a rotary sandstone cylinder or by feeding wood chips into the centre of refining rotary discs in the presence of a jet of water. These mechanical treatments damage the morphology and size of the pulp resulting in the reduction of the cellulose chains and their crystallinity [91, 95, 96]. The intense mechanical action that the cellulose fibres undergo during mechanical pulping results in the delamination and gradual peeling off of the external cell wall layers (primary and S1 layers) and the subsequent exposition of the S2 layer [76]. Although energy consuming, this process allows the fibrillation of almost whole wood material.

The dominant chemical pulping method worldwide is kraft process or sulphate process. It involves dissolving lignin with a hot solution in a mixture of sodium hydroxide and sulphite in a digester. The use of sodium chlorite promotes a higher yield of delignified fibres compared to the conventional kraft process [95]. The use of only sodium hydroxide is called soda cooking. For example, Mahardika et al. [118] reported the use of a solution of sodium hydroxide (25%) for 60 min at 60 °C and 9 bar in a digester to cook PLF as a pretreatment to further isolate CN.

Alternative chemical pulping processes are the use of sulphite pulping by sulphite acid and its salts [96]. Other pulping methods include the use of organic solvents such as ethanol, methanol and peracetic acid for delignification [96], and enzymatic methods [113].

4.1.2 Alkaline

Alkaline pretreatment of lignocellulosic fibres such as pineapple fibres aims to disrupt the lignin structure and assist to separate the structural linkages between lignin and carbohydrates. Strict control of the alkaline hydrolysis of the lignocellulosic fibre is necessary in order to avoid degradation of cellulose [164]. The efficiency of alkaline pretreatment of lignocellulosic fibres depends on the lignin content.

Alkali pretreatments are more effective for lignin solubilization. Minor cellulose and hemicellulose solubilization occurs compared to acid or hydrothermal processes, promoting less sugar degradation compared to acid pretreatment. This process can be performed at room temperature and times ranging from seconds to days.

Sodium, potassium, calcium and ammonium hydroxides are used in alkaline pretreatments. Sodium hydroxide promotes swelling, increases the inner surface of cellulose and decreases its degree of polymerization and crystallinity, resulting in rupture of the lignin structure [3].

In the literature, there are several reports about the use of alkaline pretreatment prior to the extraction of CN from pineapple fibres. Abraham et al. [1] used alkali pretreatment for PLFs which were soaked with sodium hydroxide aqueous solution (2 wt%) for 6 h at 30 °C. Cherian et al. [33] and Balakrishnan et al. [13, 15] performed pretreatment of PLF using sodium hydroxide aqueous solution (2 wt%) under 138 kPa for 1 h. Gao et al. [63] treated PLF with a sodium hydroxide aqueous solution (4 wt%) at 80 °C for 2 h under mechanical stirring. Dos Santos et al. [45] treated PLF with a sodium hydroxide aqueous solution of 2 wt% for 4 h at 100 °C under mechanical stirring. Shih et al. [159, 160] treated PLF with sodium hydroxide aqueous solution of 4 wt% for 45 min. Moreno et al. [126] treated PLF for 4 h with sodium hydroxide (2% w/v) at 80 °C. Asri et al. [9] treated PLF with sodium hydroxide aqueous solution (4 M) and stirred at 80 °C for 4 h. Prado and Spinacé [144] treated pineapple crown fibres with sodium hydroxide aqueous solution (5 wt%) at 90 °C for 1 h, under stirring.

4.1.3 Bleaching

The bleaching process removes the residual lignin of the lignocellulosic fibres resulting in a whiter material with lower impurities, without changing the cellulose crystallinity or polymorphism and improving the ageing resistance [95, 96]. The bleaching process uses various chemical agents in several stages, such as hydrogen peroxide, chlorine dioxide, oxygen, ozone, peracetic acid, sodium chlorite, chlorine and chlorine dioxide. Among them, oxygen and chlorine dioxide are the most used ones [81, 95].

Abraham et al. [1] used the bleaching treatment for PLF with a sodium chlorite ($NaClO_2$) solution (pH 2.3) for 1 h at 50 °C. Costa et al. [37] and Balakrishnan et al. [13] bleached 6 types of PLF using a mixture of sodium hydroxide aqueous solution/glacial acetic acid and a mixture of 1:3 sodium hypochlorite (NaClO) solutions. Dos Santos et al. [45] bleached PLF with a solution of acetate buffer and aqueous sodium chlorite. Shih et al. [159] bleached PLF using sodium chlorite (1.5% $NaClO_2$) at pH 4.8, 85 °C for 3 h. Shih et al. [160] bleached PLF in an acetate buffer solution containing sodium chlorite (0.3% $NaClO_2$) at pH 4.8, 60 °C for 3 h.

Balakrishnan et al. [15] bleached 6 types of PLF using hydrogen peroxide (H_2O_2) in autoclave maintained pressure of 138 kPa for 30 min. Moreno et al. [126] treated PLF for 4 h with sodium chlorite (2.5% w/v $NaClO_2$) at room temperature. Asri et al. [9] treated PLF dried sample with sodium hypochlorite (NaClO 5 wt%) and heated at 80 °C for 4 h. Prado and Spinacé [144] treated pineapple crown fibres with a mixture of hydrogen peroxide (H_2O_2, 16 v%) and sodium hydroxide aqueous solution (NaOH 5 wt%) at 55 °C for 90 min, under stirring.

4.1.4 Alkaline-Acid-Alkaline Pretreatment

Alkaline-acid-alkaline pretreatment has been used in order to solubilize lignin, hemicelluloses and pectins before mechanical isolation of CNF from lignocellulosic sources [97]. This pretreatment includes three steps:

(a) Soaking lignocellulosic fibres in sodium hydroxide (12–17.5 wt%) for 2 h, aiming to increase the surface area of the cellulosic fibres making them more susceptible to hydrolysis;
(b) Hydrolysis of lignocellulosic fibres using 1 M hydrochloric acid solution at 60–80 °C to solubilize the hemicelluloses;
(c) Treatment of the lignocellulosic fibres with sodium hydroxide (2 wt%) for 2 h at 60–80 °C, in order to break the lignin structure and the bonds between the carbohydrates and the lignin [95, 97]. Although promising, this pretreatment has not yet been reported for pineapple fibres.

4.2 Methods for the Extraction of Cellulose Nanostructures

Different methodologies can be used for CN extraction from lignocellulosic sources as pineapple fibres, such as chemical, enzymatic, mechanical and electrical procedures. Chemical methods include acid or basic hydrolysis, oxidation and treatment with ionic liquids. Mechanical methods include steam explosion, high-pressure homogenization, high shear homogenization, high-intensity ultrasound, electrospinning, among others. An example of electrical method is electrospinning. A combination of these methods is also possible to extract CN.

4.2.1 Hydrolysis

The hydrolysis of microcrystalline cellulose to extract CN can be made using acid or base, but acid is mainly used. Generally, high concentration (6–8 M) of mineral acids is used to extract CN from different sources, including lignocellulosic fibres as pineapple fibres. However, to obtain CN from plants, the hierarchical structure of the lignocellulosic fibres should be deconstructed aiming to remove hemicelluloses and lignin, than pretreatments are usually employed [7].

The hydrolysis reaction occurs under strictly controlled conditions of temperature, stirring, time and acid concentration [73, 95, 123, 150]. There is a correlation between temperature and reaction time: higher reaction temperature results in a decrease in reaction time [22].

Several acids can be used, such as sulphuric acid (H_2SO_4), hydrochloric acid (HCl), hydrobromic acid (HBr) and phosphoric acid (H_3PO_4) (Fig. 12a). However, H_2SO_4 is the most common acid used for the extraction of CNC through chemical hydrolysis [22, 95, 123, 150].

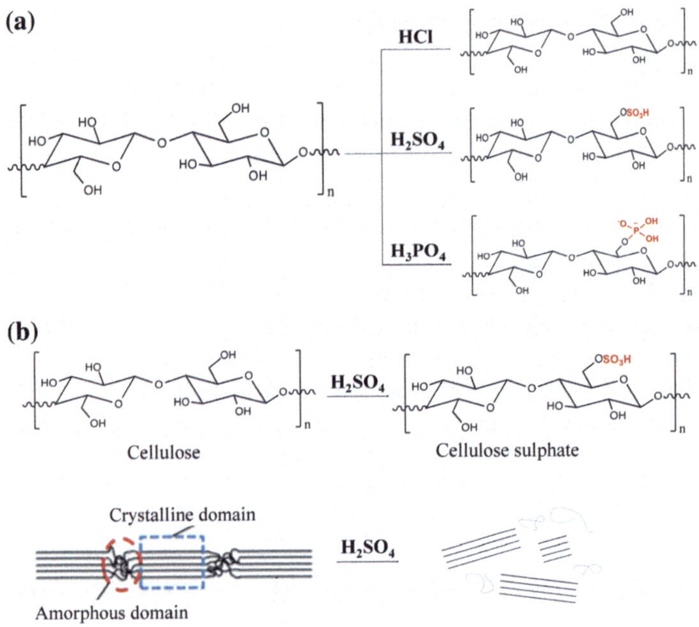

Fig. 12 **a** Reaction of cellulose with different mineral acids (HCl, H_2SO_4 and H_3PO_4) and **b** scheme of cellulose hydrolysis with sulphuric acid forming cellulose sulphate

In the acid hydrolysis of cellulose, the acid protons rapidly protonate the glycosidic oxygen or cyclic oxygen, followed by a slow cleavage of glycosidic bonds induced by the addition of water. This hydrolysis process results in two shorter chains while preserving the basic backbone structure [115]. Besides, the interfibrillar contacts of cellulose and the amorphous regions are selectively and fast hydrolysed, leaving the crystalline regions intact which remains in the acid aqueous solutions, forming CNC dispersion. The CNC particles in the dispersion are separated via consecutive centrifugations using water to remove excess of acid. Neutralization and dialysis are performed to remove any free acid molecules [22, 95, 115, 123, 150].

In the hydrolysis of cellulose with H_2SO_4, after the chain cleavage, the esterification of the hydroxyl groups occurs resulting in the so-called cellulose sulphate (Fig. 12b). The presence of sulphate groups on the CNC surfaces makes them negatively charged, favouring the anionic stabilization due to the repulsive forces of the electric double layers, which are very efficient and avoid the aggregation of the CNC resulting from the hydrogen bonds [95, 115]. The negatively charged CNC surfaces lead to a more stable cellulose suspension and form a stable colloidal system [19]. On the other hand, CNCs obtained using HCl are not charged, present a limited dispersion and tend to flocculate [73, 94].

Gao et al. [63] treated PLF with H_2SO_4 (65 wt%) for 40 min at 50 °C under mechanical stirring and produced CNC with particle size of about 30–50 nm, zeta potential value of −48.9 mV and contact angle of 30.92°. Dos Santos et al. [45]

extracted CNC from PLF by acid hydrolysis at 45 °C for 5, 30 or 60 min, using 20 mL of H_2SO_4 (9.17 M) for each gram of material. CNC obtained presented a needle-shaped nature, thermal stability of 225 °C, crystallinity degree of 73%, an average length (*L*) of 249.7 (±51.5) nm and a diameter (*D*) of 4.45 (±1.41) nm with aspect ratio (*L*/*D*) of around 60.

Moreno et al. [126] obtained CNC from PLF using sulphuric acid in two concentrations and times of reaction: (a) 30% w/v for 5 h at 50 °C and (b) 60% w/v for 30 min at 50 °C. CNC obtained using H_2SO_4 in the concentrations of 30 and 60% w/v resulted in average particle sizes of 250 and 160 nm, respectively.

Prado and Spinacé [144] isolated CNC from pineapple crown leaf fibres using H_2SO_4 (60 wt%) at 45 °C and vigorous stirring for 1 h. Obtained CNC had rod-like shape and diameter of 39 nm. CNC showed cellulose I crystalline structure and crystallinity index of 73%. The thermal degradation of CNC started at 124 °C, what was attributed to the sulphate groups in the cellulose chains.

Although acid hydrolysis is widely used to extract CNC, it has some disadvantages such as high energy consumption, generation of chemical residues, equipment corrosion, and health and environmental risks. In order to overcome these disadvantages, some alternatives have been studied such as the use of solid acids, gaseous acids and metal salt catalysts [95].

4.2.2 Oxidation

The promising oxidation treatment of lignocellulosic fibres is performed by carboxylation via 2,2,6,6-tetramethylpiperidine-1-oxyl (TEMPO)-mediated oxidation. TEMPO catalyses the conversion of primary hydroxyl groups (–OH) of cellulose into carboxyl groups (R–COOH) in the presence of a primary oxidizing agent (e.g. sodium hypochlorite) and halogen salts (e.g. sodium bromide or sodium chloride). This reaction (Fig. 13) occurs in an aqueous medium [95].

For native cellulose, the oxidation occurs only on the surface of the microfibrils, which become negatively charged. The negative charge introduced during TEMPO-mediated oxidation in cellulose results in repulsion and consequently facilitates fibrillation [164]. TEMPO-mediated oxidation can also be used as pretreatment and/or to functionalize CN.

Shih et al. [159, 160] obtained CNF with diameter of 55 nm from PLF using 2,2,6,6-tetramethylpiperidine-1-oxyl radical (TEMPO) method. PLF previously submitted to alkali and bleaching treatment was suspended in a sodium phosphate buffer

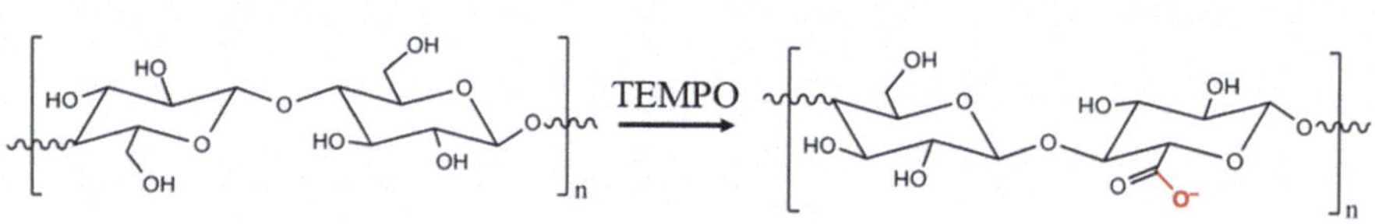

Fig. 13 Scheme of reaction of cellulose via (TEMPO)-mediated oxidation

(pH 6.8) containing TEMPO and sodium chlorite. Sodium hypochlorite solution (2 M) was diluted to 0.1 M with the same buffer, and it was added as the oxidation medium. TEMPO-mediated oxidation occurred at 60 °C for 3.5 h.

4.2.3 Ionic Liquids

Ionic liquids are composed entirely of ions (generally consisting of organic cations—namely derivatives of NN′-substituted imidazolium, N-substituted pyridinium, tetra-alkylated ammonium and tetra-alkylated phosphonium, and either organic or inorganic anions) that remain in the fluid state at temperatures below 100 °C. They are considered as green solvents, and ionic liquids containing organic cations with quaternary ammonium, phosphonium and sulphonium nuclei have been used in many fields of research in the last two decades [23]. They have interesting and valuable properties such as non-flammability, very low vapour pressure, and thermal and chemical stability [95].

The hydrophilic ionic liquids such as 1-butyl-3-methylimidazolium chloride and 1-allyl-3-methylimidazolium chloride can dissolve cellulose, disrupting its inter- and intramolecular hydrogen bonds. In the mechanism of dissolution, the ionic liquid cations attack the oxygen atoms, whereas ionic liquid anions interact with the protons of the hydroxyl groups of cellulose, resulting in disruption of the extensive network of hydrogen bonds and promoting the dissolution of cellulose (Fig. 14) [18, 95]. The calculated free solubility energy and experimental results showed the solubility of the cellulose in ionic liquid as $[OAc]^- > [Cl]^- > [BF_4]^- \sim [PF_6]^-$ [188]. After dissolution of cellulose, a wide range of nanoparticle morphologies can be produced through precipitation [150].

Ionic liquids can also be used as a pretreatment of lignocellulosic fibres to remove hemicelluloses and lignin. For example, Fu et al. [60, 61] reported the use of 1-butyl-3-methylimidazolium chloride as pretreatment of PLF to extract CN.

Similar to cellulose, the solubility of lignin in ionic liquids also seems to depend on the anion [23]. Pu et al. [145] studied the dissolution of lignin with aprotic ionic liquids and observed that up to 20 wt% of lignin could be dissolved in 1-hexyl-3-methylimidazolium trifluoromethanesulfonate, 1,3-dimethylimidazolium methylsulphate and 1-butyl-3-methylimidazolium methylsulphate. The nature of the anions influences the lignin solubility. For 1-butyl-3-methylimidazolium salts, the order of

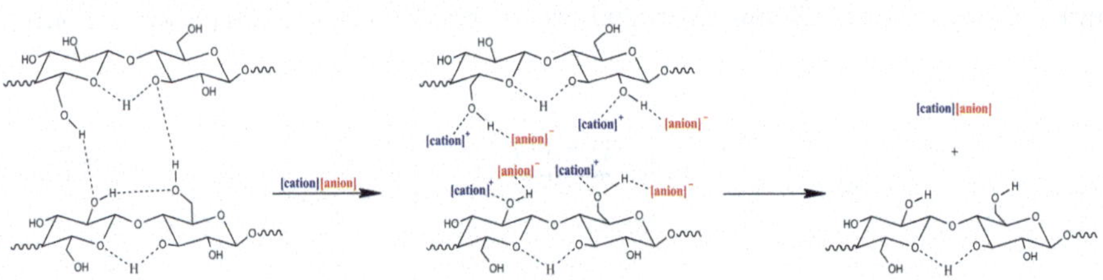

Fig. 14 Scheme of disruption of cellulose hydrogen bonds by action of ionic liquids

solubility was $[MeSO_4]^-$ > Cl^- ~ Br^- >> PF_6^-. On the other hand, ionic liquids containing large, non-coordinating anions do not act as suitable solvents for lignin dissolution [188].

4.2.4 Enzymatic Hydrolysis

Special enzymes such as cellulases, xylanases and ligninases (peroxidases and laccases) are capable of degrading the main components of lignocellulosic fibres such as cellulose, hemicelluloses and lignin [95, 140, 150, 158, 191], and produce CN from lignocellulosic sources as pineapple fibres.

Cellulases hydrolyse the β-1,4-glycosidic linkages of cellulose. They are divided into three groups referred to as (a) endoglucanases or β-1,4-endoglucanases (A- and B-type cellulases), (b) exoglucanases or cellobiohydrolases (C- and D-type cellulases) and (c) β-glucosidases [95]. Endoglucanases cleave intramolecular β-1,4-glycosidic linkages preferably in amorphous domains of cellulose, generating damaged fibres with new terminal ends. Exoglucanases cleave the accessible ends of cellulose molecules to release glucose and cellobiose. β-glucosidases hydrolyse cellobiose to glucose [95, 140, 150, 158, 191].

When used as pretreatments, lower concentrations (0.02%) of cellobiohydrolases and endoglucanases are used, since the enzymes have strong synergistic effects. The molecular weight and fibre length are preserved [95, 150, 158, 164].

Hemicellulose is mainly constituted of xylan carbohydrate. For complete degradation of xylan in the lignocellulosic fibres, it is necessary synergistic action of several hydrolytic enzymes (such as endo-1,4-β-xylanase, xylan 1,4-β-xylosidase), accessory enzymes (such as xylan esterases, ferulic and *p*-coumaric esterases, α-l-arabinofuranosidases and α-4-O-methyl glucuronosidases) and reductive enzymes, including cellobiose oxidizing enzymes, aryl alcohol oxidases and aryl alcohol dehydrogenases [140]. To degrade lignin, peroxidases and laccases are two major families of enzymes. Apparently, these enzymes act using low molecular weight mediators to carry out lignin degradation [140]. To the best of our knowledge, the literature shows no study applying enzymatic hydrolysis to pineapple fibres, but it is a viable approach that can be considered.

4.2.5 Steam Explosion

In the steam explosion method, the cellulose source is heated to 200–250 °C by a high-pressure saturated steam during a certain time [1]. After this period, the system releases the pressure and the steam can expand, separating the cellulose structures [88, 162]. Figure 15 shows a schematic representation of a steam explosion system.

The steam explosion is the most common mechanical method employed to extract CNF from the pineapple fibres [1, 13, 15, 33, 34, 37, 90, 108, 185]. Many studies reported the diameter of CNF from pineapple fibres ranging from 5 to 30 nm [13, 15, 33, 185], but it is worth to mention that Cherian et al. [34] obtained CNF with

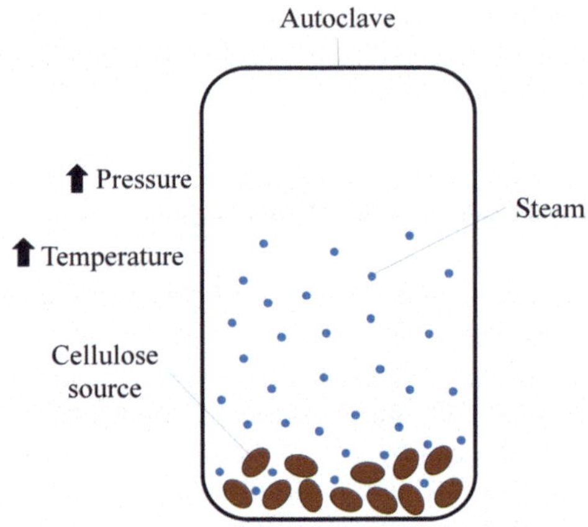

Fig. 15 Scheme of steam explosion system

a wider range (15 nm–1 μm) and an average aspect ratio of 67. Overall, the steam explosion leads to an increase from 14% [1] to 33% [13, 15] in the crystallinity index of CNF compared to pineapple fibres.

The pressure employed in these studies is usually around 138 kPa hold for different periods. While some studies focused to carry out short cycles (15–30 min) repeated several times (6–8 times) [15, 33, 34, 37], others chose only one long cycle (1 h) [1, 13].

4.2.6 High-Pressure Homogenization

In the high-pressure homogenization, the mixture of cellulose source and a fluid, such as water, is forced through a small nozzle using the high pressure and speed, and the shear forces from micro to nanoscale [97]. The generated shear forces trough the nozzle decrease the material from micro to nanoscale [97]. Figure 16 shows a schematic representation of a high-pressure homogenization system.

In the case of PLF, Fu et al. [60, 61] reported the use of high-pressure homogenization combined with the ionic liquid method to extract CN. In this method, pressure can reach 100 MPa and the process can be repeated for 45 cycles [61], resulting in crystalline spherical cellulose nanoparticles with diameters ranging from 4 to 10 nm.

4.2.7 High Shear Homogenization

Differently from the high-pressure homogenization (Sect. 4.2.6), in the high shear homogenization (or grinding method) there are two grindstones that rotate while the cellulose source is between them [97]. The shear forces created by the grindstones, Fig. 17, lead to the fibrillation of the fibrous material and the production of CN [97].

Wahyuningsih et al. [179] and Mahardika et al. [118] applied this approach as the main step or as an initial step, respectively, to extract CN from pineapple fibres in

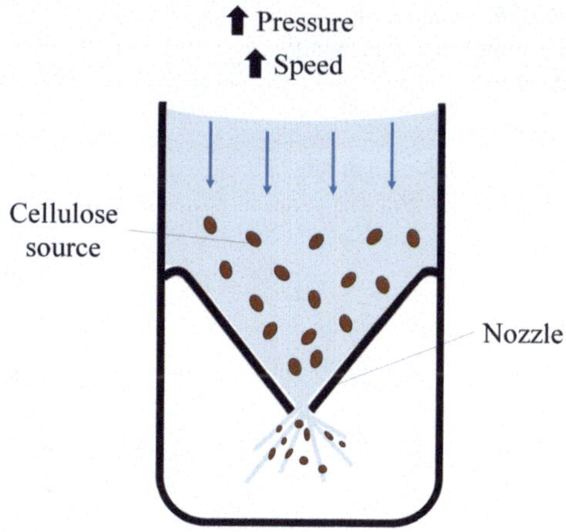

Fig. 16 Scheme of high-pressure homogenization system

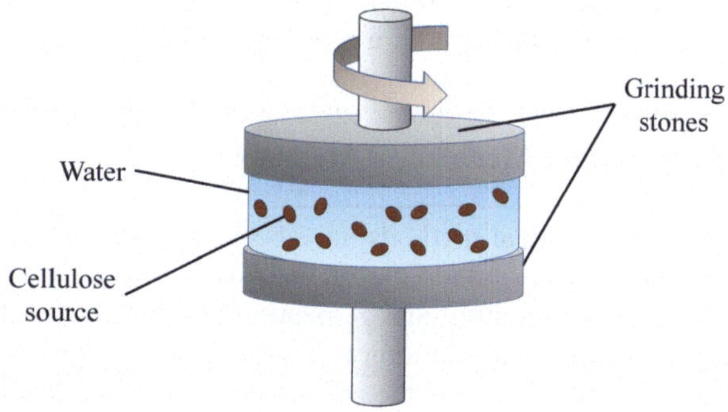

Fig. 17 Scheme of high shear homogenization system

water. The speed of rotation ranged from 1500 rpm [179] to 12,000 rpm [118]. Wahyuningsih et al. [179] reported the extraction of CNF with average size of 284.6 nm and crystallinity index of 55.4%.

4.2.8 High-Intensity Ultrasound

Another mechanical method used to extract CN from lignocellulosic sources is the high-intensity ultrasound (Fig. 18). In this method, the cellulose source is dispersed in water and high-intensity ultrasound waves are applied to the system [116, 169]. The energy of the ultrasound waves creates vapour bubbles that grow and collapse, generating micro-jets of high speed and pressure that promote the isolation of CN [104, 109, 116].

Nikmatin et al. [133] produced CN from PLF without pretreatment using only high-intensity ultrasound for 120 min. On the other hand, Mahardika et al. [118] used a combination of grinding and high-intensity ultrasound to extract CN from PLF. Ultrasonication was performed for 30–60 min, with the system heated to 60 °C

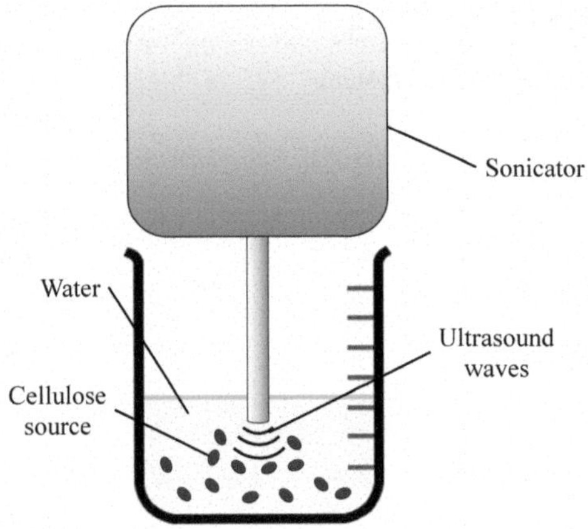

Fig. 18 Scheme of high-intensity ultrasound system

and an equipment of 400 W, resulting in CNF with diameters between 40 and 70 nm and an increased crystallinity index up to 20%.

4.2.9 Other Mechanical Methods

Other mechanical methods have been reported in the literature to extract CN from lignocellulosic fibres. One of them is the microfluidization [55, 97]. It is similar to the high-pressure homogenization (Sect. 4.2.6), but its instrumentation includes an intensifier pump aiming to increase the system pressure, besides an interaction chamber that enhances the shear forces and the impact [55, 97].

Another possibility is the cryocrushing method, in which the cellulose source is immersed in liquid nitrogen to freeze the water content in it. Then, they are crushed by a mortar and a pestle [30, 97]. Notwithstanding, to the best of our knowledge, none of these methods have been applied to PLF yet.

Mostly, mechanical methods do not need any solvent besides water, which does not generate toxic residues, neither change the surface chemistry of CN [152, 172, 190], which are advantages when compared to the chemical methods. However, these methods are generally energy and time consuming, and may lead to damages to the crystalline structure of the nanomaterial [96, 125, 172, 190].

4.2.10 Electrospinning

The electrospinning, Fig. 19, can be considered as an electrical method instead of a mechanical one [59]. In this system, the cellulose source is dissolved in a solvent, such as the N,N-dimethylacetamide (DMAc) or an ionic liquid [59, 66]. For example, Surip et al. [167, 168] reported the production of CNF from pineapple fibre through electrospinning dissolving PLF in trifluoroacetic acid.

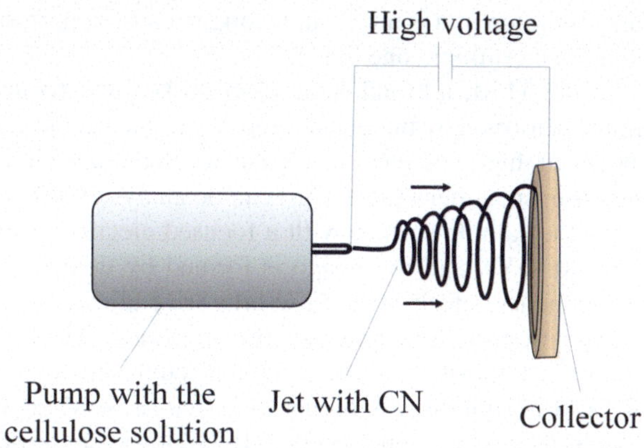

Fig. 19 Scheme of electrospinning system

After dissolution of cellulose, an electrode is immersed and a high voltage (5–30 kV) is applied to a droplet of this solution. The result is the formation of a solution jet that hits the grounded electrode which can be connected to a collector of the obtained CN [59, 114, 167].

Many factors influence the morphology and the properties of the produced CN, including the solution viscosity, surface tension, conductivity, the applied voltage and the collector distance [59, 66, 114].

5 Characterization of Cellulose Nanostructures Extracted from Pineapple Fibres

5.1 Morphology and Size Distribution

CNs present different shapes and size distributions. In addition, they can aggregate in water and agglomerate when dried, generating micrometric particles that are hard to disperse again [89]. Based on these, it is necessary to know the appropriate technique for the proper characterization of their size and morphology [56].

There are many techniques to evaluate the morphology of CN, and each one has its own benefits and limitations. Around 70% of the papers about the CN extracted from pineapple fibres characterized their morphology and/or size distribution. The most used techniques are the transmission electron microscopy (TEM), the scanning electron microscopy (SEM), the atomic force microscopy (AFM) and the dynamic light scattering (DLS).

TEM and SEM have some similarities in their instrumentation. Both of them are composed by an electron gun, the electron microscope column, magnetic lens and a vacuum system. The electron gun produces an electron beam that is focused on the specimen, and can be a thermionic emission source or a field emission source. The field emission gun is more interesting because it performs a higher spatial resolution than the thermionic one [182].

In the TEM, a broad static electron beam is focused on the specimen, and the image is formed by the transmitted electrons. As the electrons must pass through the specimen, high voltages are used to accelerate the electrons (~60 until 300 keV) and only thin specimens (>0.5 μm) can be analysed [43, 182]. On the other hand, SEM scans the sample surface with a focused electron beam of lower energy (~500 eV until 30 keV), and the image is formed by detecting the scattered (backscattered or secondary) electrons [43]. While SEM allows the study of the morphology and composition of the surface of the specimen, TEM can provide details about the internal composition of the sample through diffraction patterns. The magnification of TEM is significantly higher than that of SEM, as well as the spatial resolution, which can be even 1 nm or better [182]. However, the depth of field in TEM is smaller and the sample preparation is more complex.

TEM has been the most common technique used for the characterization of CN from pineapple fibre [9, 13–15, 33, 36, 42, 60, 118, 126, 159, 160, 179, 192]. It has been used to identify the shape, the length and the width of the nanoparticles [56, 89]. As mentioned before, the sample needs to be prepared thin for TEM, and it is usual to apply a negative staining to the sample with uranyl acetate, aiming to enhance the image contrast [105]. However, even with this resource, the contrast can be low and the lateral association of the nanostructures on the grid of TEM may influence the observed results [89, 105].

Other characterization used for CN from pineapple fibres was SEM [4, 9, 13, 15, 27, 33, 34, 63, 118, 167, 179]. Some properties such as shape, length and width of the nanostructures were determined by SEM. The main limitation for the use of SEM in the characterization of CN surface relies on their height, which can be around 5 nm. This fact hinders the imaging because the difference between the heights of the nanoparticle and the substrate is small. In addition, usually a thin layer of conductive material (e.g. platinum or gold) is needed to coat the CN to prevent charging of the specimen. Notwithstanding, it may change the observed nanoparticle, broadening its size [56, 105].

Apart from morphological features, the chemical analysis of the specimen can also be achieved with TEM or SEM. For this, the equipment must contain a system such as the energy-dispersive X-ray spectroscopy (EDS or EDX) [182]. The electron beam hits the samples and may excite the sample electrons to a free electron state or to an unoccupied level of higher energy. These excitations emit photons, such as the X-rays, which are specific to each chemical element. The detection limit for an element using EDS can reach even 1% of the composition [182].

Another technique employed to characterize the morphology of CN extracted from pineapple fibres was AFM [14, 15, 27, 33, 34, 39, 40, 42, 45, 108, 144, 177]. AFM principles are completely different from TEM and SEM. In AFM, a force-sensing cantilever scans the sample surface in the plane xy and the position z is recorded [70]. This tip can be close to the sample, and the system measures the force created by the potential energy between tip and sample, by the cantilever deflection. This operation is the contact mode or static mode [70]. The tip can also vibrate, having its amplitude or frequency modulated. Thus, the distance between the cantilever and the specimen can be higher than in contact mode. This mode is the non-contact mode [65, 70]. A combination between the contact and non-contact modes results in the *tapping* mode. Intermediary distances between the cantilever and specimen are used, and the oscillation of the cantilever results in an intermittent contact between the tip and the specimen. AFM results provide similar morphological information than TEM and SEM, but AFM allows the analyses of other dimensions, such as the height of the nanostructures [56, 89]. Further, the friction of the specimen can be estimated [70]. Among AFM limitations, there is the possible overestimation of length or height. It occurs due to the cantilever convolution effects and compression of individual particles by the cantilever too [89, 105].

DLS was also employed to characterize the size distribution of CN isolated from pineapple fibres [9, 14, 15, 60, 118, 126, 177]. It allows obtaining the hydrodynamic apparent size distribution of nanostructures. This dimension is the radius or the diameter of an equivalent sphere that shows the same diffusion coefficient of the analysed nanoparticle [21, 24]. DLS considers all particles, and it does not matter if they are individual, aggregates or agglomerates. Thus, its results alone cannot be directly related to the length or the width of CN [24, 56, 89]. Likewise, the intensity of the size distribution has no relation with the amount of the nanostructures in the samples. The intensity showed by DLS is a function of the scattered light, and it is not a percentage of the present nanoparticles [21, 24]. However, apart from its limitations, DLS shows great advantages [21, 89]. It is useful for fast evaluation of particle size, compared to the time-consuming microscopies [21, 89]. Also, the apparatus is the cheaper one among all mentioned before [21]. Moreover, the samples can be analysed just after preparation, without drying on a substrate or grid [21, 89]. In all studies characterizing CN extracted from pineapple fibres using DLS, it was used in combination with other size characterizations rather than alone [9, 14, 15, 60, 118, 126, 177]. Therefore, DLS would be a first analysis to identify the best sample to study using microscopy.

There are other techniques available to characterize the size distribution of CN. Among them, the field flow fractionation (FFF) could provide a more accurate analysis than DLS or other microscopic techniques. It was not applied to CN isolated from pineapple yet, but recent studies about the characterization of CN extracted from different sources have reported its use [80, 89, 127].

5.2 Chemical Composition

The chemical composition of the cellulose source, together with its own internal structure and extraction method, plays an essential role in the isolation of CN, e.g. the yield, geometrical dimensions and mechanical properties. For example, high extractive and lignin contents reduce the yield of CN extraction, which results in higher costs [90]. This is why pretreatments are usually employed prior to the isolation of CN in order to remove lignin and hemicelluloses.

Table 3 Standard methods used to estimate the chemical composition of pineapple fibres

Compound	ASTM standard method
Cellulose	ASTM D1103-55T[a,b,c]
	Potassium hydroxide method[d,e]
	TAPPI T9M-54[f]
Holocellulose	ASTM D1104-56[a,b,c]
	Acid chlorite method[d,e]
Lignin	ASTM D1106-56[a,b,c]
	TAPPI T13M-54[e,f]
	TAPPI T222 om-6[g]
Moisture content	ASTM D4442-92[a,b]
Ash content	Calcination for 4 h at 800 °C[e,h]

a: Abraham et al. [1]; b: Cherian et al. [33]; c: Moreno et al. [126]; d: Browning [25]; e: Dos Santos et al. [45]; f: Mahardika et al. [118]; g: Fareez et al. [53]; h: Trindade et al. [174]

There are standard methods used in order to quantify the chemical composition of the raw lignocellulosic materials before and after the chemical treatments. These methods are defined by the American Society for Testing and Materials (ASTM) or by the Technical Association of Pulp and Paper Industry (TAPPI), as shown in Table 3. Despite the importance of knowing the composition of the raw material, only 17% of the works on CN extracted from pineapple fibres determined the composition of the raw material using these methods [1, 33, 53, 118, 126].

Around 62% of the works on extraction of CN from pineapple fibres performed chemical and structural characterization of the raw material and/or the intermediates and the produced CN. Fourier transform infrared (FTIR) spectroscopy was used in 83% of the papers to determine the chemical composition of pineapple fibres and CN [1, 9, 13, 15, 42, 45, 53, 60, 63, 118, 126, 144, 160, 161, 168].

FTIR is a spectroscopic technique where the sample is exposed to radiation with wavelengths between 2.5 and 25 μm that correspond to the vibrational portion of the infrared region. If the bonds in the molecule have a dipole moment that changes as a function of time, the molecule will absorb the infrared radiation. Each molecule in the sample absorbs only specific frequencies, which match the energy range encompassing the stretching and bending vibrational frequencies of their bonds. The energy absorbed increases the amplitude of the vibrational motion of the bonds in the molecule. As the same type of bond in two different molecules is in two different environments, each molecule has a specific infrared absorption pattern, which results in its own infrared spectrum [138].

Cellulose, hemicellulose and lignin have characteristic functional groups, as shown in Table 4. It allows the differentiation between cellulose, hemicelluloses and lignin contents before and after chemical treatments, since some absorptions will be more intense than others.

Table 4 Characteristic FTIR absorptions of the main groups in the pineapple fibre

Compound	FTIR absorption (cm^{-1})	Chemical group
Cellulose	3400–3000	Hydrogen-bonded O–H stretching[a,b]
	2920–2800	C–H asymmetrical and symmetrical stretching[a,b,c]
	1640	O–H bending (absorbed water)[a,d]
	1420–1430	C–H in plane deformation[b,c]
	1200	C–OH in plane deformation[b]
	1109	C–OC symmetric stretching or ring stretching[a]
	1165	C–O–C stretching at β-glycosidic linkage[b]
	1059	C–OH stretching[a]
	1022	C–C stretching[e]
	994	C–C, C–OH and C–H ring and side group vibrations[d]
	895	COC, CCO and CCH deformation and stretching[d]
Hemicelluloses	1730–1740	C=O of carboxyl and acetyl groups[e]
Lignin	1200–1300	Aromatic ring vibration[e]
	1830–1730	– O–CH_3, C–O–C and aromatic C=C[e]
	1605	Stretching or aromatic rings in phenol groups[a]
	1506	C=C aromatic symmetric stretching[d]

a: Prado and Spinacé [144]; b: Oh et al. [134]; c: Poletto et al. [142]; d: Fan et al. [51]; e: Abraham et al. [1]

For example, Prado and Spinacé [144] reported that after alkali treatment most of the non-cellulosic compounds in the pineapple fibres were removed, as the peaks at 1735 and 1605 cm^{-1} were absent. Only small changes were verified after the bleaching step, where the absorptions at 1059, 1109 and 1316 cm^{-1} became more intense indicating the higher content of cellulose. Analogously, Dos Santos et al. [45] also reported the removal of hemicelluloses and lignin after the purification step through alkali and bleaching treatments. The content of the cellulose increased from 36.3 to 74.5% for the treated pineapple fibre, and the amounts of hemicellulose and lignin decreased from 22.9 and 27.53% to 20.4 and 8.72%, respectively. The ash content had a small decrease from 2.85 to 2.28% after the chemical treatment. In these treatments, lignin is depolymerized and hemicellulose chains break down, forming sugar and phenolic components as water-soluble materials. Further, in the acid hydrolysis, glycosidic linkages in hemicellulose and ether linkages in lignin are hydrolysed, and cellulose is defibrillated and depolymerized to form CNC [90, 132].

Abraham et al. [1] reported a combination of mechanical and chemical methods in order to purify the pineapple fibres and isolate CNF. After the alkali treatment, they exposed the mercerized fibres to steam explosion to disrupt and defibrillate the pre-treated material, and proceed with a bleaching treatment for complete removal of the remaining cementing materials from the fibres. They also verified by FTIR that the peaks at 1200–1300 cm^{-1} and 1730–1740 cm^{-1} related to lignin and hemicelluloses, respectively, were absent in the treated fibres, and the contents of cellulose, hemicellulose and lignin changed from 75.3, 13.3 and 9.8% to 97.3, 0.2 and 0%, respectively. Thus, defibrillating the fibre through steam explosion before the bleaching treatment seems to increase the efficiency of the pretreatments in order to remove the non-cellulosic materials. However, CNF extracted via pure mechanical methods may still contain small amounts of hemicelluloses and lignin. Although hemicelluloses help in the individualization of the nanofibre increasing the nanofibrillation yield, the produced CNFs are more amorphous and degrade at a lower temperature. The existence of lignin increases the mechanical properties such as tensile index, toughness and elastic modulus of films produced with CNF, as well as their barrier properties [90].

FTIR can also be used to identify surface modification of CN after functionalization processes. For instance, Gao et al. [63] reported the use of FTIR to detect the grafting of a silane coupling agent in the surface of CNC extracted from pineapple fibres. The presence of the peak at 468 cm^{-1} related to the flexural vibrations of C–O–Si showed the successful grafting of the coupling agent on the CNC surface. Another use of FTIR is to find out possible interactions between a matrix and reinforced filler [15]. This kind of interaction can be verified through frequency shifts, changes in band intensity and shape of the FTIR spectra [168].

Regarding the sample preparation to FTIR analysis, 40% of the works use potassium bromide (KBr) plates in order to characterize the raw and treated pineapple fibres as well as the extracted CN [1, 9, 45, 53, 60, 159]. Around 46% of the papers analyse their composition using an attenuated total reflectance (ATR) accessory, which is used especially when nanocomposites are characterized [13, 15, 42, 118, 126, 144, 168]. Only 7% of the works described the use of sodium chlorite (NaCl) plates [159]. Although they are cheaper, NaCl plates have a smaller spectroscopy range from 4000 to 650 cm^{-1}, while KBr plates allow measurements up to 400 cm^{-1} [138]. The remaining 7% of the works did not describe the sample preparation method [63].

Other analytical techniques can also be used in order to characterize the chemical composition of pineapple fibres and the produced CN. Fareez et al. [53] used Raman spectroscopy in order to verify that there was no significant difference in the composition of pineapple samples bleached for 1–4 h. Prado and Spinacé [144] used elemental analysis to confirm the presence of sulphur atoms in the CNC produced by acid hydrolysis with sulphuric acid. They also determined the moisture content and absorption of the isolated CNC in order to verify their possible applications. Solid-state ^{29}Si nuclear magnetic resonance (NMR) was used by Shih et al. [159] to verify the efficiency of an eco-friendly modification of CNF isolated from pineapple fibre. Shih et al. [159] used electron spectroscopy for chemical analysis (ESCA), also known as X-ray photoelectron spectroscopy (XPS), to quantify the chemical composition of the treated CNF surface.

Although not reported yet by any of the works on CN produced from pineapple fibres, another technique that could be used to evaluate the chemical composition in situ of the CN is the energy-dispersive X-ray spectroscopy (EDX or EDS). As mentioned in Sect. 5.1, it is usually coupled with electron microscopes and allows qualitative and quantitative non-destructive analysis of samples with the detection of chemical elements heavier than beryllium.

5.3 Crystalline Structure and Crystallinity Index

As mentioned in Sect. 1, cellulose elementary fibrils are formed by regions where cellulose chains are well organized called crystalline domains and by less organized regions called amorphous domains. Cellulose chains are arranged in the crystalline domains in different cellulose polymorphs (namely I, II, III_I, III_{II}, IV_I and IV_{II}), depending on the source of cellulose [74]. The interconversion between one polymorph and another can be partial, resulting in mixed polymorphic structures depending upon the chemical treatment that the fibre is subjected [110, 183]. The portion of crystalline regions in cellulose is quantified by its crystallinity index (CI), which consists in a relation between a physic-chemical parameter associated with the crystalline domain and the same physic-chemical parameter associated with all cellulose regions, including the amorphous domains.

The type of crystalline structure and CI of cellulose can be determined by several experimental techniques, such as X-ray diffraction (XRD) [41, 67, 78, 129, 155, 178], neutron diffraction [171], solid-state ^{13}C NMR [181], FTIR [29, 130, 131], Raman spectroscopy [137] and differential exploratory calorimetry [119]. Among them, XRD is the most used technique due to its versatility. It allows not only the determination of the type of crystalline structure and CI, but also the relative amount between different phases, the average size of the cellulose crystallites and their preferential orientation.

The XRD technique is based on the coherent scattering of a radiation with wavelength (λ) between 0.5 and 2.5 Å. This coherent scattering occurs when the radiation collides with scattering centres, such as atoms, molecules or chains. If these scattering centres are arranged periodically in space, constructive interference between scattered radiations results in diffraction patterns that are characteristic of the crystalline structure of the material. The relation between the wavelength (λ), the interplanar distance (d_{hkl}) and the diffraction angle (θ) is given by the Bragg's law ($\lambda = 2\, d_{hkl}$ sen θ) [67].

Several methods may be employed to characterize the CI of cellulose using the XRD data. The Segal method [155] was used in 70% of the works on CN isolated from pineapple fibres that have characterized their CI [13, 15, 42, 53, 118, 126, 144]. The deconvolution method was used in 20% of the papers [9, 45], and only 10% of the works used other methods [1].

The Segal method was developed in 1959 to estimate the CI of samples containing cellulose I [155] and was subsequently adapted for samples containing cellulose II. It is based on the relationship between the intensity of the amorphous scattering (I_a), assigned to $2\theta = 18°$ and 16° for cellulose I and II, respectively, and the maximum diffraction intensity (I_t) of the cellulose around $2\theta = 22.7°$ and 21.7°, corresponding to the planes (200) and (020) of the celluloses I and II, respectively [129]. The CI is obtained using the relation CI (%) = [1 − (I_a/I_t)] × 100.

On the other hand, the deconvolution method consists in the separation of the individual crystalline peaks composing the sample experimental diffractogram by means of a curve fitting process [137]. Since each cellulose polymorph has a specific diffraction pattern with peaks at certain 2θ positions, these peaks can be used to compose the experimental diffractogram of the sample. The amorphous fraction is also considered to influence the composition of the experimental diffractogram and forms a broad peak with area A_a. The CI is obtained by the relation between the areas of the crystalline peaks that compose the diffractogram and its total area (A_t), according to the equation CI (%) = [1 − (A_a/A_t)] × 100. Many different functions can be used to model the shape of the diffraction peaks obtained by XRD.

Other methods that may be used to calculate CI include the subtraction of the amorphous fraction. There is no consensus as to which technique and method are most suitable for the characterization of CI of cellulose. It is important to keep in mind that the different techniques are based on different principles and have different sensitivities. Accordingly, the results obtained using different techniques are generally not comparable.

XRD studies of untreated and treated pineapple fibres were done by several authors [1, 13, 33, 45, 144, 159]. The diffraction pattern of untreated and treated pineapple fibres is typical of semi-crystalline materials, having crystalline peaks and a broad hump [45]. Pineapple fibres are described as formed by cellulose I [1, 9, 13, 15, 33, 34, 45, 53, 126, 144, 159], more specifically cellulose Iβ [53], which has characteristic diffraction peaks 2θ at 14.88, 16.68 and 22.7° related to the crystallographic planes with Miller indices ($1\bar{1}0$), (110) and (200) [58]. The fibre treatment with alkali may change the topography of pineapple fibre as well as the crystallographic structure of cellulose [1].

The crystalline conversion of cellulose I to cellulose II occurs with an alkali concentration of up to 32%. The crystalline transformation of cellulose polymorphs in concentrations lower than this is limited by the reduced accessibility to the cellulose molecules [1]. Even though such transformation has been reported for the pineapple fibres due to their high cellulose content by Cherian et al. [33] and Abraham et al. [1], this transformation is not retained up to the final stage of the CN extraction since the cellulose II turns back to cellulose I upon treatment with oxalic acid [1]. The removal of the non-cellulosic compounds by alkali and bleaching treatments increases the CI of pineapple fibres from 23 to 44% [45, 126]. When these treatments are associated with steam explosion, an increase of up to 93% in CI can be observed due to the more efficient removal of non-cellulosic and amorphous compounds. However, it is worth to remember that high alkali concentrations may damage the cell wall, leading to a decrease in the CI [1].

The method employed to isolate CN also changes their CI. For example, CNCs usually exhibit sharp peaks in the XRD pattern, which indicates a higher crystallinity due to the more efficient removal of non-cellulosic polysaccharides and dissolution of amorphous zones. It demonstrates that hydrolysis takes place preferentially in the amorphous region [33]. The CI of CNC extracted by acid hydrolysis with sulphuric acid is reported to be around 73%, independently on the pretreatments employed (e.g. only chemical or combinations of chemical and mechanical treatments) and on the source of pineapple fibre (e.g. plant leaf or fruit crown) [45, 144]. This value is high compared to the CNC obtained from other lignocellulosic wastes such as garlic straw and skin, barley and sugar cane bagasse [54, 93]. Generally, CNC isolated using H_2SO_4 shows lower CI values compared to those made from HCl. In addition, the increasing of hydrolysis time promotes an increase of the CI values due to the removal of amorphous regions [124], until it starts to degrade cellulose in the crystalline regions.

Conversely, spherical cellulose nanoparticles obtained by high-pressure homogenization process presented CI almost 51% lower than the initial pineapple fibre. The authors attributed this result to the cleavage of inter- and intramolecular hydrogen bonds of cellulose, transforming the ordered crystalline region into amorphous region [60]. A similar behaviour is observed in the extraction of CN from other lignocellulosic sources using mechanical methods, since these processes usually do not have a preferential region to attack. This is why CNF generally has lower CI values than CNC [118].

XRD has also been used after the preparation of cellulose nanocomposites in order to confirm that the CNs preserve their crystalline structure [34]. Also, the crystallinity of the nanocomposites filled with different contents of CN can be evaluated [179].

5.4 *Thermal Properties*

The thermogravimetric analysis (TGA) is usually carried out to characterize the thermal properties of CN from different sources. Trache et al. [172] stated that at least 75% of all scientific publications about CN in the world considered applying it in nanocomposites. In this context, knowing some properties such as the thermal stability of CN is crucial. It allows determining the nanocomposite process temperature and the maximum temperature during its application [31, 42, 45, 139].

TGA measures the variation of a sample mass (~10 mg or less) under a controlled temperature programme. It is possible to check the mass loss as a function of time or of temperature [12, 31]. The mass loss measurements can be performed in isothermic or dynamic methods. In the first method, the sample is heated and the temperature is constant for a period. By its turn, in the dynamic methods, or non-isothermal, the temperature is increased during the time [31]. The atmosphere of the test can be nitrogen, helium, argon or even oxygen. Often, the inert atmosphere is chosen, unless the study wants to clarify the thermo-oxidative stability of the sample [12, 31, 32]. Further, not only the TGA curve is useful but also the differential thermogravimetric analysis (DTG). DTG is the mass derivative mass loss curve. TGA and DTG curves allow verifying the onset temperature of decomposition (i.e. the thermal stability), the temperature at the highest decomposition rate and the end temperature of decomposition [31].

In addition, this characterization provides the number of steps during the sample degradation. This occurs because each stage of degradation will lead to the formation of a distinct peak in the DTG curve [31]. In this manner, we can determine the presence of cellulose, hemicellulose and lignin in the lignocellulosic sources [90, 186]. The TGA curve of CN is different from the lignocellulosic source because there is only cellulose. However, it can be distinct from that of pure cellulose. It happens because the TGA of CN may be influenced by other factors, such as their high specific surface area and the extraction conditions [139]. For instance, the nanostructures obtained using the acid hydrolysis with sulphuric acid will show lower thermal stability than pure cellulose [139, 151]. The reason is that this method inserts sulphate groups on the cellulose chains, which catalyse their thermal degradation [27, 40, 45, 90, 139, 151].

Many papers have shown the characterization of the thermal stability of CN isolated from pineapple fibres [1, 15, 27, 36, 40, 42, 45, 60, 90, 118, 126, 144, 160, 177]. The thermogravimetric analysis is performed under nitrogen atmosphere, usually with a heating rate of 10 °C/min, but there are reports using 5 °C/min [1] and 20 °C/min [118, 126]. CN isolated by chemical methods showed lower thermal stability compared to the microscopic fibre. For instance, Dos Santos et al. [45] and Prado and Spinacé [144] obtained CNC from pineapple fibres using acid hydrolysis and verified a reduction in the thermal stability of CNC due to the presence of sulphate groups on the CNC chains. Likewise, Shih et al. [160] used TEMPO-mediated oxidation to extract CNF from pineapple fibres, and they also verified a reduction in the thermal stability. Meanwhile, CN extracted by mechanical methods showed similar [60] or higher [1] onset temperature of degradation than the raw material.

Finally, TGA and DTG data can be used to obtain thermodynamic parameters, such as the activation energy of each degradation process [12, 72, 135, 187]. However, to the best of our knowledge, it has not been applied to CN extracted from pineapple fibres yet.

5.5 *Other Properties*

Other properties of CN extracted from pineapple fibres have been characterized, such as the zeta potential and the contact angle. The zeta potential measurement allows determining the stability of the material in a liquid, such as water [71, 101, 176]. When CNs are in contact with a liquid, they will be surrounded by ions with opposite charge, which will create an electrical potential at the interface. The ionizable cellulose groups and the liquid phase ions form a double layer. If an electric field is applied to the system, the nanostructures and the liquid phase ions will tend to move in opposite directions. The so-called zeta potential, or electrokinetic potential, is the change in the electric potential across this double layer [71, 176]. This phenomenon is interesting because the zeta potential is related to the physical stability of the nanostructure. The nanoparticle is stable in this liquid when the zeta potential is higher than 30 mV in absolute value [57].

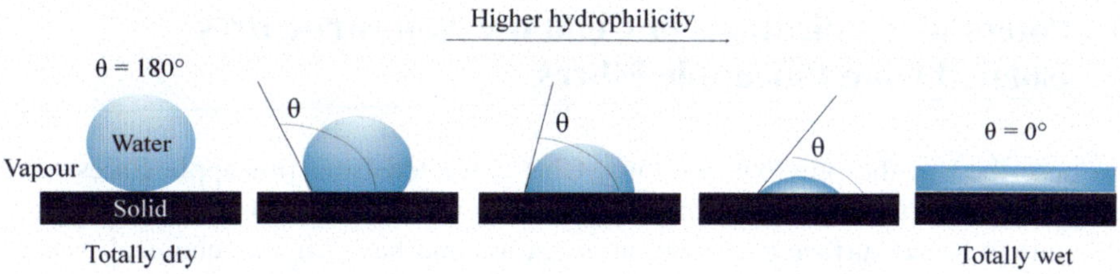

Fig. 20 Scheme of contact angle systems

The literature shows five reports about the zeta potential of CN extracted from pineapple fibres [9, 14, 15, 27, 177]. Their results reveal that CNs obtained using chemical routes produce more stable nanomaterials than those obtained by mechanical methods. However, it is important to remember that the pH, the temperature as well as the nanostructure concentration in the liquid phase may alter the electrostatic interactions in the system and consequently the zeta potential [71]. In the same manner, the groups inserted in the cellulose chains after chemical functionalization may promote an increase of the CN surface charge density and improve the absolute value of their zeta potential [71, 176].

On the other hand, the contact angle measurement indicates the hydrophilicity of the material [71, 106]. It consists in measuring the angle θ of a water droplet and a solid surface (Fig. 20). This angle θ is created by the equilibrium among the interfacial surface tensions among the liquid, the solid and the vapour, which is in contact with them [20, 71, 106].

Briefly, if the interfacial tension solid–vapour is equal or lower than the sum of the tensions solid–water and water–vapour, there is the wetting of the surface. It means that the functional groups of the solid have a strong interaction with the water molecules, leading to the droplet spreading. In other words, as shown in Fig. 20, smaller contact angles suggest higher hydrophilicity [20, 71, 106].

Gao et al. [63] and Shih et al. [158–161] studied the contact angle between a water droplet and the surface of the CN extracted from pineapple fibres. They verified that CN was hydrophilic and that this property could be changed by chemical modification of the cellulose chains.

Other characterization techniques besides those showed in this chapter are less often found in the literature. However, they have been used to study the CN isolated from pineapple. These techniques include birefringence [42, 45], inverse gas chromatography [42], nuclear magnetic resonance [159], rheological analysis [61] as well as determination of specific surface area [60].

6 Potential Applications of Cellulose Nanostructures Isolated from Pineapple Fibres

As described in the previous sections, CNs extracted from pineapple fibres have excellent properties to be used in many diverse applications. They are biodegradable, allow broad surface chemical modification and have high mechanical properties, low cost and low weight [1, 102, 111]. Currently, all studies available in the literature reported the use of CN isolated from pineapple fibres as reinforcement in nanocomposites. This occurs mainly because CNs from pineapple fibres show higher crystallinity and higher thermal stability compared to the CN obtained from other lignocellulosic wastes such as sisal, coir and banana rachis [42]. The large surface area of the CN allows that a small amount of reinforcement (usually from 0.5 to 10 wt%) promotes a significant improvement of the mechanical properties, what is an advantage compared to micro-sized reinforcements. However, this improvement only occurs when the interaction between matrix and reinforcement is favourable [33, 160].

All the papers reported in the literature on CN extracted from pineapple fibres describe the use of CN as reinforcement of polymer matrices to produce nanocomposites for different purposes. Shih et al. [160] reported the use of CNF from pineapple leaves to produce highly transparent and impact-resistant nanocomposites using poly(methyl methacrylate) as matrix. Zhou et al. [192] reported the production of lightweight biobased polyurethane (PU) nanocomposite foams reinforced with CNF from pineapple fibres. Biodegradable nanocomposites reinforced with CNF isolated from pineapple fibres have been reported by Wahyuningsih et al. [179], Shih et al. [159] and Amalia et al. [4] using polyvinyl alcohol (PVA), poly (lactic acid) (PLA) and chitosan as matrices, respectively. In addition, Balakrishnan et al. [15] reported UV-resistant transparent bio-nanocomposite films using starch as matrix. The use of biodegradable matrices such as PVA, PLA, chitosan and starch allows the production of bio-nanocomposites with several environmental benefits that can be used as packaging, where CNC can also be used to increase the barrier properties [133].

Most of the studies in the literature explore the use of CNF instead of CNC extracted from pineapple fibres as reinforcement in polymeric nanocomposites. In order to understand why, Balakrishnan et al. [14] compared the reinforcement effect of CNF and CNC isolated from pineapple fibres on thermoplastic starch. They observed that although both CNs improved the properties of the matrix, the use of CNF promotes better mechanical properties due to their higher degree of entanglement. It allows a better stress transfer to the matrix, thus resulting in a better reinforcing ability.

The nanocomposites produced with CNF isolated from pineapple fibres have potential to be used in several applications. High-volume products include automotive parts, packaging, absorbents, adsorbents and textiles. Low-volume applications include aerogels, hydrogels, cosmetics, paints, thermoset adhesives, air and water filtration, environmental remediation membranes, additive manufacturing, photocatalysts, electronics and biomedical applications [53, 133, 158].

Indeed, 23% of the studies currently available at the literature focus on the use of pineapple CNF in biomedical applications. It can be attributed to the biocompatibility, hemocompatibility and absence or low cytotoxicity of CN [111], associated with relatively low rigidity of CNF compared to CNC [28]. These features enable the use of CNF isolated from pineapple fibres in many biomedical applications such as drug excipient and delivery media, surgical wounds, scaffold for tissue and organ engineering, repair of articular cartilage and others [9, 33, 34, 53].

For instance, Cherian et al. [34] reported the use of PU nanocomposites reinforced with CNF isolated from pineapple fibres in order to produce heart valve and vascular grafts. The addition of only 5 wt% of CNF into PU matrix promoted an increase of about 300 and 2600% in the strength and the stiffness values, respectively.

The produced pineapple CNF showed unique interconnected web-like structure, which allowed the production of nanocomposites with good biological durability, fatigue resistance and hemocompatibility. The developed material has also potential to be used for the manufacturing of other products, such as non-latex condoms, surgical gloves, medical bags, organ retrieval bags and medical disposables [34].

Another example of promising application was reported by Costa et al. [37]. They produced a bio-nanocomposite constituted of PVA reinforced with pineapple CNF and containing extract of *Stryphnodendron adstringens* bark, which has antimicrobial properties and can be used in human medicine for many purposes [26]. The produced bio-nanocomposite was designed for medical implants, and it showed homogeneous distribution of pores with prospective natural antimicrobial properties.

In summary, CN extracted from pineapple fibres has potential to be used in numerous high value-added applications. This is extremely advantageous to add value to all the pineapple production chain. Moreover, the obtained CNs have interesting properties to the used in applications with many environmental and human health benefits, such as the biodegradable packaging and the biomedical applications, respectively.

7 Future Perspectives

The interest in the extraction of CN from pineapple fibres has been experiencing a significant leap in the last few years. However, despite the numerous advantages of using pineapple wastes as source to produce CN, parts of pineapple plants and fruits are still discarded owing to the unawareness of their potential economical uses [144].

In order to make pineapple fibre wastes a large-scale source to obtain CN, an economic as well as energy-efficient production is required [17]. For this purpose, fibre extraction process of pineapple leaves could be improved and automatized in order to increase the production. In addition, this automatized process should be able to work even with small leaves such as those from the pineapple crown. Another challenge lies in the scalability of the CN production, since most of the investigations are still on a laboratory scale [158].

Additionally, the main methods used in order to isolate CN from pineapple fibres involve numerous chemicals in the pretreatments and in the nanocellulose extraction, such as in the acid hydrolysis. It implies the generation of large amounts of chemical wastes that should be treated before disposal, increasing the environmental impact and the overall cost of the process. The preparation of CN from untreated fibres is an environmental-friendly tendency that reduces the production costs by decreasing the chemical and energy consumption.

Moreover, only few mechanical methods have been used in the extraction of CN from pineapple fibres, such as steam explosion [1, 33], homogenization [60, 61, 118] and ultrasonication [118], and they are usually associated with chemical treatments. Many mechanical processes that could be more sustainable and cost-effective have not yet been used for the isolation of CN from pineapple fibres, such as ball milling, extrusion, cryocrushing and others.

While CNC and CNF are the main types of CN produced from pineapple fibres, other types of CN can also be obtained such as hairy nanocrystalline cellulose and amorphous cellulose. The preparation of these CNs with different functionalities can open up new applications for these nanostructures.

An important aspect that could be more exploited is the surface modification of the prepared CN. As shown in Fig. 21, there are three main approaches for modification of the chemical surface of CN:

(1) The substitution of hydroxyl groups by small molecules, such as silanes, or by TEMPO oxidation;
(2) Polymer grafting onto the surface of CN using coupling agents as poly(ethylene glycol) (PEG), poly(ethylene oxide) (PEO) and polycaprolactone (PCL);
(3) Polymer grafting from a radical polymerization, either by ring-opening polymerization (ROP), atom transfer radical polymerization (ATRP) and single-electron transfer living radical polymerization (SET-LP) [46].

Only few studies in the literature reported the functionalization of the CN obtained from pineapple fibres, with the use of silanes [63], TEMPO oxidation [159, 160], carboxylated reactants [9], sol-gel modification and suspension polymerization on

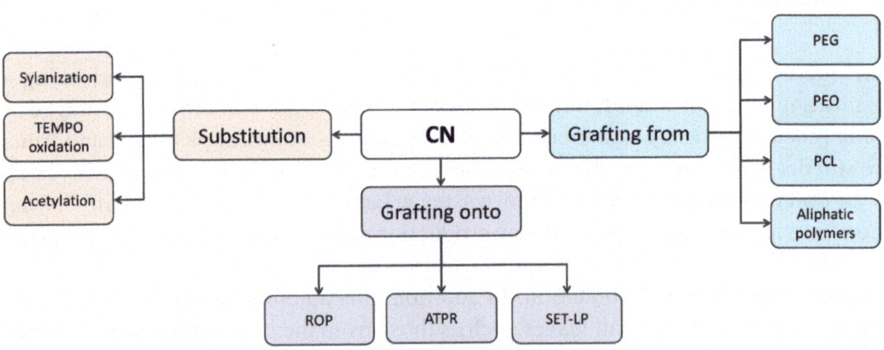

Fig. 21 Main approaches for surface covalent chemical modification of CN, where ROP is the ring-opening polymerization, ATRP is the atom transfer radical polymerization and SET-LP is the single-electron transfer living radical polymerization [112]

the surface of CN [158–161]. As can be noted in Fig. 21, there are numerous possible routes of cellulose functionalization that still can be exploited. The introduction of functional groups in the surface of CN can improve the properties of the nanoparticles as well as increase their compatibility with different matrices such as polyolefins, thus allowing their use in new applications.

The use of CN in nanocomposites with thermoplastic matrices remains a major challenge. As most of the thermoplastic matrices are hydrophobic, the interaction with the hydrophilic CN is not favourable, which impact the dispersion and the stress transfer of the reinforcement. In this case, surface treatments and coupling agents may be used to improve the adhesion between CN and the thermoplastic matrix, thus leading to the production of nanocomposites with better mechanical properties [144].

As discussed in Sect. 6, most of the studies currently available in the literature explore the use of CNF instead of CNC extracted from pineapple fibres as reinforcement in polymeric nanocomposites due to their better reinforcing ability. The properties of CNC produced from pineapple fibres, however, can enable their use in other applications that have still not been exploited. For example, due to its high hydrophilicity, potential applications of CNC rely on liquid media, such as the stabilization of water–oil emulsions [143] and the increase of the strength of cement [154]. The unique morphology, rigidity and chiral ordering of CNC lead to optical effects in aqueous media that can be used in the production of sensors and optical devices [69]. The alignment and orientation of CNC allow their use as template for inorganic and organic nanoparticles [147]. The giant permanent electric dipole of CNC allows their use in the production of piezoelectric thin films [38]. Finally, CNC can also be used in pharmaceutical and biomedical applications, such as drug excipient, drug delivery media and biosensors [28, 48].

Regarding the biomedical applications of CN isolated from pineapple fibres, although studies conducted so far reported the absence or low cytotoxicity of CN in general, the mechanisms of aggregation of these nanoparticles in the body are still unknown, as well as the long-term in vivo effects. Besides, the eco-toxicity associated with the incorporation of CN in other materials also needs deep investigations [111]. The unclear toxicology of CN may become the greatest obstacle for their application and marketability in different applications [158].

8 Conclusions

The residues of crown, stem and leaf of the pineapple plant are abundant and inexpensive and have high content of cellulose (74–83 wt%). Consequently, pineapple fibres obtained from these agro-industrial wastes are environmentally and economically viable sources for extraction of different cellulose nanostructures, including nanocrystals, nanofibres, amorphous, hairy and nanoyarns.

Usually, pretreatment steps are required prior to the isolation of CN, such as pulping process, alkaline, bleaching, alkaline-acid-alkaline and ionic liquid. However, the alkaline and bleaching are still the most used pretreatments for extracting hemicelluloses and lignin from pineapple fibres.

CN can be isolated using chemical (acid or basic hydrolysis, oxidation and ionic liquid), enzymatic, mechanical (steam explosion, high-pressure homogenization, high shear homogenization, high-intensity ultrasound, electrospinning and others) and electrical (electrospinning) methods. Among these, steam explosion and acid hydrolysis are the most used techniques for the extraction of CN from pineapple fibres. Recently, few groups published several times about CN extraction using steam explosion methodology. However, hydrolysis was the most common approach used by distinct research groups worldwide. Probably, the use of environmentally friendly mechanical methods should rise in future.

The morphological, chemical, structural and thermal characterization of CN is important to suggest potential applications, and the most reported techniques for CN produced from pineapple fibres were TEM, FTIR, XRD and TGA, respectively.

The main application of CN extracted from pineapple fibres was as reinforcement in polymer nanocomposites for diverse purposes, where the production of biomedical devices and biodegradable bio-nanocomposites for packaging stood out. Potential applications include sensors, optical devices, electronic and liquid media applications, such as the stabilization of water–oil emulsions.

References

1. Abraham E, Deepa B, Pothan LA et al (2011) Extraction of nanocellulose fibrils from lignocellulosic fibres: a novel approach. Carbohydr Polym 86:1468–1475. https://doi.org/10.1016/j.carbpol.2011.06.034
2. Alemdar A, Sain M (2008) Isolation and characterization of nanofibers from agricultural residues—wheat straw and soy hulls. Bioresour Technol 99:1664–1671. https://doi.org/10.1016/j.biortech.2007.04.029
3. Alvira P, Tomás-Pejó E, Ballesteros M et al (2010) Pretreatment technologies for an efficient bioethanol production process based on enzymatic hydrolysis: a review. Bioresour Technol 101:4851–4861. https://doi.org/10.1016/j.biortech.2009.11.093
4. Amalia B, Imawan C, Listyarini A (2018) Effect of nanofibril cellulose isolated from pineapple leaf on the mechanical properties of chitosan film. AIP Conf Proc 2023:020034/1–020034/6. https://doi.org/10.1063/1.5064031
5. Araki J, Kuga S (2001) Effect of trace electrolyte on liquid crystal type of cellulose microcrystals. Langmuir 17:4493–4496. https://doi.org/10.1021/la0102455
6. Arib RMN, Sapuan SM, Ahmad MMHM et al (2006) Mechanical properties of pineapple leaf fibre reinforced polypropylene composites. Mater Des 27:391–396. https://doi.org/10.1016/j.matdes.2004.11.009
7. Arrieta MP, Fortunati E, Burgos N et al (2016) Nanocellulose-based polymeric blends for food packaging applications. In: Puglia D, Fortunati E, Kenny JM (eds) Multifunctional polymeric nanocomposites based on cellulosic reinforcements. William Andrew Applied Science Publishers, p 394. https://doi.org/10.1016/b978-0-323-44248-0.00007-9
8. Asim M, Abdan K, Jawaid M et al (2015) A review on pineapple leaves fibre and its composites. Int J Polym Sci 2015:1–16. https://doi.org/10.1155/2015/950567

9. Asri LATW, Rahmatika A, Fahreza MZ et al (2018) Preparation and release behavior of carboxylated cellulose nanocrystals-alginate nanocomposite loaded with rutin. Mater Res Express 5:095303.1–095303.10. https://doi.org/10.1088/2053-1591/aad5d3
10. Atalla RH, Hackney JM, Uhlin I et al (1993) Hemicelluloses as structure regulators in the aggregation of native cellulose. Int J Biol Macromol 15:109–112. https://doi.org/10.1016/0141-8130(93)90007-9
11. Azizi Samir MA, Alloin F, Dufresne A (2005) Review of recent research into cellulosic whiskers, their properties and their application in nanocomposite field. Biomacromolecules 6:612–626. https://doi.org/10.1021/bm0493685
12. Bach Q-V, Chen W-H (2017) Pyrolysis characteristics and kinetics of microalgae via thermogravimetric analysis (TGA): a state-of-the-art review. Bioresour Technol 246:88–100. https://doi.org/10.1016/j.biortech.2017.06.087
13. Balakrishnan P, Sreekala MS, Kunaver M et al (2017) Morphology, transport characteristics and viscoelastic polymer chain confinement in nanocomposites based on thermoplastic potato starch and cellulose nanofibers from pineapple leaf. Carbohyd Polym 169:176–188. https://doi.org/10.1016/j.carbpol.2017.04.017
14. Balakrishnan P, Gopi S, Geethamma VG et al (2018a) Cellulose nanofiber vs nanocrystals from pineapple leaf fiber: a comparative study on reinforcing efficiency on starch nanocomposites. Macromol Symp 380:1800102–1800109. https://doi.org/10.1002/masy.201800102
15. Balakrishnan P, Gopi S, Sreekala MS et al (2018b) UV resistant transparent bionanocomposite films based on potato starch/cellulose for sustainable packaging. Starch 70:1700139.1–1700139.13. https://doi.org/10.1002/star.201700139
16. Beck-Candanedo S, Roman M, Gray DG (2005) Effect of reaction conditions on the properties and behavior of wood cellulose nanocrystal suspensions. Biomacromolecules 6:1048–1054. https://doi.org/10.1021/bm049300p
17. Berglund L, Noël M, Aitomäki Y et al (2016) Production potential of cellulose nanofibers from industrial residues: efficiency and nanofiber characteristics. Ind Crops Prod 92:84–92. https://doi.org/10.1016/j.indcrop.2016.08.003
18. Bhaumik P, Dhepe PL (2015) Conversion of biomass into sugars. In: Murzin D, Simakova O (eds) Biomass sugars for non-fuel applications. Royal Society of Chemistry, p 265. https://doi.org/10.1039/9781782622079-00001
19. Bondeson D, Mathew A, Oksman K (2006) Optimization of the isolation of nanocrystals from microcrystalline cellulose by acid hydrolysis. Cellulose 13:171–180. https://doi.org/10.1007/s10570-006-9061-4
20. Bonn D, Eggers J, Indekeu J et al (2009) Wetting and spreading. Rev Mod Phys 81:739–805. https://doi.org/10.1103/RevModPhys.81.739
21. Bootz A, Vogel V, Schubert D et al (2004) Comparison of scanning electron microscopy, dynamic light scattering and analytical ultracentrifugation for the sizing of poly(butyl cyanoacrylate) nanoparticles. Eur J Pharm Biopharm 57:369–375. https://doi.org/10.1016/S0939-6411(03)00193-0
22. Börjesson M, Westman G (2015) Crystalline nanocellulose—preparation, modification, and properties. In: Poletto M, Junior HLO (eds) Cellulose—fundamental aspects and current trends. IntechOpen, p 282. https://doi.org/10.5772/61899
23. Brandt A, Ray MJ, To TQ et al (2011) Ionic liquid pretreatment of lignocellulosic biomass with ionic liquid-water mixtures. Green Chem 13:2489–2499. https://doi.org/10.1039/c1gc15374a
24. Brar SK, Verma M (2011) Measurement of nanoparticle by light-scattering techniques. TrAC-Trend Anal Chem 30:4–17. https://doi.org/10.1016/j.trac.2010.08.008
25. Browning BL (1967) Methods of wood chemistry, 1st edn. Interscience Publishers, New York
26. Calixto JB (2000) Efficacy, safety, quality control, marketing and regulatory guidelines for herbal medicines (phytoterapeutic agents). Braz J Med Biol Res 33:79–89. https://doi.org/10.1590/S0100-879X2000000200004
27. Camacho M, Ureña YRC, Lopretti M et al (2017) Synthesis and characterization of nanocrystalline cellulose derived from pineapple peel residues. J Renew Mater 5:271–729. https://doi.org/10.7569/JRM.2017.634117

28. Cao Y (2018) Applications of cellulose nanomaterials in pharmaceutical science and pharmacology. Express Polym Lett 12:768–780. https://doi.org/10.3144/expresspolymlett.2018.66
29. Carrillo F, Colom X, Sunol JJ et al (2004) Structural FTIR analysis and thermal characterisation of lyocell and viscose-type fibres. Eur Polym J 40:2229–2234. https://doi.org/10.1016/j.eurpolymj.2004.05.003
30. Chakraborty A, Sain M, Kortschot M (2005) Cellulose microfibrils: a novel method of preparation using high shear refining and cryocrushing. Holzforschung 59:102–107. https://doi.org/10.1515/HF.2005.016
31. Chattopadhyay DK, Webster DC (2009) Thermal stability and flame retardancy of polyurethanes. Prog Polym Sci 34:1068–1133. https://doi.org/10.1016/j.progpolymsci.2009.06.002
32. Chen W-H, Peng J, Bi XT (2015) A state-of-the-art review of biomass torrefaction, densification and applications. Renew Sustain Energy Rev 44:847–866. https://doi.org/10.1016/j.rser.2014.12.039
33. Cherian BM, Leão AL, Souza SF et al (2010) Isolation of nanocellulose from pineapple leaf fibres by steam explosion. Carbohydr Polym 81:720–725. https://doi.org/10.1016/j.carbpol.2010.03.046
34. Cherian BM, Leão AL, Souza SF et al (2011) Cellulose nanocomposites with nanofibres isolated from pineapple leaf fibers for medical applications. Carbohydr Polym 86:1790–1798. https://doi.org/10.1016/j.carbpol.2011.07.009
35. Cocinero EJ, Gamblin DP, Davis BG et al (2009) The building blocks of cellulose: the intrinsic conformational structures of cellobiose, its epimer, lactose, and their singly hydrated complexes. J Am Chem Soc 131:11117–11123. https://doi.org/10.1021/ja903322w
36. Corrales-Ureña YR, Villalobos-Bermudez C, Pereira R et al (2018) Biogenic silica-based microparticles obtained as a sub-product of the nanocellulose extraction process from pineapple peels. Sci Rep-UK 8:10417–10426. https://doi.org/10.1038/s41598-018-28444-4
37. Costa LMM, Olyveira GM, Cherian BM et al (2013) Bionanocomposites from electrospun PVA/pineapple nanofibers/*Stryphnodendron adstringens* bark extract for medical applications. Ind Crop Prod 41:198–202. https://doi.org/10.1016/j.indcrop.2012.04.025
38. Csoka L, Hoeger IC, Rojas OJ et al (2012) Piezoelectric effect of cellulose nanocrystals thin films. ACS Macro Lett 1:867–870. https://doi.org/10.1021/mz300234a
39. Dai H, Huang Y, Huang H (2018a) Enhanced performances of polyvinyl alcohol films by introducing tannic acid and pineapple peel-derived cellulose nanocrystals. Cellulose 25:4623–4637. https://doi.org/10.1007/s10570-018-1873-5
40. Dai H, Ou S, Huang Y et al (2018b) Utilization of pineapple peel for production of nanocellulose and film application. Cellulose 25:1743–1756. https://doi.org/10.1007/s10570-018-1671-0
41. De Figueiredo LP, Ferreira FF (2014) The Rietveld method as a tool to quantify the amorphous amount of microcrystalline cellulose. J Pharm Sci 103:1394–1399. https://doi.org/10.1002/jps.23909
42. Deepa B, Abrahan E, Cordeiro N et al (2015) Utilization of various lignocellulosic biomass for the production of nanocellulose: a comparative study. Cellulose 22:1075–1090. https://doi.org/10.1007/s10570-015-0554-x
43. De Jonge N, Ross FM (2011) Electron microscopy of specimens in liquid. Nat Nanotechnol 6:695–704. https://doi.org/10.1038/nnano.2011.161
44. Dong XM, Kimura T, Revol JF et al (1996) Effects of ionic strength on the isotropic–chiral nematic phase transition of suspensions of cellulose crystallites. Langmuir 12:2076–2082. https://doi.org/10.1021/la950133b
45. Dos Santos RM, Neto WP, Silverio HA et al (2013) Cellulose nanocrystals from pineapple leaf, a new approach for the reuse of this agro-waste. Ind Crop Prod 50:707–714. https://doi.org/10.1016/j.indcrop.2013.08.049
46. Dufresne A (2013) Nanocellulose: a new ageless bionanomaterial. Mater Today 16:220–227. https://doi.org/10.1016/j.mattod.2013.06.004

47. Dufresne A (2017) Cellulose nanomaterial reinforced polymer nanocomposites. Curr Opin Colloid Interface Sci 29:1–8. https://doi.org/10.1016/j.cocis.2017.01.004
48. Edwards JV, Prevost NT, French AD et al (2015) Kinetic and structural analysis of fluorescent peptides on cotton cellulose nanocrystals as elastase sensors. Carbohydr Polym 116:278–285. https://doi.org/10.1016/j.carbpol.2014.04.067
49. Eichhorn SJ, Baillie CA, Zafeiropoulos N et al (2001) Review—current international research into cellulosic fibers and composites. J Mater Sci 36:19–26. https://doi.org/10.1023/A:1017512029696
50. Elazzouzi-Hafraoui S, Nishiyama Y, Putaux J-L et al (2008) The shape and size distribution of crystalline nanoparticles prepared by acid hydrolysis of native cellulose. Biomacromolecules 9:57–65. https://doi.org/10.1021/bm700769p
51. Fan M, Dai D, Huang B (2012) Fourier transform infrared spectroscopy for natural fibres. In: Salih SM (ed) Fourier transform-materials analysis. InTech. https://doi.org/10.5772/3582
52. FAO—Food and Agriculture Organization of the United Nations (2019) Crops. http://www.fao.org/faostat/en/#data/QC. Accessed 15 Jan 2019
53. Fareez IM, Ibrahim NA, Yaacob WM et al (2018) Characteristics of cellulose extracted from Josapine pineapple leaf fibre after alkali treatment followed by extensive bleaching. Cellulose 25:4407–4421. https://doi.org/10.1007/s10570-018-1878-0
54. Ferreira FV, Mariano M, Rabelo SC, Gouveia RF, Lona LMF (2018) Isolation and surface modification of cellulose nanocrystals from sugarcane bagasse waste: from a micro- to a nano-scale view. Appl Surf Sci 436:1113–1122. https://doi.org/10.1016/j.apsusc.2017.12.137
55. Ferrer A, Filpponen I, Rodríguez A et al (2012) Valorization of residual empty palm fruit bunch fibers (EPFBF) by microfluidization: production of nanofibrillated cellulose and EPFBF nanopaper. Bioresour Technol 125:249–255. https://doi.org/10.1016/j.biortech.2012.08.108
56. Foster EJ, Moon RJ, Agarwal UP et al (2018) Current characterization methods for cellulose nanomaterials. Chem Soc Rev 47:2511–3006. https://doi.org/10.1039/c6cs00895j
57. Freitas C, Müller RH (1998) Effect of light and temperature on zeta potential and physical stability in solid lipid nanoparticle (SLN™) dispersions. Int J Pharm 168:221–229. https://doi.org/10.1016/S0378-5173(98)00092-1
58. French AD (2014) Idealized powder diffraction patterns for cellulose polymorphs. Cellulose 21:885–896. https://doi.org/10.1007/s10570-013-0030-4
59. Frey MW (2008) Electrospinning cellulose and cellulose derivatives. Polym Rev 48:378–391. https://doi.org/10.1080/15583720802022281
60. Fu T-K, Li J-H, Wei X-Y et al (2016a) Preparation and characterization of pineapple leaf nanocellulose by high pressure homogenization. In: Proceedings of the 2nd annual international conference on advanced material engineering, Wuhan, China, 2016. https://doi.org/10.2991/ame-16.2016.119
61. Fu T-K, Wang Y-H, Li J-H et al (2016b) Effect of high pressure homogenization (HPH) on the rheological properties of pineapple leaf cellulose/[BMIM]Cl solution. DEStech Trans Eng Technol Res. https://doi.org/10.12783/dtetr/mdm2016/4922
62. Fukuzumi H, Saito T, Iwata T et al (2009) Transparent and high gas barrier films of cellulose nanofibers prepared by TEMPO-mediated oxidation. Biomacromolecules 10:162–165. https://doi.org/10.1021/bm801065u
63. Gao TM, Huang MF, Li PW et al (2012) Preparation and characterization nano-cellulose and its surface modification by silane coupling agent. Appl Mech Mater 217–219:260–263. https://doi.org/10.4028/www.scientific.net/AMM.217-219.260
64. García A, Gandini A, Labidi J et al (2016) Industrial and crop wastes: a new source for nanocellulose biorefinery. Ind Crop Prod 93:26–38. https://doi.org/10.1016/j.indcrop.2016.06.004
65. García R, Pérez R (2002) Dynamic atomic force microscopy methods. Surf Sci Rep 47:197–301. https://doi.org/10.1016/S0167-5729(02)00077-8
66. Gardner DJ, Oporto GS, Mills R et al (2008) Adhesion and surface issues in cellulose and nanocellulose. J Adhes Sci Technol 22:545–567. https://doi.org/10.1163/156856108x295509

67. Garvey CJ, Parker IH, Simon GP (2005) On the interpretation of X-ray diffraction powder patterns in terms of the nanostructure of cellulose I fibres. Macromol Chem Phys 206:1568–1575. https://doi.org/10.1002/macp.200500008
68. Gassan J, Bledzki AK (1999) Alkali treatment of jute fibers: relationship between structure and mechanical properties. J Appl Polym Sci 71:623–629. https://doi.org/10.1002/(SICI)1097-4628(19990124)71:4%3c623:AID-APP14%3e3.0.CO;2-K
69. George J, Sabapathi SN (2015) Cellulose nanocrystals: synthesis, functional properties, and applications. Nanotechnol Sci Appl 8:45–54. https://doi.org/10.2147/NSA.S64386
70. Giessibl FJ (2003) Advances in atomic force microscopy. Rev Mod Phys 75:949–983. https://doi.org/10.1103/RevModPhys.75.949
71. Goddard JM, Hotchkiss JH (2007) Polymer surface modification for the attachment of bioactive compounds. Prog Polym Sci 32:698–725. https://doi.org/10.1016/j.progpolymsci.2007.04.002
72. Grønli MG, Várhegyi G, Di Blasi C (2002) Thermogravimetric analysis and devolatilization kinetics of wood. Ind Eng Chem Res 41:4201–4208. https://doi.org/10.1021/ie0201157
73. Habibi Y (2014) Key advances in the chemical modification of nanocelluloses. Chem Soc Rev 43:1519–1542. https://doi.org/10.1039/c3cs60204d
74. Habibi Y, Lucia LA, Rojas OJ (2010) Cellulose nanocrystals: chemistry, self-assembly, and applications. Chem Rev 110:3479–3500. https://doi.org/10.1021/cr900339w
75. Haigler CH, Betancur L, Stiff MR et al (2012) Cotton fiber: a powerful single-cell model for cell wall and cellulose research. Front Plant Sci 3:104. https://doi.org/10.3389/fpls.2012.00104
76. Hamad WY (1997) Some microrheological aspects of wood-pulp fibres subjected to fatigue loading. Cellulose 4:51–56. https://doi.org/10.1023/A:1018463117874
77. Hariwongsanupab N, Thanawan S, Amornsakchai T et al (2017) Improving the mechanical properties of short pineapple leaf fiber reinforced natural rubber by blending with acrylonitrile butadiene rubber. Polym Test 57:94–100. https://doi.org/10.1016/j.polymertesting.2016.11.019
78. Hermans PH, Weidinger A (1948) Quantitative X-ray investigations on the crystallinity of cellulose fibers. A background analysis. J Appl Phys 19:491–506. https://doi.org/10.1063/1.1698162
79. Hoque MB, Hossain MS, Nahid AM et al (2018) Fabrication and characterization of pineapple fiber-reinforced polypropylene based composites. Nano Hybrids Compos 21:31–42. https://doi.org/10.4028/www.scientific.net/NHC.21.31
80. Hu Y, Abidi N (2016) Distinct chiral nematic self-assembling behavior caused by different size-unified cellulose nanocrystals via a multistage separation. Langmuir 32:9863–9872. https://doi.org/10.1021/acs.langmuir.6b02861
81. Hubbe MA, Rojas OJ, Lucia LA et al (2008) Cellulosic nanocomposites: a review. Bioresources 3:929–980
82. Ioelovich M (2013) Nanoparticles of amorphous cellulose and their properties. Am J Nano Res Appl 1:41–45. https://doi.org/10.11648/j.nano.20130101.18
83. Ioelovich M (2014) Peculiarities of cellulose nanoparticles. TAPPI J 13:45–51
84. Ioelovich M (2015) Recent findings and the energetic potential of plant biomass as a renewable source of biofuels—a review. Bioresources 10:1879–1914
85. Ioelovich M, Leykin A, Figovsky O (2010) Study of cellulose paracrystallinity. Bioresources 5:1393–1407
86. Jaafar J, Siregar JP, Piah MBM et al (2018) Influence of selected treatment on tensile properties of short pineapple leaf fiber reinforced tapioca resin biopolymer composites. J Polym Environ 26:4271–4281. https://doi.org/10.1007/s10924-018-1296-2
87. Jahanbaani AR, Behzad T, Borhani S et al (2016) Electrospinning of cellulose nanofibers mat for laminated epoxy composite production. Fibers Polym 17:1438–1448. https://doi.org/10.1007/s12221-016-6424-9
88. Jacque N, Vanderghem C, Danthine S et al (2012) Influence of steam explosion on physicochemical properties and hydrolysis rate of pure cellulose fibers. Bioresour Technol 121:221–227. https://doi.org/10.1016/j.biortech.2012.06.073

89. Jakubek ZJ, Chen M, Couillard M et al (2018) Characterization challenges for a cellulose nanocrystal reference material: dispersion and particle size distributions. J Nanopart Res 20(98):1–16. https://doi.org/10.1007/s11051-018-4194-6
90. Jonoobi M, Oladi R, Davoudpour Y et al (2015) Different preparation methods and properties of nanostructured cellulose from various natural resources and residues: a review. Cellulose 22:935–969. https://doi.org/10.1007/s10570-015-0551-0
91. Kalia S, Boufi S, Celli A et al (2014) Nanofibrillated cellulose: surface modification and potential applications. Colloid Polym Sci 292:5–31. https://doi.org/10.1007/s00396-013-3112-9
92. Kalia S, Dufresne A, Cherian BM et al (2011) Cellulose-based bio- and nanocomposites: a review. Int J Polym Sci 35 pp. Article ID 837875. https://doi.org/10.1155/2011/837875
93. Kallel F, Bettaieb F, Khiari R et al (2016) Isolation and structural characterization of cellulose nanocrystals extracted from garlic straw residues. Ind Crops Prod 87:287–296. https://doi.org/10.1016/j.indcrop.2016.04.060
94. Kamel S (2007) Nanotechnology and its applications in lignocellulosic composites, a mini review. Express Polym Lett 1:546–575. https://doi.org/10.3144/expresspolymlett.2007.78
95. Kargarzadeh H, Ahmad I, Thomas S et al (2017) Methods for extraction of nanocellulose from various sources. In: Kargarzadeh H, Ahamad I, Thomas S et al (eds) Handbook of nanocellulose and cellulose nanocomposites. Wiley-VCH Verlag GmbH & Co, p 849. https://doi.org/10.1002/9783527689972.ch1
96. Khalil HPSA, Bhat AH, Yusra AFI (2012) Green composites from sustainable cellulose nanofibrils: a review. Carbohyd Polym 87:963–979. https://doi.org/10.1016/j.carbpol.2011.08.078
97. Khalil HPSA, Davoudpour Y, Islam M et al (2014) Production and modification of nanofibrillated cellulose using various mechanical processes: a review. Carbohydr Polym 99:649–665. https://doi.org/10.1016/j.carbpol.2013.08.069
98. Khazraji AC, Sylvain R (2013) Self-assembly and intermolecular forces when cellulose and water interact using molecular modeling. J Nanomater 2013:48. https://doi.org/10.1155/2013/745979
99. Kim CW, Kim DS, Kang SY et al (2006) Structural studies of electrospun cellulose nanofibers. Polymer 47:5097–5107. https://doi.org/10.1016/j.polymer.2006.05.033
100. Kim DY, Nishiyama Y, Kuga S (2002) Surface acetylation of bacterial celulose. Cellulose 9:361–367. https://doi.org/10.1023/A:1021140726936
101. Kirby BJ, Hasselbrink EF Jr (2004) Zeta potential of microfluidic substrates: 1. Theory, experimental techniques, and effects on separations. Electrophoresis 25:187–202. https://doi.org/10.1002/elps.200305754
102. Klemm D, Kramer F, Moritz S et al (2011) Nanocelluloses: a new family of nature-based materials. Angew Chem Int Ed 50:5438–5466. https://doi.org/10.1002/anie.201001273
103. Kumar R, Ha SK, Verma K et al (2018) Recent progress in some selected bio-nanomaterials and their engineering applications: an overview. J Sci Adv Mater Devices 3:263–288. https://doi.org/10.1016/j.jsamd.2018.05.003
104. Kurokawa M, King P, Wu X, Joyce E, Mason T, Yamamoto K (2016) Effect of sonication frequency on the disruption of algae. Ultrason Sonochem 31:157–162. https://doi.org/10.1016/j.ultsonch.2015.12.011
105. Kvien I, Tanem BS, Oksman K (2005) Characterization of cellulose whiskers and their nanocomposites by atomic force and electron microscopy. Biomacromolecules 6:3160–3165. https://doi.org/10.1021/bm050479t
106. Kwok DY, Neumann AW (1999) Contact angle measurement and contact angle interpretation. Adv Colloid Interface Sci 81:167–249. https://doi.org/10.1016/S0001-8686(98)00087-6
107. Lee HV, Hamid SB, Zain SK (2014) Conversion of lignocellulosic biomass to nanocellulose: structure and chemical process. Sci World J 20 pp. Article ID 631013. https://doi.org/10.1155/2014/631013
108. Leão AL, Souza SF, Cherian BM et al (2010) Agro-based biocomposites for industrial applications. Mol Cryst Liq Cryst 522:18/[318]–27/[327]. https://doi.org/10.1080/15421401003719852

109. Li W, Yue J, Liu S (2012) Preparation of nanocrystalline cellulose via ultrasound and its reinforcement capability for poly(vinyl alcohol) composites. Ultrason Sonochem 19:479–485. https://doi.org/10.1016/j.ultsonch.2011.11.007
110. Li Y, Li G, Zou Y et al (2014) Preparation and characterization of cellulose nanofibers from partly mercerized cotton by mixed acid hydrolysis. Cellulose 21:301–309. https://doi.org/10.1007/s10570-013-0146-6
111. Lin N, Dufresne A (2014) Nanocellulose in biomedicine: current status and future prospect. Eur Polym J 59:302–325. https://doi.org/10.1016/j.eurpolymj.2014.07.025
112. Lin N, Huang J, Dufresne A (2012) Preparation, properties and applications of polysaccharide nanocrystals in advanced functional nanomaterials: a review. Nanoscale 4:3274–3294. https://doi.org/10.1039/c2nr30260h
113. Lin X, Wu Z, Zhang C, Liu S, Nie S (2018) Enzymatic pulping of lignocellulosic biomass. Ind Crops Prod 120:16–24. https://doi.org/10.1016/j.indcrop.2018.04.033
114. Liu H, Hsieh Y-L (2002) Ultrafine fibrous cellulose membranes from electrospinning of cellulose acetate. J Polym Sci Polym Phys 40:2119–2129. https://doi.org/10.1002/polb.10261
115. Lu P, Hsieh Y-L (2010) Preparation and properties of cellulose nanocrystals: rods, spheres, and network. Carbohydr Polym 82:329–336. https://doi.org/10.1016/j.carbpol.2010.04.073
116. Luo J, Fang Z, Smith RL Jr (2014) Ultrasound-enhanced conversion of biomass to biofuels. Prog Energy Combust Sci 41:56–93. https://doi.org/10.1016/j.pecs.2013.11.001
117. Madsen B, Gamstedt EK (2013) Wood versus plant fibers: similarities and differences in composite applications. Adv Mater Sci Eng 14 pp. Article ID 564346. https://doi.org/10.1155/2013/564346
118. Mahardika M, Abral H, Kasim A et al (2018) Production of nanocellulose from pineapple leaf fibers via high-shear homogenization and ultrasonication. Fibers 6:28–40. https://doi.org/10.3390/fib6020028
119. Mahato DN, Mathur BK, Bhattacherjee S (2013) DSC and IR methods for determination of accessibility of cellulosic coir fibre and thermal degradation under mercerization. Indian J Fibre Text Res 38:96–100. http://nopr.niscair.res.in/handle/123456789/16380. Accessed 22 Jan 2019
120. Milanez DH, Amaral RM, Faria LIL et al (2013) Assessing nanocellulose developments using science and technology indicators. Mater Res-Ibero-Am J 16:635–641. https://doi.org/10.1590/S1516-14392013005000033
121. Milanez DH, Amaral RM, Faria LIL et al (2014) Technological indicators of nanocellulose advances obtained from data and text mining applied to patent documents. Mater Res-Ibero-Am J 17:1513–1522. https://doi.org/10.1590/1516-1439.266314
122. Milanez DH, Noyons E, Faria LIL (2016) A delineating procedure to retrieve relevant publication data in research areas: the case of nanocellulose. Scientometrics 107:627–643. https://doi.org/10.1007/s11192-016-1922-5
123. Mishra RK, Ha SK, Verma K et al (2018a) Recent progress in selected bio-nanomaterials and their engineering applications: an overview. J Sci Adv Mater Devices 3:263–288. https://doi.org/10.1016/j.jsamd.2018.05.003
124. Mishra RK, Sabu A, Tiwari SK (2018b) Materials chemistry and the futurist eco-friendly applications of nanocellulose: status and prospect. J Saudi Chem Soc 22:949–978. https://doi.org/10.1016/j.jscs.2018.02.005
125. Moon RJ, Martini A, Nairn J et al (2011) Cellulose nanomaterials review: structure, properties and nanocomposites. Chem Soc Rev 40:3941–3994. https://doi.org/10.1039/c0cs00108b
126. Moreno G, Ramirez K, Esquivel M et al (2018) Isolation and characterization of nanocellulose obtained from industrial crop waste resources by using mild acid hydrolysis. J Renew Mater 6:362–369. https://doi.org/10.7569/jrm.2017.634167
127. Mukherjee A, Hackley VA (2018) Separation and characterization of cellulose nanocrystals by multi-detector asymmetric flow field-flow fractionation. Analyst 143:731–740. https://doi.org/10.1039/c7an01739a
128. Nachtkamp K, Krüger C, Engelhardt J et al (2018) Nanoparticles made of amorphous cellulose. US patent 8722092B2, 10 Oct 2007

129. Nam S, French AD, Condon BD et al (2016) Segal crystallinity index revisited by the simulation of X-ray diffraction patterns of cotton cellulose Iβ and cellulose II. Carbohydr Polym 135:1–9. https://doi.org/10.1016/j.carbpol.2015.08.035
130. Nelson ML, O'Connor RT (1964a) Relation of certain infrared bands to cellulose crystallinity and crystal latticed type. Part I. Spectra of lattice types I, II, III and of amorphous cellulose. J Appl Polym Sci 8:1311–1324. https://doi.org/10.1002/app.1964.070080322
131. Nelson ML, O'Connor RT (1964b) Relation of certain infrared bands to cellulose crystallinity and crystal lattice type. Part II. A new infrared ratio for estimation of crystallinity in celluloses I and II. J Appl Polym Sci 8:1325–1341. https://doi.org/10.1002/app.1964.070080323
132. Nguyen HD, Mai TTT, Nguyen NB et al (2013) A novel method for preparing microfibrillated cellulose from bamboo fibers. Adv Nat Sci Nanosci Nanotechnol 4:015016. https://doi.org/10.1088/2043-6262/4/1/015016
133. Nikmatin S, Rudwiyanti JR, Prasetyo KW et al (2015) Mechanical and optical characterization of bio-nanocomposite from pineapple leaf fiber material for food packaging. In: Proceedings of the international seminar on photonics, optics, and its applications, Sanur, Indonesia, 2014. https://doi.org/10.1117/12.2081112
134. Oh SY, Yoo DI, Shin Y, Kim HC, Kim HY, Chung YS, Park WH, Youk JH (2005) Crystalline structure analysis of cellulose treated with sodium hydroxide and carbon dioxide by means of X-ray diffraction and FTIR spectroscopy. Carbohydr Res 340:2376–2391. https://doi.org/10.1016/j.carres.2005.08.007
135. Ounas A, Aboulkas A, El Harfi K et al (2011) Pyrolysis of olive residue and sugar cane bagasse: non-isothermal thermogravimetric kinetic analysis. Bioresour Technol 102:11234–11238. https://doi.org/10.1016/j.biortech.2011.09.010
136. Pakzad A, Simonsen J, Heiden PA et al (2012) Size effects on the nanomechanical properties of cellulose I nanocrystals. J Mater Res 27:528–536. https://doi.org/10.1557/jmr.2011.288
137. Park S, Baker JO, Himmel ME et al (2010) Cellulose crystallinity index: measurement techniques and their impact on interpreting cellulase performance. Biotechnol Biofuels 3:10. https://doi.org/10.1186/1754-6834-3-10
138. Pavia DL, Lampman GM, Kriz GS et al (2009) Introduction to spectroscopy, 4th edn. Cengage Learning, Belmont, pp 15–26
139. Peng Y, Gardner DJ, Han Y et al (2013) Influence of drying method on the material properties of nanocellulose I: thermostability and crystallinity. Cellulose 20:2379–2392. https://doi.org/10.1007/s10570-013-0019-z
140. Pérez J, Munõz-Dorado J, de la Rubia T et al (2002) Biodegradation and biological treatments of cellulose, hemicellulose and lignin: an overview. Int Microbiol 5:53–63. https://doi.org/10.1007/s10123-002-0062-3
141. Phanthong P, Reubroycharoen P, Hao X et al (2018) Nanocellulose: extraction and application. Carbon Resour Convers 1:32–43. https://doi.org/10.1016/j.crcon.2018.05.004
142. Poletto M, Zattera AJ, Santana RM (2012) Structural differences between wood species: evidence from chemical composition, FTIR spectroscopy, and thermogravimetric analysis. J Appl Polym Sci 126:E337–E344. https://doi.org/10.1002/app.36991
143. Prado KS, Gonzales D, Spinacé MAS (2019) Recycling of viscose yarn waste through one-step extraction of nanocellulose. Int J Biol Macromol. https://doi.org/10.1016/j.ijbiomac.2019.06.124
144. Prado KS, Spinacé MA (2019) Isolation and characterization of cellulose nanocrystals from pineapple crown waste and their potential uses. Int J Biol Macromol 122:410–416. https://doi.org/10.1016/j.ijbiomac.2018.10.187
145. Pu Y, Jiang N, Ragauskas AJ (2007) Ionic liquid as a green solvent for lignin. J Wood Chem Technol 27:23–33. https://doi.org/10.1080/02773810701282330
146. Quan SL, Kang SG, Chin IJ (2010) Characterization of cellulose fibers electrospun using ionic liquid. Cellulose 17:223–230. https://doi.org/10.1007/s10570-009-9386-x
147. Querejeta-Fernández A, Kopera B, Prado KS et al (2015) Circular dichroism of chiral nematic films of cellulose nanocrystals loaded with plasmonic nanoparticles. ACS Nano 9:10377–10385. https://doi.org/10.1021/acsnano.5b04552

148. Reddy N, Yang Y (2005) Biofibers from agricultural byproducts for industrial applications. Trends Biotechnol 23:22–27. https://doi.org/10.1016/j.tibtech.2004.11.002
149. Revol JF, Godbout L, Dong XM et al (1994) Chiral nematic suspensions of cellulose crystallites; phase separation and magnetic field orientation. Liq Cryst 16:127–134. https://doi.org/10.1080/02678299408036525
150. Rojas J, Bedoya M, Ciro Y (2015) Current trends in the production of cellulose nanoparticles and nanocomposites for biomedical applications. In: Poletto M, Junior HLO (eds) Cellulose—fundamental aspects and current trends. IntechOpen, p 282. https://doi.org/10.5772/61899
151. Roman M, Winter WT (2004) Effect of sulfate groups from sulfuric acid hydrolysis on the thermal degradation behavior of bacterial cellulose. Biomacromolecules 5:1671–1677. https://doi.org/10.1021/bm034519+
152. Sacui IA, Nieuwendaal RC, Burnett DJ et al (2014) Comparison of the properties of cellulose nanocrystals and cellulose nanofibrils isolated from bacteria, tunicate, and wood processed using acid, enzymatic, mechanical, and oxidative methods. ACS Appl Mater Interfaces 6:6127–6138. https://doi.org/10.1021/am500359f
153. Saito T, Kimura S, Nishiyama Y et al (2007) Cellulose nanofibers prepared by TEMPO-mediated oxidation of native cellulose. Biomacromolecules 8:2485–2491. https://doi.org/10.1021/bm0703970
154. Salas C, Nypelö T, Rodriguez-Abreu C et al (2014) Nanocellulose properties and applications in colloids and interfaces. Curr Opin Colloid Interface Sci 19:383–396. https://doi.org/10.1016/j.cocis.2014.10.003
155. Segal LG, Creely JJ, Martin AE Jr et al (1959) An empirical method for estimating the degree of crystallinity of native cellulose using the X-ray diffractometer. Text Res J 29:786–794. https://doi.org/10.1177/004051755902901003
156. Sena Neto AR, Araujo MAM, Souza FVD et al (2013) Characterization and comparative evaluation of thermal, structural, chemical, mechanical and morphological properties of six pineapple leaf fiber varieties for use in composites. Ind Crop Prod 43:529–537. https://doi.org/10.1016/j.indcrop.2012.08.001
157. Sena Neto AR, Araujo MAM, Barboza RMP et al (2015) Comparative study of 12 pineapple leaf fiber varieties for use as mechanical reinforcement in polymer composites. Ind Crop Prod 64:68–78. https://doi.org/10.1016/j.indcrop.2014.10.042
158. Shak KP, Pang YL, Mah SK (2018) Nanocellulose: recent advances and its prospects in environmental remediation. Beilstein J Nanotechnol 9:2479–2498. https://doi.org/10.3762/bjnano.9.232
159. Shih Y-F, Chou M-Y, Chang W-C et al (2017) Completely biodegradable composites reinforced by the cellulose nanofibers of pineapple leaves modified by eco-friendly methods. J Polym Res 24:209–221. https://doi.org/10.1007/s10965-017-1367-4
160. Shih Y-F, Chou M-Y, Lian H-Y et al (2018a) Highly transparent and impact-resistant PMMA nanocomposites reinforced by cellulose nanofibers of pineapple leaves modified by eco-friendly methods. Express Polym Lett 12:844–854. https://doi.org/10.3144/expresspolymlett.2018.72
161. Shih Y-F, Tsou Z-G, Wang C-H et al (2018b) Eco-friendly modification for the cellulose nanofibers derived from pineapple leaves for high-performance nanocomposite. IOP Conf Ser Earth Environ Sci 171:1–7. https://doi.org/10.1088/1755-1315/171/1/012041
162. Sing J, Suhag M, Dhaka A (2015) Augmented digestion of lignocellulose by steam explosion, acid and alkaline pretreatment methods: a review. Carbohydr Polym 117:624–631. https://doi.org/10.1016/j.carbpol.2014.10.012
163. Siqueira G, Bras J, Dufresne A (2010) Cellulosic bionanocomposites: a review of preparation, properties and applications. Polymers 2:728–765. https://doi.org/10.3390/polym2040728
164. Siró I, Plackett D (2010) Microfibrillated cellulose and new nanocomposite materials: a review. Cellulose 17:459–494. https://doi.org/10.1007/s10570-010-9405-y
165. Smith CW (1995) Crop production: evolution, history, and technology. Wiley, United States, p 320

166. Stylianopoulos T, Kokonou M, Michael S et al (2012) Tensile mechanical properties and hydraulic permeabilities of electrospun cellulose acetate fiber meshes. J Biomed Mater Res B 100:2222–2230. https://doi.org/10.1002/jbm.b.32791
167. Surip SN, Aziz FMA, Bonnia NN et al (2017) Process parameter and surface morphology of pineapple leaf electrospun nanofibers (PALF). AIP Conf Proc 1885:020269/1–020269/5. https://doi.org/10.1063/1.5002463
168. Surip SN, Aziz FMA, Bonnia NN et al (2018) Effect of pineapple leaf fibers (PALF) concentration on nanofibers formation by electrospinning. IOP Conf Ser Mater Sci Eng 290:1–7. https://doi.org/10.1088/1757-899x/290/1/012003
169. Sutkar VS, Gogate PR, Csoka L (2010) Theoretical prediction of cavitational activity distribution in sonochemical reactors. Chem Eng J 158:290–295. https://doi.org/10.1016/j.cej.2010.01.049
170. TAPPI WI 3021. Standard terms and their definition for cellulose nanomaterial. https://www.tappi.org/content/hide/draft3.pdf. Accessed 25/11/2019
171. Thygesen A, Oddershede J, Lilholt H et al (2005) On the determination of crystallinity and cellulose content in plant fibres. Cellulose 12:563. https://doi.org/10.1007/s10570-005-9001-8
172. Trache D, Hussin MH, Haafiz MKM et al (2017) Recent progress in cellulose nanocrystals: sources and production. Nanoscale 9:1763–1786. https://doi.org/10.1039/c6nr09494e
173. Tran AV (2006) Chemical analysis and pulping study of pineapple crown leaves. Ind Crop Prod 24:66–74. https://doi.org/10.1016/j.indcrop.2006.03.003
174. Trindade WG, Hoareau W, Megiatto JD, Razera AT, Castellan A, Frollini E (2005) Thermoset phenolic matrices reinforced with unmodified and surface grafted furfuryl alcohol sugar cane bagasse and curaua fibers: properties of fibers and composites. Biomacromolecules 6:2485–2496. https://doi.org/10.1021/bm058006+
175. Van de Ven TG, Sheikhi A (2016) Hairy cellulose nanocrystalloids: a novel class of nanocellulose. Nanoscale 8:15101–15114. https://doi.org/10.1039/c6nr01570k
176. Van der Biest O, Vandeperre LJ (1999) Electrophoretic deposition of materials. Annu Rev Mater Sci 29:327–352. https://doi.org/10.1146/annurev.matsci.29.1.327
177. Vega-Baudrit JR, Camacho M, Ureña YRC et al (2018) Synthesis and characterization of nanocellulose obtained from pineapple peel wastes. In: Abstracts of the 255th national meeting & exposition of the American Chemical Society—division of cellulose & renewable materials, New Orleans, 18–22 Mar 2018
178. Wada M, Chanzy H, Nishiyama Y et al (2004) Cellulose III_I crystal structure and hydrogen bonding by synchrotron X-ray and neutron fiber diffraction. Macromolecules 37:8548–8555. https://doi.org/10.1021/ma0485585
179. Wahyuningsih K, Iriani ES, Fahma F (2016) Utilization of cellulose from pineapple leaf fibers as nanofiller in polyvinyl alcohol-based film. Indones J Chem 16:181–189. https://doi.org/10.22146/ijc.1086
180. Wan Nadirah WO, Jawaid M, Al Masri AA, Khalil HPSA, Suhaily SS, Mohamed AR (2012) Cell wall morphology, chemical and thermal analysis of cultivated pineapple leaf fibres for industrial applications. J Polym Environ 20:404–411. https://doi.org/10.1007/s10924-011-0380-7
181. Wang T, Yang H, Kubicki JD et al (2016) Cellulose structural polymorphism in plant primary cell walls investigated by high-field 2D solid-state NMR spectroscopy and density functional theory calculations. Biomacromolecules 17:2210–2222. https://doi.org/10.1021/acs.biomac.6b00441
182. Wang ZL (2000) Transmission electron microscopy of shape-controlled nanocrystals and their assemblies. J Phys Chem B 104:1153–1175. https://doi.org/10.1021/jp993593c
183. Whitmore RE, Atalla RH (1985) Factors influencing the regeneration of cellulose I from phosphoric acid. Int J Biol Macromol 7:182–186. https://doi.org/10.1016/0141-8130(85)90022-4
184. Xu X, Liu F, Jiang L et al (2013) Cellulose nanocrystals vs. cellulose nanofibrils: a comparative study on their microstructures and effects as polymer reinforcing agents. ACS Appl Mater Interfaces 5:2999–3009. https://doi.org/10.1021/am302624t

185. Xue Y, Mou Z, Xiao H (2017) Nanocellulose as sustainable biomass material: structure, properties, present status and future prospects in biomedical applications. Nanoscale 9:14758–14781. https://doi.org/10.1039/c7nr04994c
186. Yang H, Yan R, Chen H et al (2006) In-depth investigation of biomass pyrolysis based on three major components: hemicellulose, cellulose and lignin. Energy Fuel 20:388–393. https://doi.org/10.1021/ef0580117
187. Yao F, Wu Q, Lei Y et al (2008) Thermal decomposition kinetics of natural fibers: activation energy with dynamic thermogravimetric analysis. Polym Degrad Stabil 93:90–98. https://doi.org/10.1016/j.polymdegradstab.2007.10.012
188. Yoo CG, Pu Y, Ragauskas AJ (2017) Ionic liquids: promising green solvents for lignocellulosic biomass utilization. Curr Opin Green Sustain Chem 5:5–11. https://doi.org/10.1016/j.cogsc.2017.03.003
189. Zamil MS, Geitmann A (2017) The middle lamella—more than a glue. Phys Biol 14:015004. https://doi.org/10.1088/1478-3975/aa5ba5
190. Zhang Q, Benoit M, Vigier KO et al (2013) Pretreatment of microcrystalline cellulose by ultrasounds: effect of particle size in the heterogeneously-catalyzed hydrolysis of cellulose to glucose. Green Chem 15:963–969. https://doi.org/10.1039/c3gc36643j
191. Zhang YHP, Himmel ME, Mielenz JR (2006) Outlook for cellulase improvement: screening and selection strategies. Biotechnol Adv 24:452–481. https://doi.org/10.1016/j.biotechadv.2006.03.003
192. Zhou X, Wang H, Zhang J et al (2018) Lightweight biobased polyurethane nanocomposite foams reinforced with pineapple leaf nanofibers (PLNFs). J Renew Mater 6:68–74. https://doi.org/10.7569/JRM.2017.634150

Tensile Behaviour of Centrally Holed Pineapple Fibre Reinforced Vinyl Ester Composites

Nadendla Srinivasababu

Abstract The composite parts having holes need to experimentally examined for understanding their behaviour under mechanical loading conditions. So, an initial attempt was made to reinforce the locally available pineapple leaf fibre in as-is condition and after chemical treatment into vinyl ester matrix for preparation of the composites according to ASTM D5766/5766M—07 standard by rolling cum hand lay-up technique. Drilling holes of 3, 6 and 8 mm in diameter was performed slowly and carefully without disturbing the fibres in matrix. Fibres were examined under SEM and its diameter is in the range of 3.12–16.6 μm. Unwanted impurities cum waxy materials washed away from the fibre after alkali treatment and were confirmed from the SEM image. Plain, untreated pineapple leaf fibre composites tensile strength was decreased up to 6 mm hole and thereafter, it was increased. Similar trend was observed after determination of modulus of the composites. However, treated fibre composites tensile strength and modulus were improved beyond the 3 mm hole. Tensile fractured specimens revealed the fibre–matrix interactions.

Keywords Pineapple leaf fibre (PALF) · Open-hole tensile test · Scanning electron microscope (SEM) · Tensile strength · Tensile modulus · Fractured surface

1 Introduction

Pineapple is botanically called *Ananas comosus* and is abundantly grown in tropical parts of the country in India. Retted pineapple leaf fibre consists of holocellulose (91.94%), alpha cellulose (87.36%), hemicellulose (4.58%), lignin (3.62%), alcohol benzene extractives (2.72%), ash (0.54%), moisture (11.61%) and moisture regain (13.15%). Mukherjee et al. described that the fibres were multicellular with chemical constituents of cellulose—70–82%, lignin 5–12%, ash 1.1% with an ultimate tensile strength and modulus of 362–748 MN/m^2, 25–36 GN/m^2, respectively. Yu found that the pineapple leaf consists of nearly 56–62% cellulose, 16–19% hemicellulose,

N. Srinivasababu (✉)
Fibrous Composites Research Lab, Department of Mechanical Engineering, Vignan's Lara Institute of Technology & Science, Vadlamudi 522213, Andhra Pradesh, India
e-mail: drnsbabu_vlits@vignan.ac.in

M. Jawaid et al. (eds.), *Pineapple Leaf Fibers*, Green Energy and Technology,
https://doi.org/10.1007/978-981-15-1416-6_11

2–2.5% pectin, 9–13% lignin, 1–1–1.5% water-soluble materials, 4–7% fat and wax, 2–3% ash, 3–8 mm diameter, 7–18 μm diameter and density 1.543 g/cm^3 [3].

Pineapple leaf fibre was easily extracted by using a newly developed machine that consists of feed, leaf scratching and serrated roller which facilitates retting process [5]. Further, fibre extraction methods like manual, mechanical with subsequent degumming of them through chemical, enzymatic from an agro-waste like pineapple leaf were reviewed [9].

The quality of pineapple fibre extracted from different age leaf, i.e. 4, 8 and 12 months was studied. Tensile test was conducted on famous Malaysian Joseapine/Johor Sarawak Pineapple, Morris leaf fibres which have different physical properties [10]. From Kok Kwai subdistrict of Thailand, pineapple leaves were collected from the cultivation area. Different parts of the fibre like bottom, middle and top were used to examine its morphology, size and mechanical properties [14].

Hydrogen peroxide bleached pineapple leaf fibre collected from north, south Bengal and Maharashtra and their yarn tensile behaviour was experimentally studied [8]. Pineapple leaf fibres collected from Medellin were subjected to delignification in an autoclave and subjected to thermal treatments for studying tensile strength [6].

Khadi and Village Industries Commission, Trivandrum, supplied pineapple leaf fibre ultimate tensile strength, Young's modulus, average modulus and elongation was determined with respect to fibre diameter, length and test speed [11]. An easy decorative machine extracts the pineapple leaves received from Muzium Nanas and was subjected to alkali, heat and combination treatments for fibre degumming. Using video analyzer, fibre diameter was measured and its tensile properties determined [16].

Thirty days pond water retted pineapple leaf taken from two districts of Bangladesh, i.e. Khagrachari, Jhenidah districts, was used to estimate the constituents of it. The fibre has mainly α-cellulose: 74.44%, hemicellulose: 13.39%, ligin: 7.12%, pectic matter: 2.89%, aqueous extract: 0.58% [1].

Pineapple leaf fibre of different lengths, with constant, varying content was reinforced into ETERSET 2504APT unsaturated resin to make and test the composites for determining tensile strength, modulus [13]. Unidirectional pineapple leaf fibre mat and different fillers like alumina, silicon carbide, fly ash and red mud were reinforced into vinyl ester resin to prepare the composites for determining density, microhardness, tensile, flexural and inter-laminar shear and impact strength [15].

Alkali treated 5 mm length pineapple fibre and fly ash (0–1.2%) was added to geopolymer paste to make composites and was tested for compressive, flexural properties [2]. Oko Oba village, Nigeria sourced pineapple leaf fibre reinforced polyester, epoxy composites were tested for tensile, flexural and impact properties. The pineapple epoxy composites were suitable for development of prosthetic socket [12].

Ananas comosus fibres available in Ayer Hitam, Johor, were analyzed for chemical composition according to TAPPI and morphology for their use in paper making industry as alternative pulp [7]. Useful furnishing, decorative fabrics can be made from blends of pineapple leaf fibre in wool, waste from polyester, viscose rayon fibres [4].

From the literature review conducted on pineapple leaf fibre, mechanical behaviours of its composites the following points are noted.

- Lot of efforts was taken by the researchers in order to find the constituents like cellulose, hemicellulose, lignin, pectin and wax cum other materials of pineapple leaf fibre. The obtained values were varied with respect to origin/location of plant growth.
- Using retting, chemical, mechanical peeling and in diversified ways with innovative equipment, pineapple fibre was extracted.
- Fibre was used in polyester, epoxy matrices for making composites and to determine their mechanical behaviour.

When mechanical fasteners, screws were in use to assemble different components, holes made in the composite structures need to be studied experimentally. The effects of such holes on composite strength must be determined. Such cut-outs and holes create the stress concentration in the laminates, which reduces the load-carrying capacity. The works done in this area are limited.

So, in the present work, commercially available pineapple leaf fibre was procured and was subjected to alkali treatment for enhancing the bonding between fibre–matrix. Treated fibres were also examined under SEM. Untreated and chemically treated fibres were reinforced up to maximum volume of the mould. Using rolling cum hand lay-up, pineapple leaf fibre reinforced vinyl ester composites were made. The prepared composites were machined with different hole sizes in order to assess its tensile behaviour as per ASTM procedures. Finally, the fractured treated pineapple fibre composites were seen under SEM to understand the fibre–matrix interaction.

2 Materials and Methods

2.1 Materials

Pineapple leaf fibre was procured from Cherukupalli village in the state of Andhra Pradesh, India. Ecmalon 9921 vinyl ester resin was purchased from Ecmas Resins Pvt. Ltd., Hyderabad, and Telangana, India. The properties of the liquid resin given by the supplier are given in Table 1. It is an epoxy novalac-based resin resistant to several chemicals, oxidizing acids, solvents and is especially recommended in chloralkali industries.

Table 1 Vinyl ester liquid resin properties from manufacturer datasheet

Appearance	Liquid of amber to light brown in colour
Viscosity	210 Brookfield at 25 °C
Specific gravity	1.07
Gel time	30 min at 25 °C

2.2 Fibre Chemical Treatment

Pineapple leaf fibres were initially dried in an atmosphere and soaked in a tub containing alkaline solution of 0.7025 M concentration up to 49 h 20 min. Then the fibres were washed with a huge quantity of drinking water till the complete chemical goes away from them. Ambient dried fibres were kept in an NSW 143 oven at 80, 100 °C up to 1 h 10 min and 2 h 20 min respectively. Fibres received in as-is condition and chemically treated fibres were reinforced into the resin one after another for making composites.

2.3 Composites Fabrication, Hole Drilling and Testing

Vinyl ester resin, catalyst and accelerator were simultaneously added one after another into borosilicate beaker and were stirred manually using glass stirring rod. Here onwards, this is called as the resin mixture or simply resin invariably. Pineapple fibres are silky in nature, try to come out of the mould quickly and settle over its bank when the low viscous vinyl ester resin mixture was poured in the hand lay-up method. So, nearly 200 ml of resin was poured over the pre-placed fibres in the mould and by using 25 mm steel rod, rolling was performed up to 2 min. After observation of proper fibres settlement, smash of air bubbles in the mould, the remaining resin was poured over them. Excess resin was squeezed out with steel rule and pressure of 1350 Pa was applied over the mould up to 24 h. This procedure was adopted for making the composites reinforced with untreated and chemically treated fibres up to maximum permissible level in the mould.

Then the specimens were taken out from the mould and ground using Bosch grinding machine in order to obtain flat straight edges. Using soft cloth dust during finishing was wiped off and dimensions of the specimen were taken to measure the cross section. In order to test the ability of continuous fibre composites behaviour at crack, open-hole tensile test according to ASTM D5766 standard was conducted.

With hand drill and bit, various size holes viz. 3, 6 and 8 mm were made at the centre of the plain, untreated (Figs. 1, 2 and 3) and alkali treated (Figs. 4, 5 and 6) pineapple leaf fibre composite specimens. Using Mitutoyo Japan make digital vernier caliper having 0.02 mm readability machined hole of all the specimens is measured at three locations on top and bottom surface. Then the average values of all the specimens were calculated and given in Tables 2, 3, 4, 5, 6 and 7.

All the drilled specimens were tensile tested at a cross-head speed of 5 mm/min on PC 2000 Electronic Tensometer having 20 KN load cell. After tests, the data was transferred to the computer interface for further analysis.

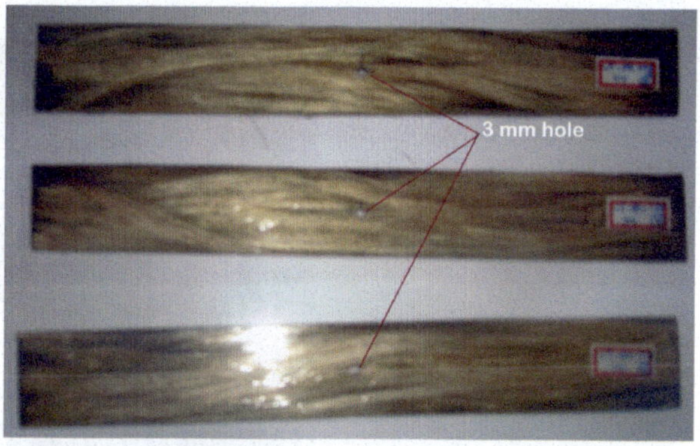

Fig. 1 Pineapple leaf fibre reinforced vinyl ester composites—3 mm hole drilled

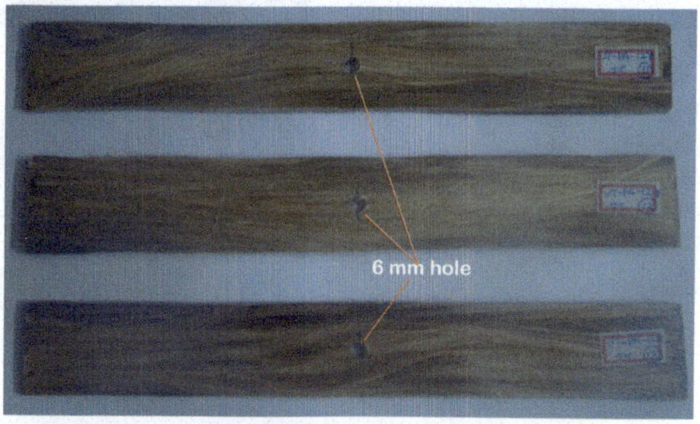

Fig. 2 Pineapple leaf fibre reinforced vinyl ester composites—6 mm hole drilled

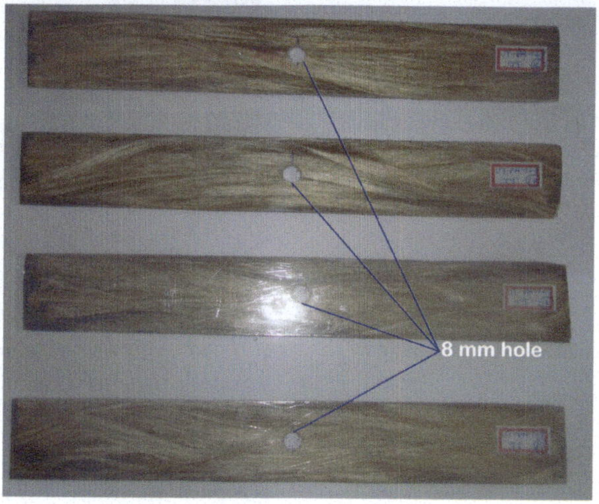

Fig. 3 Pineapple leaf fibre reinforced vinyl ester composites—8 mm hole drilled

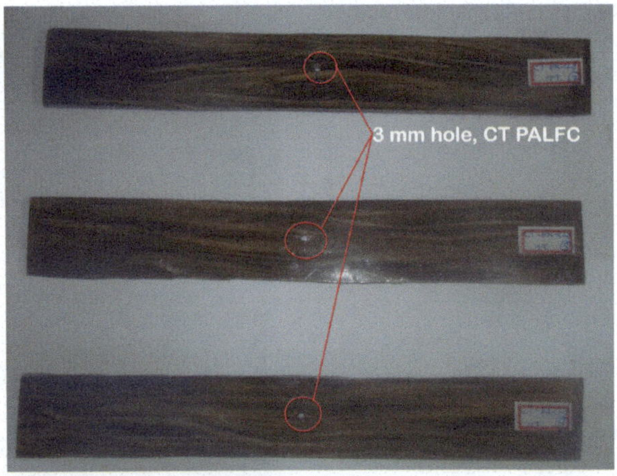

Fig. 4 Chemically treated pineapple leaf fibre reinforced vinyl ester composites—3 mm hole drilled

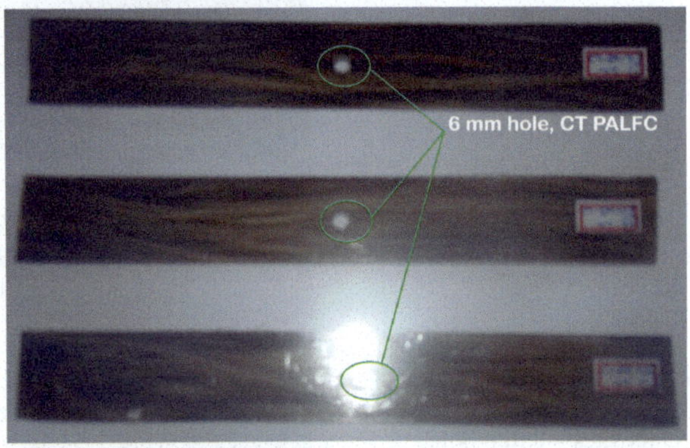

Fig. 5 Chemically treated pineapple leaf fibre reinforced vinyl ester composites—6 mm hole drilled

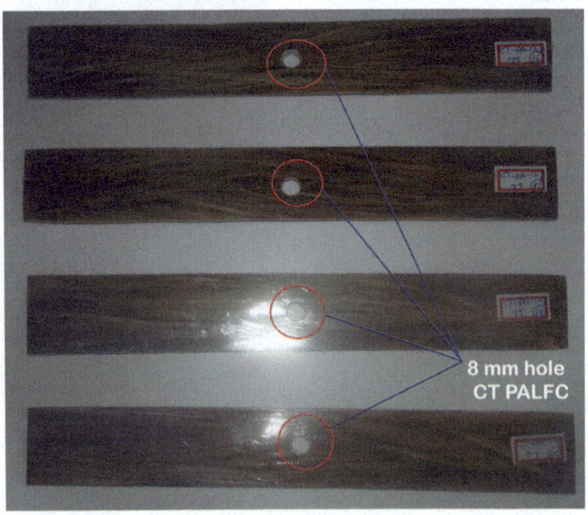

Fig. 6 Chemically treated pineapple leaf fibre reinforced vinyl ester composites—8 mm hole drilled

Table 2 Diameter of 3 mm hole machined on PALF vinyl ester composites

Specimen no.	Diameter of hole after machining (mm)						Average diameter (mm)
	Top surface			Bottom surface			
	Position 1	Position 2	Position 3	Position 1	Position 2	Position 3	
1	3.05	3.06	3.02	3.01	3.04	3.07	3.04
2	3.05	3.07	3.03	3.05	3.02	3.06	3.05
3	3.05	3.02	3.03	3.09	3	2.95	3.02
4	3.07	3.02	2.99	3.05	3.04	3.06	3.04
5	3.09	3.02	3.02	3.03	2.96	2.97	3.02

Table 3 Diameter of 6 mm hole machined on PALF vinyl ester composites

Specimen no.	Diameter of hole after machining (mm)						Average diameter (mm)
	Top surface			Bottom surface			
	Position 1	Position 2	Position 3	Position 1	Position 2	Position 3	
1	6.05	6.03	6.09	6.02	6.07	6.08	6.06
2	6.12	6.05	6.06	6.03	6.07	6.01	6.06
3	6.18	6.21	6.08	5.91	5.95	5.96	6.05
4	6.27	5.99	6.14	6.08	6.15	5.23	5.98
5	5.99	6.02	6.11	6.15	5.84	6.03	6.02

Table 4 Diameter of 8 mm hole machined on PALF vinyl ester composites

Specimen no.	Diameter of hole after machining (mm)						Average diameter (mm)
	Top surface			Bottom surface			
	Position 1	Position 2	Position 3	Position 1	Position 2	Position 3	
1	8.31	8.27	8.21	8.18	8.22	8.29	8.25
2	8.30	8.38	8.29	8.00	8.02	8.10	8.18
3	8.32	8.25	8.37	8.04	7.98	7.94	8.15
4	8.38	8.02	8.02	7.60	8.11	7.82	7.99
5	8.33	8.24	8.30	8.04	8.00	8.05	8.16

3 Results and Discussion

Pineapple leaf in the received conditioned was tested under Scanning Electron Microscope (SEM) and the images were shown in Fig. 7a–d. Multicellular fibrils were observed and their diameter was measured which was ranges from 3.12 to 15.3 μm, Fig. 3c. Unwanted pulp was visible in the fibre, Fig. 7d, and was washed away

Table 5 Diameter of 3 mm hole machined on chemically treated PALF vinyl ester composites

Specimen no.	Diameter of hole after machining (mm)						Average diameter (mm)
	Top surface			Bottom surface			
	Position 1	Position 2	Position 3	Position 1	Position 2	Position 3	
1	3.11	3.09	3.06	3.02	3.09	3.08	3.08
2	3.12	3.15	3.13	3.05	3.08	3.06	3.10
3	3.15	3.08	3.14	3.11	3.08	3.07	3.11
4	3.08	3.17	3.07	2.99	2.89	3.12	3.05
5	2.92	2.86	2.94	3	2.86	2.98	2.93

Table 6 Diameter of 6 mm hole machined on chemically treated PALF vinyl ester composites

Specimen no.	Diameter of hole after machining (mm)						Average diameter (mm)
	Top surface			Bottom surface			
	Position 1	Position 2	Position 3	Position 1	Position 2	Position 3	
1	6.19	6.13	6.17	6.19	6.13	6.16	6.16
2	6.22	6.13	6.18	6.19	6.11	6.18	6.17
3	6.23	6.13	6.24	6.13	6.12	6.02	6.15
4	6.05	6.18	6.23	5.94	6.02	5.95	6.06
5	6.22	6.24	6.19	5.64	5.98	5.87	6.02

Table 7 Diameter of 8 mm hole machined on chemically treated PALF vinyl ester composites

Specimen no.	Diameter of hole after machining (mm)						Average diameter (mm)
	Top surface			Bottom surface			
	Position 1	Position 2	Position 3	Position 1	Position 2	Position 3	
1	8.12	8.04	8.09	8.11	8.13	8.16	8.11
2	8.15	8.04	8.24	8.06	8.12	8.09	8.12
3	8.48	8.34	8.47	8.05	8.03	8.08	8.24
4	8.21	7.99	8.1	7.62	7.59	7.38	7.82
5	8.16	8.09	8.2	8.01	7.9	7.83	8.03

after alkali treatment. SEM images of chemically treated pineapple leaf fibres were shown in Fig. 8a–d where the fibrils are clearly recognizable. Diameter of treated fibres (Fig. 8d) starts from 5.26 μm and ends with 16.6 μm.

Plain, untreated and chemically treated PALF reinforced vinyl ester composites tensile strength and modulus were determined and the average of five specimens

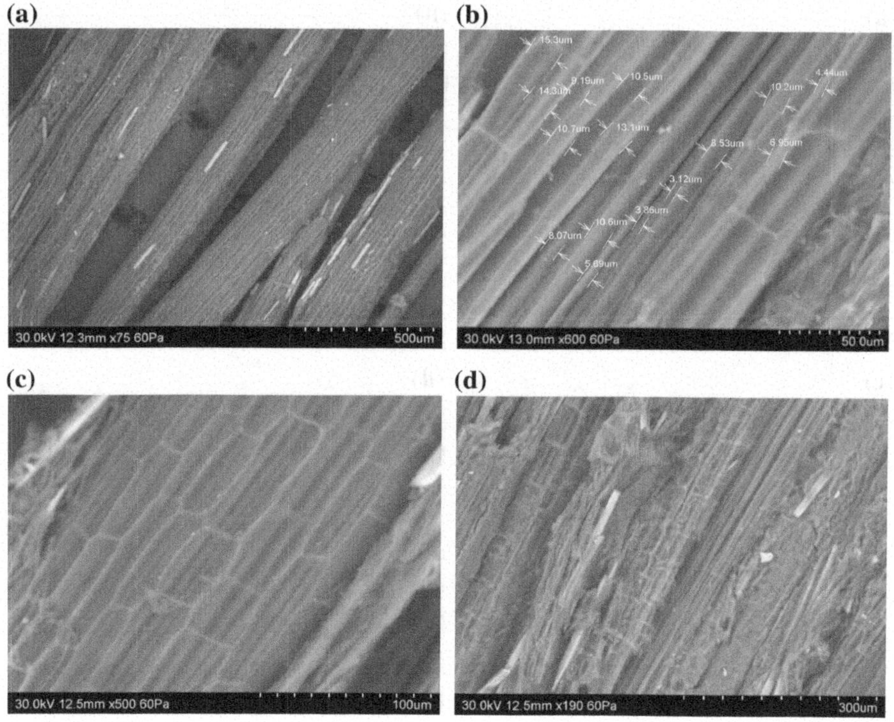

Fig. 7 SEM images of pineapple leaf fibre

calculated. Figures 9 and 10 show the graphs indicate the tensile strength and modulus against diameter of the hole machined on the specimen.

With increase in diameter of the hole from 3 to 6 mm, the tensile strength of plain specimens decreased from 49.88 to 18.03 MPa, and thereafter, it was increased to 27.92 MPa at 8 mm hole. A clear decreasing trend in tensile strength was shown by pineapple leaf fibre reinforced vinyl ester composites with increase in hole size from 0 to 8 mm. Chemically treated PALF reinforced vinyl ester composites expected to exhibit good tensile strength when compared with untreated fibre reinforced composites. But the untreated composites with 0, 3 mm hole have exhibited 33.92, 45.34% more tensile strength than the treated fibre composites, respectively. An enhanced tensile strength of 7.7 and 15.61% was visualized from graph of the treated PALF composites having 6, 8 mm when compared with untreated fibre composites.

Tensile modulus of chemically treated pineapple leaf fibre reinforced composites are higher than all the composites experimentally tested in this work at all the hole sizes except at 3 mm. Plain specimens had shown more or less similar value of modulus at all hole sizes and its value varies as 0.59–0.58 GPa. The highest tensile modulus of 0.96 GPa was achieved at 6 mm hole treated PALF composites. Untreated pineapple fibre composites tensile modulus followed the trend of tensile strength. But the alkali treated pineapple leaf fibre composites modulus was decreased from

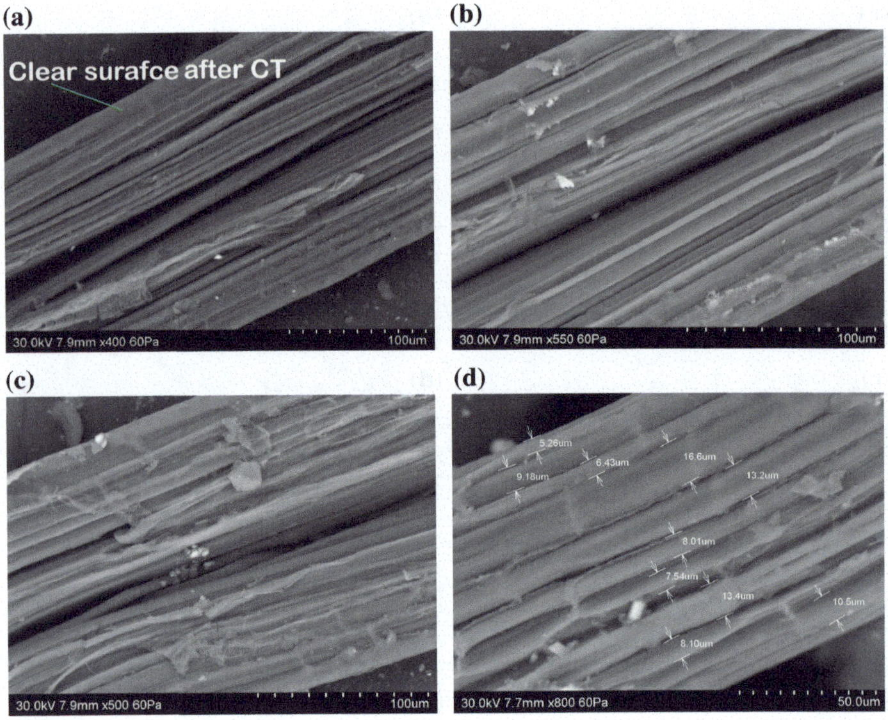

Fig. 8 SEM images of alkali treated pineapple leaf fibre

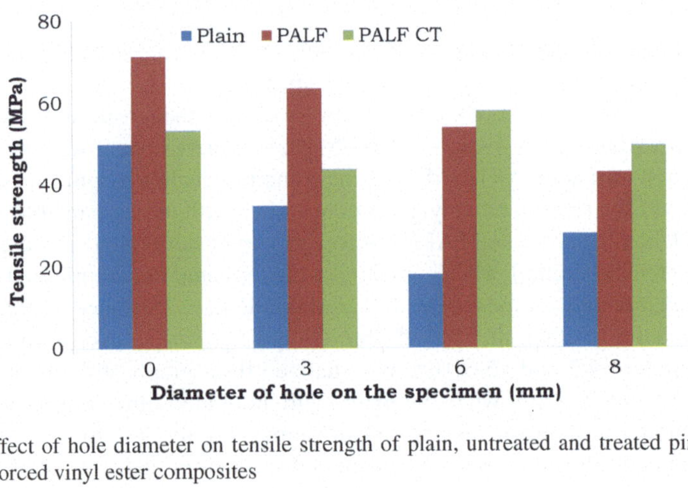

Fig. 9 Effect of hole diameter on tensile strength of plain, untreated and treated pineapple leaf fibre reinforced vinyl ester composites

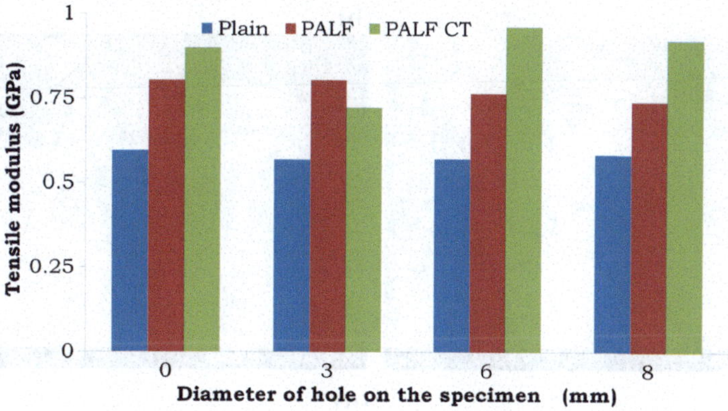

Fig. 10 Effect of hole diameter on tensile modulus of plain, untreated and treated pineapple leaf fibre reinforced vinyl ester composites

0 to 3 mm hole diameter, and thereafter, it was increased up to maximum hole size. The tensile modulus of chemically treated PALF composites at 0, 3, 8 mm size hole was 11.25, 26.31 and 24.32% higher than untreated composites.

An SEM examination was conducted on tensile fractured treated PALF vinyl ester composites with 6 mm hole and was given in Fig. 11a–d. Fibres were failed due to tensile only and pull out was observed from Fig. 11b. But minimum to large space around the fibres was observed in Fig. 11c image at some locations. The length of the pulled out fibre varied among the locations on the fractured surface. When the electron beam further focuses on a very small area, the multicellular fractured fibrils were identified, Fig. 11d.

4 Conclusions

Pineapple leaf fibre (untreated and treated) reinforced composites were prepared successfully by rolling cum hand lay-up method. The composite specimens were drilled up to the required diameter holes for open-hole tensile test without disturbing the fibres. Chemical treatment has resulted in good fibre–matrix interface locking and evidenced after tensile test of composites. Majority of the open-hole composite specimens were failed at the location of hole and few of them were failed at other locations. There is a clear decrease in load-carrying capacity in the specimens with varying the hole diameter. After a certain value of the hole size, tensile strength and modulus were increased.

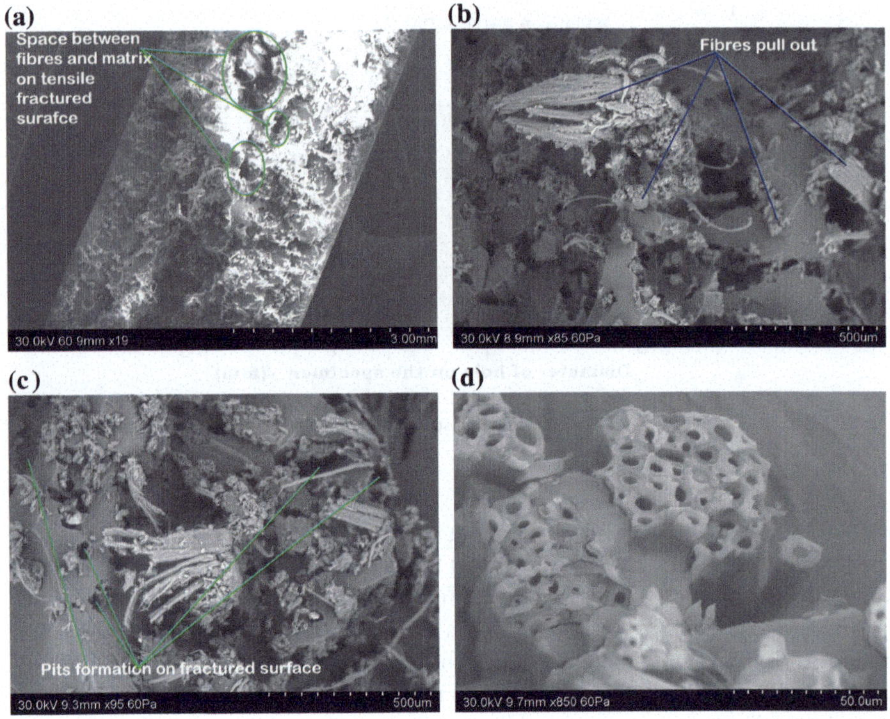

Fig. 11 SEM images tensile fractured pineapple fibre reinforced vinyl ester composites—6 mm hole size

References

1. Alam MS, Khan GMA, Razzaque SMA (2009) Estimation of main constituents of ananus comosus (pineapple) leaf fiber and its photo-oxidative degradation. J Nat Fib 6:138–150
2. Amalia N, Hidayatullah S, Nurfadilla Subaer (2017) The mechanical properties and microstructure characters of hybrid composite geopolymers-pineapple fiber leaves (PFL). IOP Conf Ser: Mat Sci Eng 180:1–8
3. Arib RMN, Sapuan SM, Hamdan MAMM, Paridah MT, Zaman HMDK (2004) A literature review of pineapple fibre reinforced polymer composites. Polym Polym Comp 12:341–348
4. Arora RK, Gupta NP, Patni PC (1985) Characteristics and processing performance of pineapple leaf fibre. Ind J Fib Text Res 10:125–126
5. Banik S, Nag D, Debnath S (2011) Utilization of pineapple leaf agro-waste for extraction of fibre and the residual biomass for vermicomposting. Ind J Fib Text Res 36:172–177
6. Buitragoa B, Jaramilloa F, Gómez M (2015) Some properties of natural fibers (sisal, pineapple, and banana) in comparison to man-made technical fibers (aramide, glass, carbon). J Nat Fib 12:357–367
7. Daud Z, Hatta MZM, Kassim ASM, Aripin AM (2014) Analysis of the chemical compositions and fiber morphology of pineapple (Annas comosus) leaves in Malaysia. J App Sci 14:1355–1358
8. Ghosh SK, Day A, Dey SK (1988) Tensile behaviour and processing of bleached yarn from pineapple leaf fibre. Ind J Fib Text Res 13:17–20

9. Jose S, Salim R, Ammayappan L (2016) An overview on production, properties, and value addition of pineapple leaf fibers (PALF). J Nat Fib 13:362–373
10. Mazalan MF, Yusof Y 00043 (2017) Natural pineapple leaf fibre extraction on Josephine and Morris. In: MATEC web of conferences ICME'17 135:1–6
11. Mukherjee PS, Satyanarayana KG (1986) Structure and properties of some vegetable fibres part 2 pineapple fibre (anannus comosus). J Mat Sci 21:51–56
12. Odusote JK, Oyewo AT (2016) Mechanical properties of pineapple leaf fiber reinforced polymer composites for application as a prosthetic socket. J Eng Tech 7:125–139
13. Siregar JP, Cionita T, Bachtiar D, Rejab MRM (2015) Tensile properties of pineapple leaf fibre reinforced unsaturated polyester composites. Appl Mech Mat 695:159–162
14. Surajarusarn B, Traiperm P, Amornsakchai T (2019) Revisiting the morphology, microstructure, and properties of cellulose fibre from pineapple leaf so as to expand its utilization. Sains Malaysiana 48:145–154
15. Yogesh M, Rao ANH (1943) Fabrication, mechanical characterization of pineapple leaf fiber (PALF) reinforced vinylester hybrid composites. In: AIP conference proceedings, pp 020062-1–020062-12
16. Yusof Y, Yahya SA, Adam A (2015) Novel technology for sustainable pineapple leaf fibers productions. Proc CIRP 26:756–760

Micromechanical Modelling and Evaluation of Pineapple Leaves Fibre (PALF) Composites Through Representative Volume Element Method

Yashwant S. Munde, Ravindra B. Ingle, Avinash S. Shinde and Siva Irulappasamy

Abstract Owing to the present scenario of industries, a massive demand for sustainable green materials made of natural fibre is provoking. Besides, the cost involved in experimental trails could be reduced. Perhaps, experimental never reflects the ideal conditions of any materials system due to their natural heterogeneity. In the present study, an attempt is made to develop a representative volume element (RVE)-based micromechanical model to evaluate mechanical properties of pineapple leaf fibre (PALF) composites numerically before being fabricated really. A 3D model of RVE is prepared using finite element analysis software ANSYS® 15 in the unit cell. To model the perfect fibre–matrix bonding, RVE modelled with both the square and hexagonal array of packaging. Results on longitudinal modulus, transverse modulus, in-plane Poisson's ratio and shear modulus of PALF composites as a function of varying fibre loading (10–50 wt% in steps of 10) have been done. Present numerical prediction (RVE) for PALF composites is compared with different analytical models like parallel and series model, Hirsah's model and Halpin–Tsai model and concluded with proper agreements.

Keywords Micromechanical modelling · PALF composites · Analytical models

Y. S. Munde (✉)
Department of Mechanical Engineering, Sinhgad College of Engineering, Savitribai Phule Pune University, Pune, Maharashtra 411041, India
e-mail: yashwant.munde@gmail.com

Y. S. Munde · R. B. Ingle · A. S. Shinde
Department of Mechanical Engineering, MKSSS'S Cummins College of Engineering for Women, Savitribai Phule Pune University, Pune, Maharashtra 411052, India

A. S. Shinde · S. Irulappasamy
Center for Composite Materials, Department of Mechanical Engineering, Kalasalingam Academy of Research and Education, Krishnankoil 626126, Tamil Nadu, India

M. Jawaid et al. (eds.), *Pineapple Leaf Fibers*, Green Energy and Technology,
https://doi.org/10.1007/978-981-15-1416-6_12

1 Introduction

In the recent past, a good deal of work has dedicated to natural fibres to replace human-made fibres. There are many reasons to select these natural fibres as reinforcement; easy renewability, availability, specific weight far less than glass and good biodegradability are promoting factors [1]. The perspective application suggested by different researchers for natural fibre reinforced composites are automotive, aerospace engineering [2], packaging, renewable energy industries [3]. The natural fibres have minimal effect on the environment because of their biodegradable properties. Among, pineapple leaf fibre (PALF) is one of the most extensively available economic and renewable resources and has limited use as ropes in the marine and agricultural industries [4].

Lopattananon et al. [5] used three different surface treatments to modify the interfacial bonding of pineapple leaf fibre (PALF). Improvement in the interfacial strength because of surface treatment leads to enhanced flexural and impact properties of modified PLAFs. Jaafar et al. [6] added inclusions of maleic anhydride polyethylene (MAPE) and studied the effect of fibre loading on mechanical properties of PALF. As the fibre percentages increased, the impact and flexural properties are found to decrease, and at 10% fibre weight, tensile strength is at peak. Moreover, the MAPE inclusions found to reduce the tensile strength because of the incompatibility between matrix and fibre. Jaafar et al. [7] investigated the mechanical properties of short PALF reinforced tapioca biopolymer with a variation of fibre length and fibre content; mechanical properties enhanced up to 30% fibre content.

Siakeng et al. [8] reviewed the mechanical and thermal properties towards the biodegradable food packaging applications, a PALF and coir fibre with PLA composite found increased as compared to neat PLA. Also, coir and PALF fibre in 1:1 ratio showed excellent properties suitable for targeted application. Asim et al. [9] also studied same compositions for comparing physical properties like density, water absorption (WA) and thickness swelling (TS) of untreated CF/PALF reinforced PLA composites and hybrid composites. Water absorption and thickness swelling depends on the fibre percentage and soaking time and found to increase with coir fibre percentage. Rihayat et al. [10] added bamboo fibre to the previous configuration and studied the mechanical properties. Different fibre–matrix volume fractions evaluated for tensile and flexural strength. The 45% volume fraction is found to be giving the highest mechanical strength. Another study on PALF composites [11] revealed that fibre loading increases the strength and flexural properties considerably. The SEM images supported these results, as less fibre pull-out and breakage observed due to proper fibre–matrix bonding.

Pratumshat et al. [12] were treated PALF with different silane solutions. The treated PALF showed more roughness than untreated. The tensile strength was found to decrease with fibre loading while tensile modulus showed improvement as compared to PLA. Glass transition temperature and melting temperature remained unchanged while percentage crystallinity increased. Asim et al. [13] studied PALF/Kenaf and their hybridisation with phenolic for characterising mechanical properties. 3P7K Hybrid showed improved tensile/flexural strength and flexural modulus. SEM images showed good interfacial bonding and addition of KF showed enhanced strength. Another study [14] on the hybridization of kenaf/PALF focused on the storage/loss modulus and damping parameter. The effect of fibre loading and fibre length investigated. Initial storage modulus is found to be improved drastically. Also, the lower percentage of PALF is sufficient to achieve improved properties. At higher fibre loading, dynamic modulus showed an increase in storage modulus with temperature. It also raised the damping peak. Up to 65 °C, fibre length played a significant role in increasing the storage modulus, but after that, it has no influence. Marginal difference in loss modulus and no difference in the tan delta is an observer with a change in fibre length.

Huda et al. [15] prepared a biodegradable composite of PALF/PLA by using film stacking method. Their study focused on the effect of surface treatment on mechanical properties. Two different surface treatments viz. silane and alkali are done, and both showed improvement in the mechanical properties. Effect of temperature is also studied by performing DMA tests, which exhibited an increase in storage modulus. In SEM images, an excellent interfacial bonding found which is important for the performance of the composite. Glóriaa et al. [16] study revealed the improvement in flexural strength of composite with the addition of the PALF fibre with 30% volume fraction. In another work [17], author has investigated tensile properties with the same fibre volume fraction. The result shows a considerable increase in tensile strength and elastic modulus with fibre loading. Mechanism of crack arrest can observe in SEM images due to long fibres. The DMA [18] of 30% PALF reinforced composite with is carried out to analyse the parameters viz. loss modulus, storage modulus and tan delta. The test temperature range was 25–195 °C with 1 Hz frequency under nitrogen flow. Continuous and aligned PALF tends to improve the viscoelastic stiffness, whereas the glass transition temperature and damping remained unaffected.

Uma Devi et al. [19] study DMA of hybrid glass/PALF in polyester with different volume ratios. The overall fibre percentage is maintained at 40% by weight. With the addition of glass fibre, the dynamic modulus increased. The intimately mixed hybrid composite properties compared with the layered composition. Intimately, mixed and glass skin layered structures found more effective than PALF skin composite. Nasir et al. [20] analysed hybrid composite of PALF and kenaf fibre with and without silane treatment for thermal stability, mechanical stability, dynamic mechanical properties and phase behaviour. Treated composite has shown improved storage modulus as compared to the untreated one. The flammability does not show any changes for treated and untreated. Motaleb et al. [21] evaluated tensile strength (TS), tensile modulus (TM), elongation at break (Eb%), bending strength (BS), bending modulus (BM) and impact strength (IS) for different fibre content. The 45 wt% composite exhibited drastic improvement in all the properties. The effect of NaOH treatment was observed to increase TS, TN, BS and BM but IS decreased as compared to untreated composite.

Nagarajan et al. [22] evaluated different characterisations like Fourier transform infrared spectra, DMA, thermogravimetric analysis, SEM and mechanical properties of alkali treated long PALF composites. The mechanical properties found to improve for treated composites, but no change in thermal properties observed. Mittal et al. [23] study the effect of fibre content and chemical treatment on the biodegradability and mechanical properties of PALF/coir epoxy composites. Specimens are prepared using hand lay-up and tests are performed as per ASTM standards. Hybrid composite with alkali treatment showed a rapid loss of strength as compared to untreated composite. The mechanical strength and biodegradability of an epoxy thermoset increased with the incorporation of PALF and coir fibres. Luz et al. [24] investigated the dynamic mechanical properties of coir and PALF epoxy composites. The storage and loss modulus is found improved as compared to neat epoxy composite. Also, the interfacial properties of PALF/epoxy are superior to the coir/epoxy composite.

Ghassemieh et al. [25] developed a finite element model to simulation of fillers and fibres reinforced polymer composites. The modulus and Poisson's ratio results are compared with available experimental literature, which shows good accuracy. Ionita et al. [26] executed a computational model to investigate the effect of inhomogeneity of randomly reinforced polymer composites on mechanical properties and compared with experimental results. Kari et al. [27] used a random sequential adsorption algorithm to model RVE of particle reinforced composites. The effect of RVE size, spherical particles size and volume fraction of filler on mechanical properties is evaluated. Devireddy et al. [28] developed a micromechanical model of short banana/jute fibre reinforced epoxy hybrid composites; authors assessed the influence of total fibre loading and the relative weight ratio of fibres on thermal conductivity. The model result shows good agreement with the error of 6–11% in comparison with experimental values.

Directional-based elastic properties required for polymer matrix composites and determination of these properties by an experimental method is a deadly and costly process. Also, the elastic properties varied with a change in the type of reinforcement, type of matrix, the volume fraction of fibre. Homogenisation techniques/micromechanics models can use as a substitute or accompaniment to experimentation to predict the elastic properties PMC with input as elastic properties of the reinforcement and matrix. Empirical and semi-empirical-based micromechanics models require adjusting factors so numerical methods as finite element method (FEM) can be an excellent alternative to evaluate elastic properties.

In the past, a great deal of work is devoted to experimental studies to evaluate the mechanical properties of bio-composites by various researchers but limited work to make use of RVE-based micromechanical modelling. An attempt is made to model PLAF composites based on RVE-based micromechanical modelling to evaluate static mechanical elastic properties.

2 Properties and Assumptions

The composite material considered for present work is unidirectional PALF reinforced epoxy composites. Physical properties of the PALF fibre are given in Table 1. The input for this analysis, elastic properties of PALF fibre and epoxy matrix are as shown in Table 2. Figure 1 shows the schematic diagram of the packaging of fibre arrangement in a matrix. The fibre volume fraction of PALF/epoxy composite is varied from 0.1 to 0.5 to evaluate elastic properties.

2.1 Numerical Method

The FEM is an extensively known numerical method which helps to evaluate the engineering constants of material as such as Young's modulus, Poisson's ratios and

Table 1 Physical properties of the PALF fibre

Physical properties	PALF fibre [8]
Cellulose content (%)	70–82
Hemicellulose content (%)	18.8
Lignin content (%)	5–12
Microfibrillar angle (°)	8
Diameter (um)	20–80

Table 2 Mechanical properties of the PALF fibre and epoxy matrix

Mechanical properties	PALF fibre [29]	Epoxy matrix
Density (g/cc)	1.526	1.15
Tensile strength (MPa)	413	–
Tensile modulus (GPa)	34.5	3.76
Poisson's ratio	0.2	0.39
Elongation at break (%)	1.6	3.5

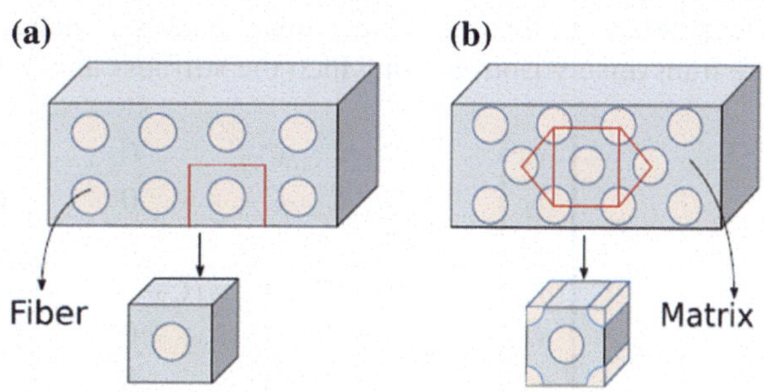

Fig. 1 Fibre packaging arrangement as a square array (**a**) and hexagonal array (**b**)

Fig. 2 Modelled unit cell of RVE with square array (**a**), and hexagonal array (**b**)

shear modulus. In FEM, the micromechanical modelling method based on the representative volume element (RVE) implemented to investigate elastic properties. It assumed that constituents are homogeneous and isotropic material having perfect binding at the interface. The 3D model of RVE prepared using finite element analysis software ANSYS 18 by considering square array and hexagonal array packing of fibre and matrix in the unit cell, as shown in Fig. 2. Further, this geometrical model of RVE discretised into finite element mesh, and appropriate boundary conditions will apply to determine stress and strain field.

2.2 *Constitutive Equations*

For an anisotropic material, Hook's law gives the constitutive equation as a stress–strain relationship as shown in Eq. (1)

$$\{\sigma_{ij}\} = [C_{ij}]\{\varepsilon_{ij}\} \tag{1}$$

where σ_{ij} and ε_{ij} are stress and strain components, respectively, (normal and shear type) and C_{ij}—stiffness matrix with a total 36 elastic constants of which 21 are independent. In this micromechanical analysis, material characteristics considered are transversely isotropic in which the stiffness tensor C_{ij} is represented by Eq. (2).

$$\begin{Bmatrix} \sigma_1 \\ \sigma_2 \\ \sigma_3 \\ \sigma_4 \\ \sigma_5 \\ \sigma_6 \end{Bmatrix} = \begin{bmatrix} C_{11} & C_{12} & C_{12} & 0 & 0 & 0 \\ C_{12} & C_{22} & C_{23} & 0 & 0 & 0 \\ C_{12} & C_{23} & C_{22} & 0 & 0 & 0 \\ 0 & 0 & 0 & \frac{1}{2}(C_{22} - C_{23}) & 0 & 0 \\ 0 & 0 & 0 & 0 & C_{66} & 0 \\ 0 & 0 & 0 & 0 & 0 & C_{66} \end{bmatrix} \begin{Bmatrix} \varepsilon_1 \\ \varepsilon_2 \\ \varepsilon_3 \\ \varepsilon_4 \\ \varepsilon_5 \\ \varepsilon_6 \end{Bmatrix} \tag{2}$$

These values of stiffness tensor used to calculate elastic property as longitudinal modulus (E_1), transverse modulus (E_2), Poisson's ratio (γ_{12}) and shear modulus (G_{12}) are given in Eqs. (3)–(6)

$$E_1 = C_{11} - \frac{2C_{12}^2}{(C_{22} + C_{23})} \tag{3}$$

$$E_2 = \frac{[2C_{11}(C_{22} + C_{23}) - 2C_{12}^2](C_{22} - C_{23})}{\left(C_{11}C_{22} - C_{12}^2\right)} \tag{4}$$

$$\gamma_{12} = \frac{C_{12}}{(C_{22} + C_{23})} \tag{5}$$

$$G_{12} = \frac{1}{2}(C_{22} - C_{23}) \tag{6}$$

2.3 Preparation of RVE

For square RVE, dimensions of the unit cell as length (a_1), width (a_2) and height (a_3) are equal. Based on these values and different volume fraction of fibre ($_f$), the diameter of fibre (d_f) is calculated using Eq. (7) for the preparation of RVE. The maximum fibre volume fraction can attain with square RVE is 78%, whereas, for hexagonal RVE, it can be 90.6%. It shows that the hexagonal array of packing offers more compactness to composites. The d_f for hexagonal array calculated by using Eq. (8)

$$V_f = \frac{a_1\left(\frac{\pi}{4}\right)d_f^2}{4a_1a_2a_3} \tag{7}$$

$$V_f = \frac{2a_1\left(\frac{\pi}{4}\right)d_f^2}{a_1a_2a_3} \tag{8}$$

where $a_3 = a_2 \tan(60°)$ and $a_2 = 4\,a_1$.

2.4 Finite Element Modelling with Boundary Conditions

The modelling of RVE handled with ANSYS APDL program. Here, we considered an *XYZ* as an orthogonal coordinate with *z* coordinate which is parallel to fibre directions, and *x* and *y* are the coordinates perpendicular to fibre directions. Geometrical dimensions for square RVE are $a_1 = a_2 = a_3 = 1$ mm, whereas, for hexagonal RVE, $a_1 = 1$ mm and a_2, a_3 calculated using relation given in Eq. (8). In RVE model, the radius of fibre used based on the corresponding variation of volume fraction of fibre from 0.1 to 0.5. The element type of SOLID 186, which has 20 nodes with three degrees of freedom per node, used for determining elastic properties of the PALF composites. Meshed model for PALF30 composites with square and hexagonal is shown in Fig. 3.

RVE model of composites is a pattern of a periodic array of the unit cell; the intermittent type of boundary conditions is applied. All the RVE has similar displacements and perfect bonding between them without overlap. Table 3 gives detail of boundary conditions used for three different load cases with deformation for the constituents is assumed the same. Separately for each surface, the displacement along X, Y and Z represented by U, V and W, respectively, on the distinct surface. On the application of displacement boundary conditions, the average stresses and average strain are calculated using equation. The average stress and average strain calculated using Eqs. (9) and (10), respectively, and which are used to calculate coefficients of stiffness matrix (C_{ij})

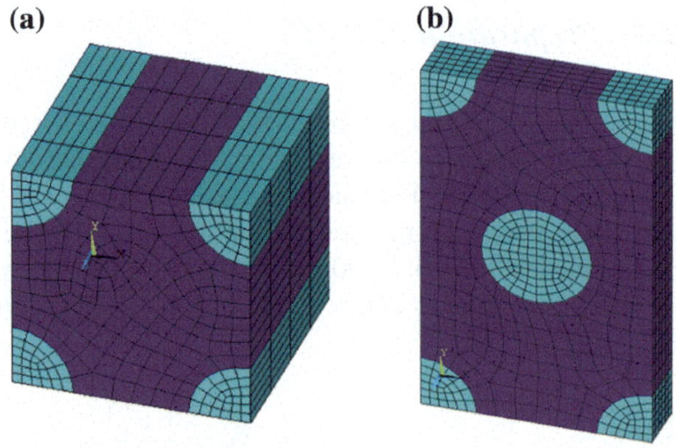

Fig. 3 Meshed model of PALF30 composites with square array (**a**) and hexagonal array (**b**)

Table 3 Boundary condition for RVE models along the X, Y and Z directions

Load case	Displacement direction	Surface					
		+X	−X	+Y	−Y	+Z	−Z
Load XX	U	1	0				
	V			0	0		
	W					0	0
Load YY	U	0	0				
	V			1	0		
	W					0	0
Load ZZ	U	0	0				
	V			0	0		
	W					1	0

$$\bar{\sigma}_{ij} = \frac{1}{V}\int_V \sigma_{ij}\mathrm{d}V \tag{9}$$

$$\bar{\varepsilon}_{ij} = \frac{1}{V}\int_V \varepsilon_{ij}\mathrm{d}V \tag{10}$$

Table 3 shows boundary condition for RVE models along the *X*, *Y*, and *Z* directions

2.5 Analytical Modelling

To verify the results of elastic properties obtained by RVE-based finite element method, well-established analytical model as a parallel and series model, Hirsch's model and Halpin–Tsai model used in the present work.

2.5.1 Series and Parallel Model

Tensile modulus of composite by series and parallel model is given by Eqs. (11) and (12), respectively,

$$E_{\mathrm{C}} = E_{\mathrm{f}}V_{\mathrm{f}} + E_{\mathrm{m}}V_{\mathrm{m}} \tag{11}$$

$$E_{\mathrm{C}} = \frac{E_{\mathrm{f}}E_{\mathrm{m}}}{E_{\mathrm{f}}V_{\mathrm{f}} + E_{\mathrm{m}}V_{\mathrm{m}}} \tag{12}$$

where E_{c}, E_{f} and E_{m} are the tensile moduli of composite, matrix and fibre, respectively. For a parallel model, the assumption would be "uniform strain throughout the lamina" and for series model "uniform stress throughout the lamina".

2.5.2 Hirsch's Model

It is a combination of series and parallel model. Equation (13) used for calculation of tensile modulus is as follows

$$E_{\mathrm{C}} = x(E_{\mathrm{f}}V_{\mathrm{f}} + E_{\mathrm{m}}V_{\mathrm{m}}) + (1 - x)\frac{E_{\mathrm{f}}E_{\mathrm{m}}}{E_{\mathrm{f}}V_{\mathrm{f}} + E_{\mathrm{m}}V_{\mathrm{m}}} \tag{13}$$

where the value of x in the above equation is 0.1 for randomly oriented composites. The x is a parameter which determines the stress transfer between fibre and matrix.

2.5.3 Halpin–Tsai

According to Halpin–Tsai, tensile modulus composite is given by

$$E_C = E_m\left(\frac{1 + A\eta V_f}{1 - \eta V_f}\right) \tag{14}$$

$$\eta = E_m\left(\frac{E_f/E_m + 1}{E_f/E_m + A}\right) \tag{15}$$

where η is given by Eq. (15) used for the relative module of fibre and matrix, and A is the measure of fibre geometry, fibre distribution and fibre loading conditions.

3 Results and Discussion

3.1 Longitudinal Modulus

Stress–strain counterplot for PALF composites is shown in Figs. 4 and 5, respectively. This response of RVE model obtained for this is by applying displacement parallel to the fibre, i.e. along the Z-axis. The longitudinal modulus of PALF composites is calculated by varying weight fraction of fibre, as shown in Fig. 6. It observed that the longitudinal modulus rises with an increase in fibre loading. Stiffer fibre reinforcement with effective stress transfer attributes the property improvements. Longitudinal modulus predicted by hexagonal RVE model is less that of square RVE model. It observed that up to 20% of fibre weight fraction, the square and the hexagonal RVE models predicted the longitudinal modulus similarly. At higher values of weight fraction, no such similar predictions noted between square and hexagonal models. Compact packaging of fibre in hexagonal array compares to the square array caused such variations at higher fibre weight fractions. The longitudinal modulus values are obtained by the numerical method compared with the results of analytical models, and it shows the best agreement with parallel and Halpin–Tsai models compare to other models.

3.2 Transverse Modulus

On the application of displacements along X- and Y-axis means perpendicular to fibre direction received the response of RVE model. This response as a ratio of transverse stress to transverse strain gives transverse modulus. Figure 7 shows the effect of fibre loading on transverse modulus. Transverse modulus rises with an increase in fibre loading, but the values are less than that of longitudinal modulus; low load

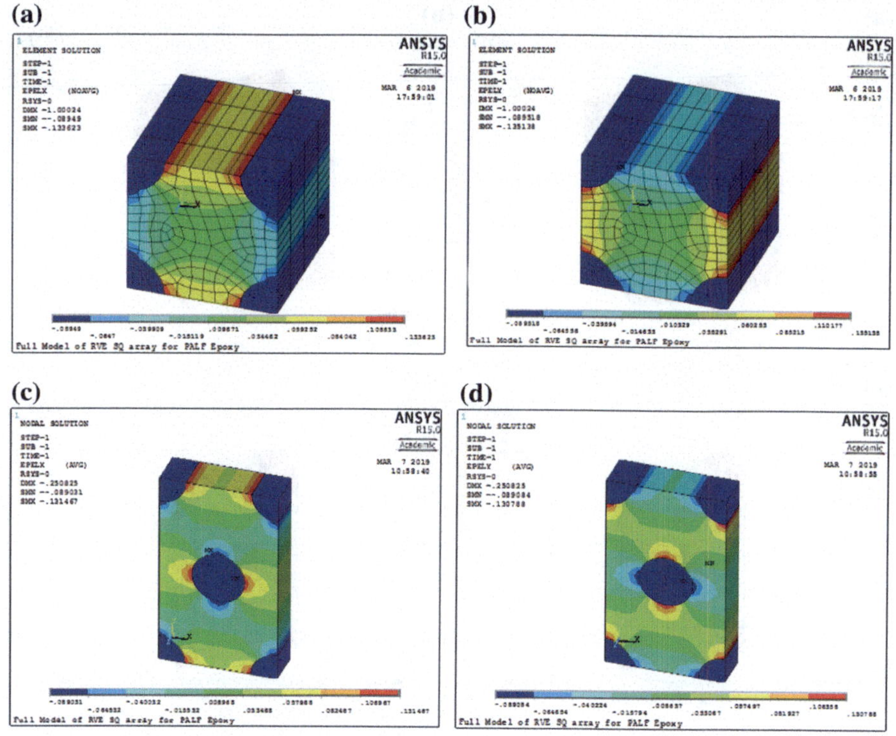

Fig. 4 Strain counterplot of PALF30 composites in **a** square RVE (strain X); **b** square RVE (strain Y); **c** hexagonal RVE (strain X); **d** hexagonal RVE (strain Y)

capability of fibre during the load orientation is perpendicular to the fibre direction caused such effect. Transverse modulus predicted by hexagonal RVE and square RVE shown significant deviation above 30% weight fraction of fibre. Transverse modulus obtained numerical method shown close agreement to series model compared to other analytical models.

3.3 In-Plane Poisson's Ratio

In-plane Poisson's ratio of PALF composite for different fibre volume fractions is shown in Fig. 8. It is apparent from the figure that the Poisson's ratio declines with a rise in PALF fibre content. The high resistance in transverse deformation of fibre compared to pure epoxy matrix attributes the finding. The results of the numerical method have shown the best agreement with the parallel model and Halpin–Tsai model for Poisson's ratio.

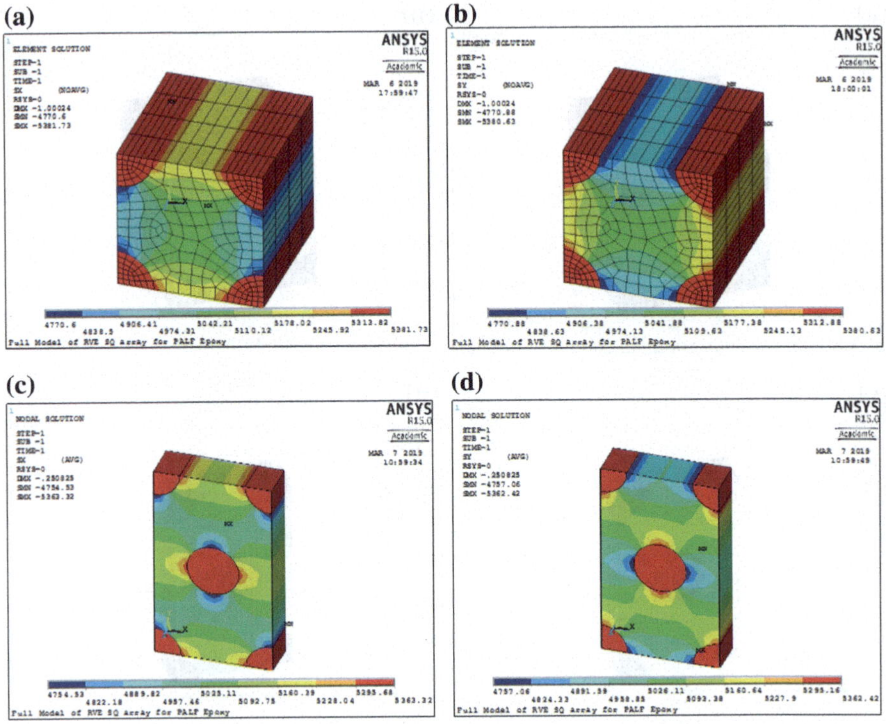

Fig. 5 Stress counterplot of PALF30 composites in **a** square RVE (stress X); **b** square RVE (stress Y); **c** hexagonal RVE (stress X); **d** hexagonal RVE (stress Y)

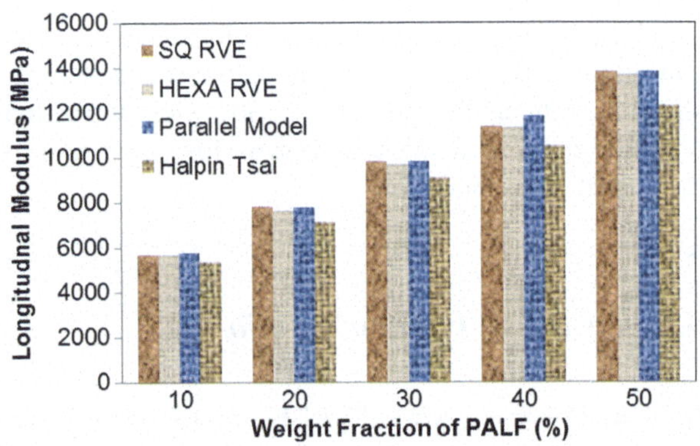

Fig. 6 Variation of longitudinal modulus with weight fraction of fibre

3.4 Shear Modulus

For the loading along the longitudinal direction, the ratio of shear stress to shear strain is evaluated as in-plane shear modulus of PALF composite. Figure 9 shows the

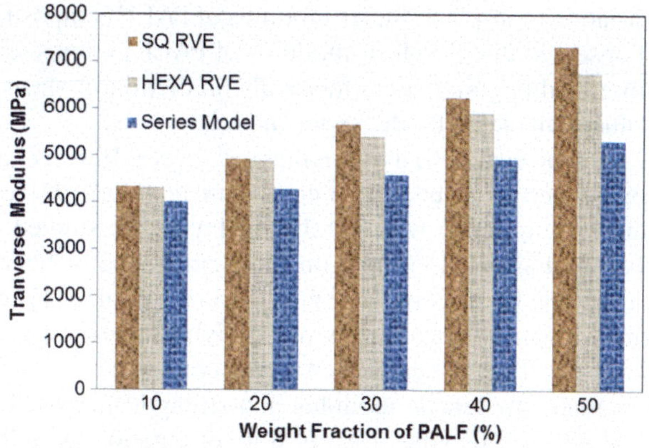

Fig. 7 Variation of transverse modulus with weight fraction of fibre

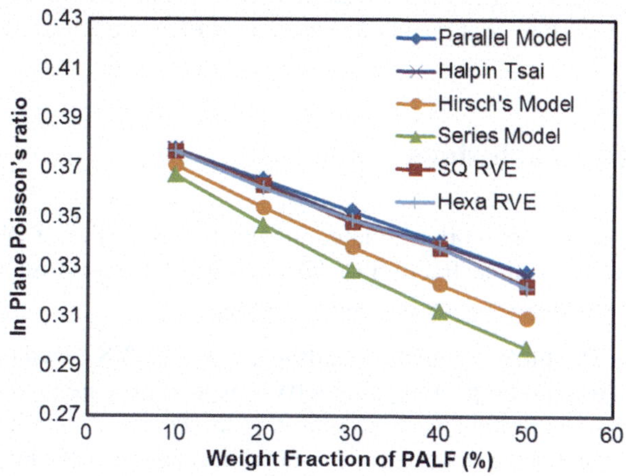

Fig. 8 Variation of in-plane Poisson's ratio with weight fraction of fibre

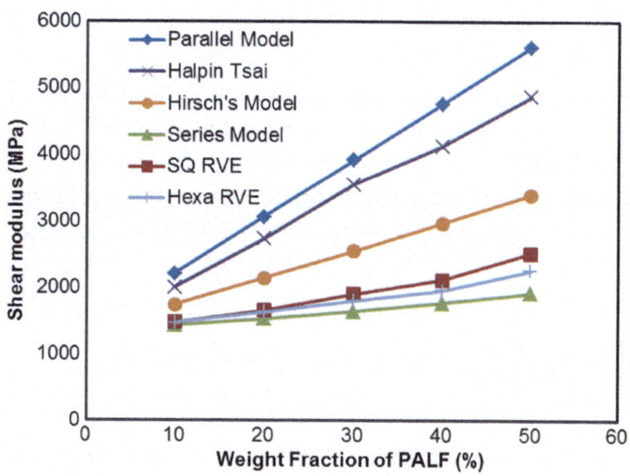

Fig. 9 Variation of shear modulus with weight fraction of fibre

variation of in-plane shears modulus of PALF composite with fibre volume fractions. It observed that the shear modulus of PALF composite increases with an increase in fibre loading. Numerical method's prediction of shear modulus firmly agreed with values calculated by the series model.

Ahmad et al. [30] used Siemens PLM NX 10.0, a FEM tool to predict the mechanical properties of jute/epoxy composite materials. The effect of orientation, fibre type and loading along with number of layers are studied for stiffness of the laminate. Reported simulation trend on the significance of fiber volume on modulus similar to the results of current article. Nirbhay et al. [31] used FEA tool ABAQUS® to study the mechanical behaviour of hybrid jute–coir/epoxy composite plate and box structure. They observed that as the content of coir fibres grew in hybrid composite upsurges, the tensile modulus and strength increased up to 50% volume. Ramesh et al. [32] studied the effect of fibre orientation by finite element analysis technique to predict mechanical strength properties of flax/epoxy and glass/epoxy composites. They show that results forecasted by FEA are adjacent to the experimental values.

4 Conclusions

The present work evaluated the elastic properties of PALF composites with the different volume fraction of fibre using RVE-based micromechanical approach. The following conclusions can be drawn.

- The finite element modelling using ANSYS based on 3D RVE with a square and hexagonal packing geometry was well executed to calculate elastic properties.
- Longitudinal modulus and in-plane Poisson's ratio predicted by RVE-based finite element analysis shown a good agreement between the parallel and Halpin–Tsai models.
- RVE-based results of transverse modulus and shear modulus shown best agreement with the series model.
- In RVE-based micromechanical modelling, the elastic properties of PALF composites are affected mainly by weight fraction of the fibres compared to the packaging of RVE.
- FEM-based micromechanical modelling technique can be adapted to predict vibration damping properties of PALF composites.

References

1. Pandey JK, Nagarajan V, Mohanty AK, Misra M (2015) Commercial potential and competitiveness of natural fibre composites. Fourteenth, Elsevier Ltd
2. Balakrishnan P, John MJ, Pothen L, (2016) Natural fibre and polymer matrix composites and their applications in aerospace engineering. Elsevier Ltd

3. Sanyang ML, Sapuan SM, Jawaid M (2016) Recent developments in sugar palm (Arenga pinnata) based biocomposites and their potential industrial applications: a review. Renew Sustain Energy Rev 54:533–549. https://doi.org/10.1016/j.rser.2015.10.037
4. Sahu P, Gupta MK (2017) Sisal (Agave sisalana) fibre and its polymer-based composites: a review on current developments. J Reinf Plast Compos 36:1759–1780. https://doi.org/10.1177/0731684417725584
5. Lopattananon N, Payae Y, Seadan M (2008) Influence of fiber modification on interfacial adhesion and mechanical properties of pineapple leaf fiber-epoxy composites. J Appl Polym Sci 110:433–443. https://doi.org/10.1002/app
6. Jaafar J, Rejab MRM, Jie CC (2019) The effect of maleic anhydride polyethylene on mechanical properties of pineapple leaf fibre reinforced polylactic acid composites. Int J Precis Eng Manuf Technol. https://doi.org/10.1007/s40684-019-00018-3
7. Jaafar J, Siregar JP, Piah MBM (2018) Influence of selected treatment on tensile properties of short pineapple leaf fiber reinforced tapioca resin biopolymer composites. J Polym Environ 26:4271–4281. https://doi.org/10.1007/s10924-018-1296-2
8. Siakeng R, Jawaid M, Ariffin H, Sapuan SM (2018) Mechanical, dynamic, and thermomechanical properties of coir/pineapple leaf fiber reinforced polylactic acid hybrid biocomposites. Polym Compos 1–12. https://doi.org/10.1002/pc.24978
9. Asim M, Paridah MT, Jawaid M, (2018) Physical and flammability properties of kenaf and pineapple leaf fibre hybrid composites. IOP Conf Ser Mater Sci Eng 368. https://doi.org/10.1088/1757-899X/368/1/012018
10. Rihayat T, Agusnar H, Wirjosentono B (2018) Mechanical properties evaluation of single and hybrid composites polyester reinforced bamboo, PALF and coir fiber. IOP Conf Ser Mater Sci Eng 334:12081. https://doi.org/10.1088/1757-899x/334/1/012081
11. Rajini N, Senthilkumar K, Siengchin S (2018) Evaluation of mechanical and free vibration properties of the pineapple leaf fibre reinforced polyester composites. Constr Build Mater 195:423–431. https://doi.org/10.1016/j.conbuildmat.2018.11.081
12. Pratumshat S, Soison P, Ross S (2015) Mechanical and thermal properties of silane treated pineapple leaf fiber reinforced polylactic acid composites. Key Eng Mater 659:446–452. https://doi.org/10.4028/www.scientific.net/kem.659.446
13. Asim M, Jawaid M, Nasir M, Saba N (2017) Effect of fiber loadings and treatment on dynamic mechanical, thermal and flammability properties of pineapple leaf fiber and kenaf phenolic composites. J Renew Mater 6:383–393. https://doi.org/10.7569/jrm.2017.634162
14. Ishak MR, Asim M, Abdan K (2017) Effect of hybridization on the mechanical properties of pineapple leaf fiber/kenaf phenolic hybrid composites. J Renew Mater 6:38–46. https://doi.org/10.7569/jrm.2017.634148
15. Huda MS, Drzal LT, Mohanty AK, Misra M (2008) Effect of chemical modifications of the pineapple leaf fiber surfaces on the interfacial and mechanical properties of laminated biocomposites. Compos Interfaces 15:169–191. https://doi.org/10.1163/156855408783810920
16. Glória GO, Teles MCA, Neves ACC (2017) Bending test in epoxy composites reinforced with continuous and aligned PALF fibers. J Mater Res Technol 6:411–416. https://doi.org/10.1016/j.jmrt.2017.09.003
17. Glória GO, Teles MCA, Lopes FPD (2017) Tensile strength of polyester composites reinforced with PALF. J Mater Res Technol 6:401–405. https://doi.org/10.1016/j.jmrt.2017.08.006
18. Margem FM, Monteiro SN, Margem JI (2018) Dynamic-mechanical analysis of epoxy composites reinforced with giant bamboo fiber 492–497
19. Uma Devi L, Bhagawan SS, Thomas S (2010) Dynamic mechanical analysis of pineapple leaf/glass hybrid fiber reinforced polyester composites. Polym Compos 31:956–965. https://doi.org/10.1002/pc.20880
20. Nasir M, Saba N, Paridah MT (2018) Thermal, physical properties and flammability of silane treated kenaf/pineapple leaf fibres phenolic hybrid composites. Compos Struct 202:1330–1338. https://doi.org/10.1016/j.compstruct.2018.06.068
21. Motaleb KZMA, Islam MS, Hoque MB (2018) Improvement of physicomechanical properties of pineapple leaf fiber reinforced composite. Int J Biomater. https://doi.org/10.1155/2018/7384360

22. Nagarajan TT, Babu AS, Palanivelu K, Nayak SK (2016) Mechanical and thermal properties of PALF reinforced epoxy composites. In: Macromolecular Symposia, pp 57–63
23. Mittal M, Chaudhary R (2018) Biodegradability and mechanical properties of pineapple leaf/coir fiber reinforced hybrid epoxy composites. Mater Res Express 6:45301. https://doi.org/10.1088/2053-1591/aaf8d6
24. da Luz FS, Monteiro SN, Tommasini FJ (2018) Evaluation of dynamic mechanical properties of PALF and coir fiber reinforcing epoxy composites. Mater Res 21:1–5. https://doi.org/10.1590/1980-5373-mr-2017-1108
25. Ghassemieh E, Nassehi V, Le L, Kingdom U (2000) Stiffness analysis of polymeric composites using the finite element method. Polymer (Guildf) 20:42–57
26. Ionita A, Weitsman YJ (2006) Randomly reinforced composites: properties, failure and aspects of material design. Probabilistic Eng Mech 21:64–72. https://doi.org/10.1016/j.probengmech.2005.07.006
27. Kari S, Berger H, Rodriguez-Ramos R, Gabbert U (2007) Computational evaluation of effective material properties of composites reinforced by randomly distributed spherical particles. Compos Struct 77:223–231. https://doi.org/10.1016/j.compstruct.2005.07.003
28. Devireddy SBR, Biswas S (2016) Thermo-physical properties of short banana-jute fiber-reinforced epoxy-based hybrid composites. Proc Inst Mech Eng Part L J Mater Des Appl 0:1–13. https://doi.org/10.1177/1464420716656883
29. Zin MH, Abdan K, Norizan MN (2018) The effect of different fiber loading on flexural and thermal properties of banana/pineapple leaf (PALF)/glass hybrid composite. Elsevier
30. Ahmad F, Bajpai PK (2018) Evaluation of stiffness in a cellulose fiber reinforced epoxy laminates for structural applications: experimental and finite element analysis. Def Technol 14:278–286. https://doi.org/10.1016/j.dt.2018.05.006
31. Nirbhay M, Misra RK, Dixit A (2015) Finite-element analysis of jute- and coir-fiber-reinforced hybrid composite multipanel plates. Mech Compos Mater 51:505–520. https://doi.org/10.1007/s11029-015-9521-8
32. Ramesh M, Sudharsan P (2018) Experimental investigation of mechanical and morphological properties of flax-glass fiber reinforced hybrid composite using finite element analysis. Silicon 10:747–757. https://doi.org/10.1007/s12633-016-9526-5

Fabrication of Pineapple Leaf Fibers Reinforced Composites

I. Cesarino, M. B. Carnietto, G. R. F. Bronzato and A. L. Leao

Abstract Consumers are more aware of environmental impacts and climatic problems, which leads to a greater demand for products with technological innovations. Research has the aim to replace and reduce raw materials from fossil sources to renewable sources, such as the natural fibers. Natural fiber composites result from the blending of two materials: one is the plastic and the other a fiber, from agricultural waste in most of the cases. Compared to polymers from fossil sources, this new material has three main advantages: they have an environmental approved; low cost and its physical and mechanical properties are superior. The cultivation of this fruit is large in many tropical countries. After harvesting, the fruit and shoots are removed, and the rest needs to be cut and removed from the soil. This material, most leaves, becomes waste and goes to disposal. However, the use of pineapple leaf fibers as a raw material for natural fiber composites production helps to reduce the pollution caused by these residues and can increase the income of pineapple producers making a channel to new business. To have success in producing NFC, it is necessary to understand process techniques; to the adhesion between fiber and the polymer; the ratio of polymer and natural fiber; and the market (automotive, construction, etc.). But, after reading this chapter, it will be possible to conclude that there is a huge opportunity to improve the natural fibers market in front of the other reinforcements because of their properties.

Keywords Pineapple fibers · Natural fiber composites · Polymers · PALF properties

1 Introduction

The environmental problems are pushing the development of products more sustainable [1]. The mix of natural fibers and polymers as natural fibers composites (NFC) is an example. The natural fibers composites materials are in deck, facades,

I. Cesarino · M. B. Carnietto · G. R. F. Bronzato · A. L. Leao (✉)
School of Agriculture, Sao Paulo State University (UNESP), Botucatu, Brazil
e-mail: alcides.leao@unesp.br

M. Jawaid et al. (eds.), *Pineapple Leaf Fibers*, Green Energy and Technology,
https://doi.org/10.1007/978-981-15-1416-6_13

pergolas, fences, automotive sectors, etc. The usage of natural fibers as reinforcement in polymeric material has increased, due to their advantage as coming from renewable source, low cost, biodegradable, recyclable, and non-toxic. Natural fibers have competitive mechanical properties—when compared to fibers such as glass, carbon, and aramid—such as stiffness, impact strength, flexibility, and elasticity [2]. According to Asim [3], the main properties of natural fibers are biodegradability and non-carcinogenic characteristic, besides that the low cost.

Composites are two or more different materials to form and chemical composition mixed resulting in a third product with superior quality to the materials individually. The development of polymeric composites involving the use of lignocellulosic residues as reinforcement has been increasing in response to environmental conservation, which has become more frequent in the polymer industry. The addition of natural fibers into a plastic matrix aims to improve the mechanical properties of these materials, in particular reducing the costs of the polymer composition, the carbon footprint, and the generation of effluents and pollutant residues. Among the advantages obtained with these composites are resources from renewable energy source; low cost; low density; non-toxic; greater ease of recycling, composting of material at the end of its useful life, reducing environmental impact; avoid the increase of the greenhouse effect; reuse of agroindustrial waste, thereby reducing its quantity in the environment.

2 Natural Fibers—Context and Applications

Natural fibers can be found in nature and can be used "in natura" or with some beneficiation. Always, the fibers vegetable were used, mainly in the textile sector; however, in the actual days, it has been gaining space and importance in other sectors, like the construction industry and automotive. For the industrial, the material more interesting is the unused or those leftovers in the sector agroforestry, the residues. Cellulose, hemicelluloses, and lignin are the main constituents of plants, together represent more than 50% of its macrocomponents; because of this, the natural fibers can be called from lignocelluloses fibers. For the industrial process, the use of natural fibers is many benefits like [4–6]:

- A renewable resource without limits disponibility;
- Biodegradables;
- For agroforestry residues, the prices are less than synthetic material;
- Represents a new font income for a small farmer;
- Least abrasives than artificial fibers.

2.1 Natural Fiber-Based Composites

In recent studies, the use of natural fibers as reinforcement has shown that replacing conventional wood with composites is a viable alternative to valorize the waste, with numerous advantages such as increased moisture resistance and deterioration; resistance to pests and insects; better dimensional stability; resistance to warping and cracking; less need for maintenance, eliminating the use of surface protection such as paints and varnishes [7].

According to Nova Institute [8], the volume of NFC production was 92,000 tons. The segment of a major market, with 97.8% in the Union European, is the automotive sector (mainly, the products in automotive are passenger cars, using fibers from flax, kenaf, hemp, jute, coir, sisal, and others). At that time, the forecast projected that the natural fiber composite materials market would grow to 531.3 million dollars in 2016. According to Grand View Research [9], natural fiber composites are used in the automotive industry to produce parts with lightweight and better mechanical property to achieve an improvement in fuel efficiency and reduce CO_2 emissions. Chandramohan and Marimuthu [10] said composites from natural fibers are from 30 to 40% lighter than aluminum structures designed to the same functional objectives, for example. It contributes to weight reduction by 30.0% and cost reduction by 20.0% during manufacturing of a vehicle. The production is based on compression molding and injection molding, where 55% using the thermoplastics as matrix. The forecast of Nova for 2020 of NFC in Europe is 120,000 tons, but in a positive scenario, it can reach 350,000 tons. Some of the key players identified in the global natural fiber composites are: Greengran B. V FlexForm Technologies LLC., FiberGran GmbH & Co. KG Tecnaro GmbH, Kafus Bio-Composites Inc. Stemergy, Procotex Corporation SA, and Bast Fiber LLC [11].

The composites made by the mixture of natural fibers with different plastics, such as polypropylene (PP), polyvinylchloride (PVC), and polyethylene (PE), are composites with a huge application, opening the opportunity to develop new business using wastes and local crops, besides of reducing the demand for tropical woods that goes to houses, furniture, and also reduce the use of plastics. Nowadays, more resins are being under study in natural fibers composites production specially using engineer plastics. That is the case of ABS, which is largely blended at 50% ratios with several natural fibers [12]. The prices offered to end users may vary depending on material used. The products with PVC are more expensive because of their attributes and advantage over other polymers used in the composites. The PVC does not depend as much as other polymers of crude oil or natural gas, that means PVC has a lower carbon footprint than PP, PE, and PEAD. The WPC produced from PVC keeps the color for a long time in comparison with others, being UV resistant and fire self-extinguish.

Some automotive companies like Mercedes Benz use natural fibers in luxury car seats and backrests. The natural fibers guarantee the softness of the piece and improved passenger health and comfort. The parts with natural fibers also guarantee greater safety in relation to the synthetic ones; because in case of fire release toxic gases and in the event of an accident, material rupture does not produce tips [13]. The disadvantages of natural fibers for composites are highly variable quality; because it depends on the agricultural conditions, the moisture absorption for the fiber, which influences the external use and their maximum temperature when processed, is restricted. Regard to the disadvantages of natural fibers, [14] the main one is the variability in mechanical properties, such as variation in plant age, geographical area, and climatic harvesting methods. However, Zah et al. [15] declare that the low cost of the fibers and their application is still very interesting economically.

3 Properties of PALF

The properties of PALF put the fiber as one of the best to produce composites to construction materials, automotive parts, and many furniture. Studies have been done and proved for many researchers that PALF is an excellent option of reinforcement to composite because of their mechanic resistance [16]. Natural fibers are composed of cellulose fibrils held together by a matrix formed by lignin and hemicelluloses, which serve as a natural barrier against microbial degradation and have excellent mechanical properties. The natural fibers are excellent reinforcements for polymer matrix composites due to their excellent mechanical strength characteristics [17]. The best fiber to reinforce the composites is the fibers obtained from the leaves because they are also longer than the stem. Chollakup et al. [18] observed that the natural fibers composites containing long fiber PALF were stronger than the short ones as determined by greater tensile strength. According to Leao [19], the pineapple nanofibers are already being compared to polyamide (PA), aramid, and carbon fibers. Compared to glass fibers, nanofibers are up to 30 times lighter and 3–4 times stronger in polypropylene (PP) matrices. The use of 0.2% by weight of these fibers can increase the mechanical properties of materials by 50%.

Alexandre [20], studies of pineapple fiber composites, found in his work that the mechanical properties of tensile strength, tensile strength flexion, strength, and modulus of elasticity presented better results for composites with 30% fiber volume and 55 mm fiber length. Some of the previous works on PALF as reinforcement to composites utilized fine bundles with diameters less than 100 mm [21]. The superior properties of pineapple fiber are associated with the high cellulose content and low microfibrillary angle [22]. Another advantage of pineapple leaf fibers is a weight reduction compared to glass fiber reinforced materials, with the possibility of improving or maintaining mechanical properties [23]. PALF has every fiber that has cellulose, lignin, and hemicelluloses. Its cellulose percentage is about 70 and 82%, and this gives to PALF the good mechanical properties [13]. Cellulose is a biopolymer formed by repeating cellobiose units and classified as a linear polymer, joined by β 1,4 glycosides bonds and hydrogen bonds. These components contain hydroxyl groups that establish intra and intermolecular hydrogen bond interactions. These hydrogen bonds allow cellulose to have a water-insoluble crystalline structure and most organic solvents [6]. The efficiency of natural fiber reinforcement is related

Table 1 Chemical composition

Chemical composition (%)		
	Leao et al. [24]	Mohanty et al. [41]
Cellulose	73.4	70–82
Hemicellulose	7.1	–
Lignin	10.5	5–12.7

to your cellulose and its crystallinity; in Table 1, there is a research from Leao et al. [24] according to PALF chemical composition, which proves the high quantity of cellulose.

The interfacial contact between natural fibers and the polymeric matrix to be effective surface modifications may be required. Studies indicate that the surface modification of the fiber decreases its hydrophilic character, increases adhesion with the polymer matrix, and reduces the polarity difference between fiber and matrix. According to Chollakup et al. [18], the incorporation of natural cellulosic fibers into composites can cause poor dispersion in the matrix because of the strong hydrogen bonds that keep the fibers bonded.

The cellulose needs to be free to be bound to the polymer, therefore a fiber treatment that separates hemicellulose and lignin from the main biopolymer is required. The method used for pretreatment depends on each biomass and the proportions of the lignin–cellulose–hemicellulose complex, so there are several possible methods that can be classified into: physical, chemical, and biological. Chemical treatment is the most effective for this situation.

Chemical treatment of fibers to increase adhesion between the hydrophilic surface of the fibers and the hydrophobic surface of the polymer is a great solution to this problem, for example, the treatment with sodium hydroxide (NaOH). Mishra et al. [22] observed an important increase after the treatment of the PALF in the strength of the composites. The chemical treatment also has changed the natural fiber composite, resulting in a reduction of water absorption because there was a better interfacial bonding. Natural fibers composites may have a higher susceptible to water absorption that will cause a negative effect damaging the final product [3]. When the natural fiber composite passes through a pretreatment of the fibers, occurs an improvement of mechanical properties, moisture resistance, and biodegradation. Natural fiber composites made from fiber treated with NaOH showed an improvement of 3% increase in resistance tensile strength, 24% in tensile modulus, 30% bending strength, and 12% impact strength compared to composites with fiber non-treated [25].

The modification of the fiber is a key area of research at present to obtain optimum fiber–matrix properties, according to Mishra et al. [22]. One of the most commonly used chemical treatments is alkaline treatment, or mercerization, where wax, lignin, and oils are removed from the fiber in this treatment. Lignin is removed because it makes it difficult for the fiber to adhere to the matrix. Alkaline treatment also breaks hydrogen bonds on the surface of the chain, increasing roughness, resulting in better mechanical properties [13]. It increases surface roughness resulting in better mechanical interlocking, and it increases the amount of cellulose exposed on the fiber surface, thus increasing the number of possible reaction sites [26]. Adding NaOH to natural fiber causes the ionization of the hydroxyl group to the alkoxide, as shown in Eq. (1):

$$\text{Fiber} - \text{OH} + \text{NaOH} \rightarrow \text{Fiber} - \text{O} - \text{Na} + \text{H}_2\text{O} \quad (1)$$

One of the challenges about work with PALF is to avoid the thermal degradation during the process. About that, there is a loss of weight in two moments. In the first moment, from 60 to 100 °C, because of dehydration—loss of 1.6%, and from 250 to 294 °C, loss of 7%, due to thermal degradation of lignin and dehydrocellulose. In the second moment, 364 °C, losing 56%, corresponding also the thermal degradation. The fiber has a thermal stability up to a temperature of 250 °C and after 450 °C, we have the formation of ashes, which is around 7%. The temperature variation at which the peaks occur is related to the percentage of cellulose in vegetable fiber, which in the case of pineapple, leaf fibers are high. This information is important to select the temperature of the composite production process [18]. When working with a natural fiber composite, the polymer used as matrix will be the responsible for distributing the stress put on the composite, and it is possible to select it for the composite because of its temperature, as can be possible to see in Table 2. The temperature required for the processing of the mixture is very important because it needs to be adequate to have homogeneous mixtures without fiber degradation, which could interfere in the mechanical properties of the composite [27].

The physicomechanical properties of PALF involve tensile strength between 400 and 627 MPa; Young's modulus 1,44 GPa; elongation at break 14.5%; and density about 0.8–1.6 g/cm^3, while the main polymers used in the natural fiber composites

Table 2 Properties of thermoplastics to produce NFC

Properties of thermoplastics polymers used in NFC fabrication			
Property	PP	LDPE	HDPE
Density (g/cm^3)	0.899–0.920	0.910–0.925	0.94–0.96
Water absorption-24 h (%)	0.01–0.02	<0.015	0.01–0.2
T_g (°C)	−10 to −23	−125	−133 to −100
T_m (°C)	160–176	105–116	120–140
Heat deflection temp (°C)	50–63	32–50	43–60
Coefficient of thermal expansion (mm/mm/°C × 10^5)	6.8–13.5	10	12.0–13.0
Tensile strength (Mpa)	26–41.4	40–78	14.5–38
Elastic modulus (Gpa)	0.95–1.77	0.055–0.38	0.4–1.5
Elongation (%)	15–700	90–800	2–130
Izod impact strength (J/m)	21.4–267	>854	26.7–1068

Source Adapted from Ku et al. [42]

have the properties according to Table 2. PALF fibers are reinforced with PP in the main natural fiber composites, but it is possible to work with all polymers.

Another opportunity to pineapple leaves is to produce green composites. Green composites have matrix and reinforcement (polymer and fiber) taken from renewable resources, for example, a natural fiber and PLA [14]. However, according to Siakeng et al. [28], for a better PLA-based NFC composite, a cost decrease will be necessary to get the place of synthetic polymer composites in the market.

4 Pineapple Leaf Fibers Composites

Anannus comosus, belongs to the *Bromeliaceae* family, commonly known as pineapple (Fig. 1). World pineapple production was 27,402.956 tons in 2017, according to Graphic 1. Costa Rica, Philippines, Brazil, China, Thailand, India, and Indonesia are leading this tropical fruit production, as Table 3. As possible to see, with the increase of the planted area of pineapple in the world, it becomes necessary to develop new alternatives to pineapple leaves. These leaves have been wasted after harvest either going to be composted or burnt [21]. Burning or composting these agricultural wastes will cause environmental pollution and will lose opportunities to make money once the countries involved in produce pineapple are in development and need to increase

Fig. 1 Pineapple production. *Source* Embrapa (author: Davi Theodoro Junghans) 2019

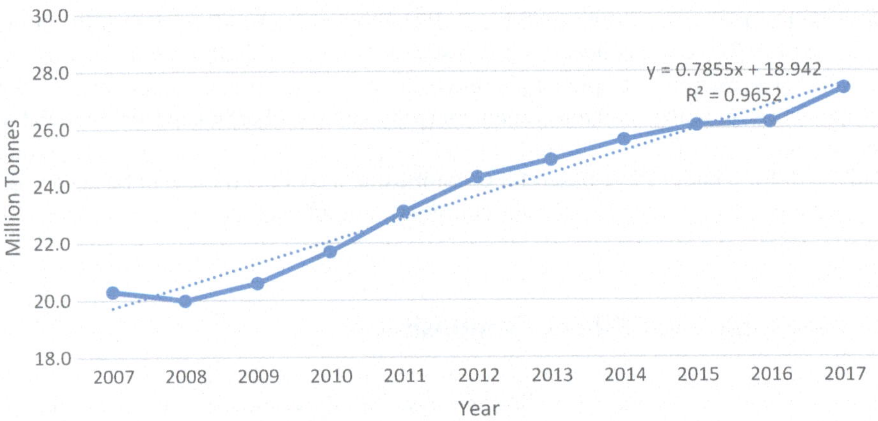

Graphic 1 World pineapple production. *Source* FAO [40]

Table 3 Countries leaders in pineapple production—2017

	Country	Tons
1st	Costa Rica	3056.445
2nd	Philippines	2671.711
3rd	Brazil	2253.897
4th	China	2129.936
5th	Thailand	2123.177
6th	India	1861.000
7th	Indonesia	1795.986

Source FAO [40]

their business. For example, in an e-commerce called Juch in France, one shoe made from PALF is sold from 50 euros [29]. To value the agricultural disposal, it is necessary to use new technologies to produce high value materials. This fiber has already used in some countries to produce dresses, clothing items, bags, shoes, etc. (Fig. 2).

Depending on which part of the plant the fiber is extracted, the fiber is categorized as: bast or stem fiber (as jute, flax, hemp, ramie, kenaf, etc.); Leaf fibers (as sisal, banana, pineapple, etc.); and seed fibers (as cotton, coir, oil palm, etc.) [30]. Natural fibers, when incorporated into plastics, can be processed by virtually all conventional plastic processing methods (extrusion, calendaring, and pressing) and have a lower density than inorganic fibers such as glass fibers. The world consumption of natural fibers totaled $4.3 billion in 2018, and China is leading in this production, as shown in Graphic 2 [31].

Fig. 2 World production of natural fibers. *Source* Adapted from ITFN [43]

Graphic 2 World productivity of natural fiber by country

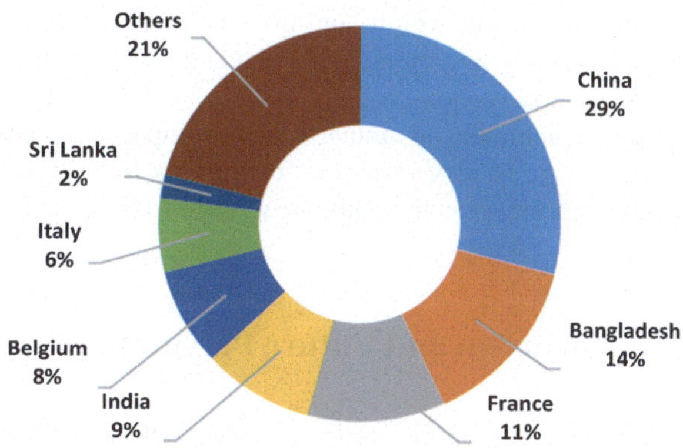

5 Applications

The lignocellulosic materials made from natural fibers as reinforcement is taking the place of synthetic fibers worldwide, and one of the best fibers to produce the composites that are growing in the market is pineapple leaf fibers (PALFs). The PALFs have good mechanical, thermal, and acoustic properties when used as reinforcement and are presented as an important raw material to produce composites [15]. PALF has the best impact properties of the composites comparing to other fibers [32]. It can be extracted manually or mechanically, and the manual process involves soaking the fibers in water—for approximately, 18 days—and then manually scraping them with the aid of a small knife or piece of ceramics. The mechanical process is performed with a defibrating machine that works according to the principles of a sisal machine [33]. According to Paul [34], the manual process can extract a minimum of 3–4% of fibers from the leaves.

The production of pineapple fiber per hectare is around 15 tons, depending on the varieties. This productivity is similar to softwood that reaches a maximum of 15 tons per hectare per year [35]. Pineapple fibers come from agricultural waste that would otherwise be discarded, since the main purpose of the crop is to produce food (juice, fruit, etc.), pharmaceutical (bromelain) and pulp [3]. Therefore, this waste has a low value, about $10 a ton in the field. Estimated by Leao et al. [36] that leaves can represent 6 tons/ha of pulp year and, for 1 ton of PALF, there are about 30 kg of dry fiber. In addition to the fruit, which is already sold "*in natura,*" farmers could have an increase in their income from fiber production, a postharvest use. Another benefit of the usage of PALF as reinforcement is the social impact, such as an increase of jobs in the sector [15].

About the environment item for NFC, some studies comparing glass fibers and natural fiber composites life cycle environmental performance figured out NFC are environmentally superior. The life cycle assessment (LCA) is an important tool to evaluate the environmental impact of the fiber for its entire life cycle and used to compare two or more elements and evaluate which one is more durable and preferable under certain environmental conditions [37]. That is because natural fibers composites have lower weight and give a better fuel efficiency, reducing the emissions when in auto-applications. Also, the end of life incineration of natural fibers results in energy and carbon credits [38]. Besides that, the natural fibers are cheaper and have environmental advantages while compared to glass and carbon fibers [39]. This characteristic is very important once glass fiber is the major competitor of natural fiber composites for the automotive sector (Fig. 3).

6 Conclusion and Future Perspective

There are many opportunities for pineapple leaf fibers (PALFs). It is a renewable filler from agriculture, while wood takes much longer time; most crops are near populated areas, by-product of food/feed, or use of marginal areas, and increase value to farming and a lightweight filler. The emerging trends in the global natural fiber industry are about to increase emphasis on recyclability, price-performance balance of natural fiber composites, and a global concern toward global warming. The natural fiber composites market has challenges to overcome: to make the NFC well-known; players of this sector make more investments in marketing and P and A; to do more research to present to consumer; to standard the market about raw material—the consumer needs to compare the products; to do more studies about economics that can support investors and more researches about technical information—pretreatment of the fibers and the ratio to some polymer along another. The success of NFC as reviewed in this chapter will vary according to the techniques used to produce it; to the adhesion between fiber and the polymer; the ratio of polymer and natural

Fig. 3 Automotive components from natural fibre composites [44]

fiber; etc. Fiber ratio will be influenced by the way in which fibers are extracted and processed. But it has a huge growth perspective and many opportunities. Producing natural fiber composites from PALF is a great opportunity to valorize the countries that produce the pineapple plant and are the most countries in development with agricultural potential. The composites made from pineapple fiber can be used in many sectors like automotive, construction, furniture, packaging, consumer goods, etc. The future is "green" and consumer wants day-by-day more bio-based materials, so at this moment, natural fibers will be as green as the future.

References

1. Faruk O et al (2014) Progress report on natural fiber reinforced composites. Macromol Mater Eng 299:9–26. https://doi.org/10.1002/mame.201300008
2. Goulart SAS et al (2011) Mechanical behavior of polypropylene reinforced palm fibers composites. Procedia Eng 10:2034–2039
3. Asim M et al (2015) Review article a review on pineapple leaves fibre and its composites. Int J Polym Sci
4. Dermibas A (2008) Biofuels sources, biofuel policy, biofuel economy and global biofuel projections. Energy Convers Manag 49(8):2106–2116. https://doi.org/10.1016/j.enconman.2008.02.020

5. Marinelli AL et al (2008) Desenvolvimento de Compósitos Poliméricos com Fibras Vegetais Naturais da Biodiversidade: Uma Contribuição para a Sustentabilidade Amazônica. Polímeros: Ciência e Tecnol 18(2):92–99
6. Silva R et al (2009) Applications of lignocellulosic fibers in polymer chemistry and in composites. Quím Nova 32(3). https://doi.org/10.1590/s0100-40422009000300010
7. Eckert CH (2000) Opportunities for natural fibers in plastic composites. In: Proceedings of the conference on progress in woodfibre-plastic composites, University of Toronto, Canada, pp 87–106
8. Nova Institute (2014) Wood-plastic composites (WPC) and natural fibre composites (NFC): European and global markets 2012 and future trends in automotive and construction. Nova Institute
9. Grand View Research (2016) Natural fiber composites (NFC) market size, share & trends report natural fiber composites (NFC) market size, share & trends analysis report by raw material, by matrix, by technology, by application, and segment forecasts, 2018–2024. 2016. Available in: https://www.grandviewresearch.com/industry-analysis/natural-fiber-composites-market. Accessed in: 15 July 2019
10. Chandramohan D, Marimuthu K (2011) A review on natural fibers. IJRRAS 8(2)
11. Honest Version, Available in: https://honestversion.com/natural-fiber-composites-market-2022-top-key-players-kg-tecnaro-gmbh-kafus-biocomposites-inc/. Accessed in: 20 Aug 2019
12. Pupo HFF (2017) Viabilidade técnica da produção de compósitos fibra polímero à base de resíduos. Doctoral thesis. Universidade Estadual Paulista Júlio de Mesquita Filho
13. Marcon JS et al (2009) Estudo da modificação da fibra proveniente da coroa de abacaxi para a formação de compósitos poliméricos. Anais do 10º Congresso Brasileiro de Polímeros, Foz do Iguaçu, PR
14. Scandola EZM (2011) Green composites: an overview. Polym Compos, 1906–1915
15. Zah R et al (2007) Curauá fibers in the automobile industry—a sustainability assessment. J Cleaner Prod 15(11):1032–1040. ISSN 0959-6526
16. Leao AL et al (2000) Curaua fiber–a tropical natural fibers from Amazon potential and applications in composites. Nat Polym Agrofibers Bases Compos Embrapa Instrumentacao Agropecuaria, 257–272
17. Santosha PVChRK et al (2018) Effect of fiber loading on thermal properties of banana and pineapple leaf fibers reinforced polyester composites. Sci Direct Mater Today: Proc 5:5631–5635
18. Chollakup R et al (2011) Pineapple leaf fibers reinforced thermoplastic composites: effects of fiber length and fiber content on their characteristics. J Appl Polym Sci 119:1952–1960
19. Leao AL, Available in: http://moldesinjecaoplasticos.com.br/estudo-desenvolve-fibras-de-banana-e-abacaxi-para-reforco-de-termoplastico/. Accessed in: 06 Aug 2019
20. Alexandre MEO (2005) Compósitos de Matriz Poliéster Reforçados com Fibra da Folha do Abacaxi. Doctoral thesis. Universidade Federal do Rio Grande do Norte, Natal
21. Sapuan SM et al (2011) Pineapple leaf fibers and PALF-reinforced polymer composites. Cellul Fibers Bio- Nano-Polym Compos, 325–343. https://doi.org/10.1007/978-3-642-17370-7_12
22. Mishra S et al (2004) A review on pineapple leaf fibers, sisal fibers and their biocomposites. Macromol Mater Eng 289:955–974. https://doi.org/10.1002/mame.200400132
23. Kleba I, Zabold J (2004) Poliuretano com fibras naturais ganha espaço na indústria automotiva. Plástico Ind, 88–99
24. Leao AL et al (2010) Agro-based biocomposites for industrial applications. Mol Cryst Liq Cryst, 318–327. https://doi.org/10.1080/15421401003719852
25. Beltrami LVR et al (2014) Efeito do tratamento alcalino de fibras de curauá sobre as propriedades de compósitos de matriz biodegradável. Polímeros 24(3):388–394
26. Li X et al (2007) Chemical treatments of natural fiber for use in natural fiber-reinforced composites: a review. J Polym Environ 15:25–33. https://doi.org/10.1007/s10924-006-0042-3
27. Ishizaki MH et al (2006) Caracterização mecânica e morfológica de compósitos de polipropileno e fibras de coco verde: influência do teor de fibra e das condições de mistura. Polímeros: Ciência e Tecnologia, São Carlos 16(3):182–186

28. Siakeng R et al (2019) Natural fiber reinforced polylactic acid composites: a review. Polym Compos 40:446–463. https://doi.org/10.1002/pc.24747
29. JUCH, Available in: https://juch.fr/en/product/pineapple-fiber-natural-men/. Accessed in: 22 Aug 2019
30. Jawaid M, Khalil HPSA (2011) Cellulosic synthetic fibre reinforced polymer hybrid composites: a review. Carbohydr Polym 86:1–18
31. Composite World, Available in https://www.compositesworld.com/blog/po st/natural-fiber-composites-whats-holding-them-back. Accessed in: 22 Aug 2019
32. Mohanty AK et al (2002) Sustainable bio-composites from renewable resources: opportunities and challenges in the green materials world. J Polym Environ 10(1/2)
33. Aquino MS (2006) Desenvolvimento de uma desfribadeira para obtenção da fibra da folha do abacaxi. Master dissertation. Universidade Federal do Rio Grande do Norte, Natal
34. Paul NG (1980) Some methods for the utilisation of waste from fibre crops and fibre waste from other crops. Agric Waste 2:313–318
35. Leao AL et al (2014) The use of pineapple leaf fibers (PALFs) as reinforcements in composites. In: Biofiber reinforcements in composite materials, vol 1(1), pp 211–235
36. Leao AL et al (2007) Production of curaua (*Ananas Erectifolius* LB SMITH) fibers for industrial applications: characterization and micropropagation. In: VI international pineapple symposium, vol 822, pp 227–238
37. Ahmad F et al (2015) A review: natural fiber composites selection in view of mechanical, light weight, and economic properties. Macromol Mater Eng 300:10–24. https://doi.org/10.1002/mame.201400089
38. Joshi SV et al (2004) Are natural fiber composites environmentally superior to glass fiber reinforced composites? Compos Part A 35:371–376
39. Bongarde US, Shinde VD (2014) Review on natural fiber reinforcement polymer composites. Int J Eng Sci Innov Technol (IJESIT). 3(2)
40. FAO, Accessible online at http://www.fao.org/faostat/en/#data/QC. Accessed in: 01 Aug 2019
41. Mohanty AK et al (2000) Surface modification of jute and its influence on performance of biodegradable jute-fabric/biopol composites. Compos Sci Technol 60:1115–1124
42. Ku H et al (2011) A review on the tensile properties of natural fiber reinforced polymer composites. Compos Part B 42:856–873
43. ITFN, Available in: https://www.itfnet.org/v1/2015/12/canada-pineapple-now-a-leather-alternative-for-shoes/. Accessed in: 22 Aug 2019
44. Natural fiber for automotive, Avaiable in: https://www.naturalfibersforautomotive.com/wp-content/uploads/2014/04/04-mercedes-s-class-11.jpg Accessed 08 Jan 2019

Pineapple Leaf Fibres for Automotive Applications

Beyanagari Sudheer Reddy, M. Rajesh, Edwin Sudhakar, Ariful Rahaman, Jayakrishna Kandasamy and M. T. H. Sultan

Abstract Fibre-reinforced polymer composites (FRPCs) are playing a significant role in manufacturing of goods/products in service for lightweight applications. Among FRPCs, natural fibre-reinforced polymer composites (NFRPCs) are one in forefront and replacing both the conventional and unconventional reinforced composites since they are eco friendly in nature and have several benefits like low price, ease of manufacturing, denseness, biodegradability, etc. In this chapter, a solemn attempt is made to study the pineapple leaf fibre (PALF) bolstered with polymer matrix composites (PMCs). PALFs are rich in cellulose, comparatively cheap and extravagantly available. PALFs reinforced with polymers such as thermoplastic/thermoset matrices are widely used in automotive sectors. PALF-reinforced polymer matrix composites have a wide range of applications in automotive industries, manufacturing of dashboards, package trays, door panels, headliners, seat backs, interior parts and many other parts. This chapter also explores the type of NFRPCs used by several automotive organizations.

Keywords Natural fibre · Pineapple leaf fibre · Polymer composites · Automotive · Hybrid-electric vehicles

Abbreviations

BMC	Bulk moulding compound
CMC	Ceramic matrix composites
ESEM	Environmental scanning electron microscopy
FRPC	Fibre-reinforced polymer composites

B. S. Reddy · M. Rajesh · E. Sudhakar · A. Rahaman · J. Kandasamy (✉)
School of Mechanical Engineering, Vellore Institute of Technology, Vellore, Tamil Nadu 632014, India
e-mail: mail2jaikrish@gmail.com

M. T. H. Sultan
Laboratory of Bio-composite Technology, Institute of Tropical Forestry and Forest Products (INTROP), Universiti Putra Malaysia, 43400 Serdang, Selangor, Malaysia

M. Jawaid et al. (eds.), *Pineapple Leaf Fibers*, Green Energy and Technology,
https://doi.org/10.1007/978-981-15-1416-6_14

MMC	Metal matrix composites
NFM	Non-fibrous material
NFRPC	Natural fibre-reinforced polymer composites
LDPE	Low-density polyethylene
PALF	Pineapple leaf fibre
PHBV	Poly-hydroxybutyrate-co-valerate
PLA	Polylactide
PM	Polymer matrix
PMC	Polymer matrix composites
PP	Poly-propelene
SMC	Sheet moulding compound
WGL	Whole ground pineapple leaf

1 Introduction

Materials play a crucial role in any manufacturing sector for producing goods with desired shape and size [1]. In today's scenario, the researchers are concentrating on the production of new materials. The fields of material science and technology have rapidly developed, and in-numerous changes have taken place, resulting in new materials with the composition of base metal and reinforcement of other metals leading to the composite materials [2].

For the past few decades, composite materials are used widely in the global market to produce sophisticated and qualitative products to meet the customer needs and demand. When compared to metals, the composite materials utilization is huge/widely due to their high strength to weight ratios and high modulus to weight ratio [3, 4]. They also offer new opportunities for designing lightweight, strong and inexpensive products [5]. At present, composite materials are being utilized in few engineering and industries' applications like automotive, aircraft and manufacture of spaceships, marine applications, sporting goods, wind energy, electronics and so forth [6].

The composite materials are arranged significantly into three kinds to be specific metal matrix composites (MMCs), polymer matrix composites (PMCs) and ceramic matrix composites (CMCs) [7, 8]. This chapter focuses on composites made using polymers. PMCs are ubiquitous, and the major constituents of PMCs are reinforcements in the form of fibres, fillers or particles that are implanted into a matrix. The strength and stiffness of the matrix material depend on the reinforcement type (short or continuous) and the matrix (polymer or natural fibre-reinforced polymer matrix). This chapter deals with the combination of pineapple leaf fibres (PALFs) reinforced with the polymer matrix (PM) for automotive applications. PALF-reinforced PM has a wide scope of applications and preferred primarily as alternatives for lightweight automotive applications as they offer better manufacturing properties such as less-in-weight, easy moldability, improved aesthetics and strength and relatively in expensive as compared to other conventional automotive components [9].

In developing countries, different approaches are adopted to manage organic/natural waste. In fact, the word waste is often an improper term for organic matter, which is regularly put to good use [10, 11]. The economies of most creating countries deal with that materials and resources should be accustomed their maximum capacity [12]. This is being propagated as a culture of recycle, repair and utilization [13, 14]. Realizing the importance of fibre composites, the European automotive manufacturing industries utilize the natural fibres like kenaf, hemp, flax, jute and sisal for creating the thermoplastic and thermosetting matrices for door panels, dashboards, headliners, seat backs, package trays and interior components. The natural fibre composites used widely in automotive applications in spite of their wetness stability and fibre epoxy bonding [5].

1.1 Fibre Types

The natural fibres generally originate from the base stem, leaf, seed, fruit, wood and grasses are organized into bundles, whereas the fibres originate from seeds lead to single cells and are referred as fibres, and these bundles are called as fibre bundles [15, 16]. The classifications of fibres are divided into natural and synthetic (artificial) fibres. Further, the natural fibres are partitioned into blast, plant fibre, grasses, seeds and fruit fibres [17, 18]. The synthetic fibres are also classified as organic and inorganic fibres as shown in Fig. 1 [18–21].

1.1.1 Pineapple Leaf Fibres (PALF)

Fibres are normally divided into natural fibres and synthetic fibres. Irrespective of the fibre type, the manufacturing techniques remain the same [15]. An attempt is made in this chapter to present the usage of PALF composites in automotive manufacturing industries.

Pineapple plant's scientific name is Ananascomosus, belonging to the Bromeliaceae family [22, 23]. PALF is extracted from leaves of the plant. PALF is also known as pina-fibre, and the word pina is originated from Spanish. PALF is one of the lingocellulosic fibres and has good potential in terms of yarn production [24]. Pineapple fibre can be used as reinforcement due to its rich cellulose content, abundant availability and cheap cost [11]. The Ananascomosus contains exceptionally short stem that initially produces a rosette of leaves, however, that lengthens and bears varied spirally organized fibrous leaves. Generally, leaves are 90 cm long, 5–8 cm wide sword moulded, dull green in colour and bear spines of claws on their edges [12]. The leaves of the plant yield robust, white fine silky-smooth fibres.

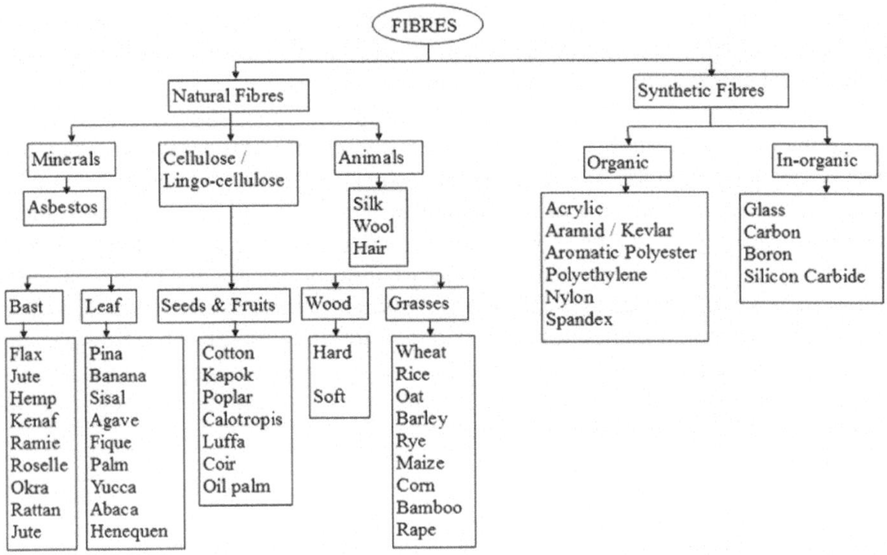

Fig. 1 Classification of fibres

Pineapple leaf fibres are extracted by manual and mechanical processes [23, 25]. The traditional method of scraping is by painstaking and requires skilful labour. The initial step in manual process is mixing of layered fibres in water for nearly 20 days to become saturated, before they are manually scratched. The manual procedure starts with shredding through beating, scraping and husking the leaves [24, 26, 27, 41]. Microorganisms play a major role in removing the unwanted material/gummy substance and separating the fibres. After this procedure, fibres are cleaned and then naturally dried.

The mechanical method is carried out because the leaves area unit fed through the feed rollers and gone through a series of scratching rollers [10, 11]. The side of the leaves is scraped by scratching roller skates to dispose of the waxy layer. Later, it passed through the toothed roller where the intimately fitted cutting edges of roller macerates. The leaf delivers with numerous breaks on its surface for easy/simple passage of retting microbes [28].

Pineapple leaves contain fibres with low density and widely preferred for fabrication of reinforced polyethylene composites. The influence of fibre length, fibre loading and orientation is studied by George et al. [27]. Further, the scanning electron microscope analysis shows that fibres are well oriented during the composite fabrication addition to fibre damage and breakage during melt mixing. Such identification of damages requires further attention. Therefore, the stress–strain behaviour in tension reports that pure polyester is brittle and with addition of fillers makes the matrix more ductile [26]. Further studies on the effect of environment of temperature and chemical treatment are studied on the short pineapple leaf fibres-reinforced polyethylene composites (PALF/LDPE), [41]. It proved that the tensile properties are decreased with the plasticization/immersion of water. Such limitations can be overtaken by "green" composites of pineapple fibres and poly-hydroxybutyrate-co-valerate (PHBV) resin, and pineapple fibres are arranged in longitudinal directions. Such process resulted in the increase of tensile and flexural strengths but decreases in transverse directions [29]. Further studies show that there is no impact on fibre content at thermal properties of the resin, and then, thermo-gravimetric analysis shows that thermal decomposition of PHBV is same in air and nitrogen atmospheres. Pineapple leaf composites at different weight percentages were studied and observed that thermal properties of composites decrease with increase in fibre content. Therefore, suitable fibre content and its concentrations are necessary for arrangement in the longitudinal directions, [31].

The isolation and the effectiveness of the high-pressure hydrothermal process of PALF have been investigated for nano-celluose of PALF [30]. The usage of PAL waste for polymer reinforcement by mechanical milling was studied by Nanthaya and Taweechai [10]. They proposed mechanical grinding methods like ball mill and disc mill can be utilized to remove fibres from slashed fresh/new pineapple leaf. Further, they extended the study for performance and cost effectiveness, with another way to deal with "Greening" plastic composites utilizing PAL squander [11]. Fresh pineapple leaves contain nearly 85% of water. These leaves are shredded into little pieces and ground into glue/paste is named as whole ground pineapple leaf (WGL). PALF restrains nearly 2.8% by weight of excellent dry fibres moreover as an oversized fraction of non-fibrous material (NFM) of roughly 10% by weight.

Munawar et al. [25] have done the experimental investigation on the PALF strengthened polylactide (PLA) and also done a comparison of material properties over the PALF bolstered polypropelene (PP). The influences of various fibres characterizations were also studied. The fibres were removed from differing kinds of pineapples particularly Moris Gajah, Maspine, Jasopine and N36.

1.1.2 Advantages of Pineapple Leaf Fibre

Pineapple leaf fibre has ecological and economic benefits, eco friendly, non-toxic/non-poisonous, completely biodegradable, easy to handle, and separation of fibres is free from hazard. These fibres are non-abrasive throughout process and use. PALF has comparatively light-in-weight, low density, low cost, is widely abundant, making certain continuous supply of raw materials, enhanced energy recovery, acceptable specific strength properties, high toughness and also possesses sensible thermal properties [25, 31, 29]. PALF is supply of financial gain for rural agricultural community in several countries.

1.1.3 Disadvantages of Pineapple Leaf Fibre

Pineapple leaf fibres are removed from pineapple plant leaves. Polymer matrix composites incorporated with PALFs are environmentally friendly but also got some disadvantages. Some of the disadvantages are high moisture consumption and low thermal stability [32]. The high moisture at times weakens the interfacial bonding among the polymer matrix and fibres, thereby reducing the mechanical properties [4].

1.2 *Natural Fibre Polymer Matrix Composites*

The NFPMCs are the combination of polymer matrix with resin (thermosets and thermoplastics) and reinforcement material (natural fibre). The main objective of the matrix/resins is to transfer the stresses and loads among the fibres, in order to act as adhesives that bond structural fibres firmly in place, and to shield the fibres from ecological and mechanical damage [8]. Bio-based resins refer to thermoset or thermoplastic resins that are acquired from natural sources [13]. NF-reinforced PMCs can be fabricated as different kinds of shapes and sizes with better quality, strength, stiffness and corrosive resistance with low price. Basically, PMCs are divided into two categories: Thermosetting and Thermoplastics [8, 33]. The use of lingo-cellulosic pine fibres as fortifications in thermosetting and thermoplastic resins for developing biodegradable, ease and lightweight composites in automotive field of research is exceptionally getting [22].

1.2.1 Thermosetting

Thermosetting resins are a tough and hard cross-connected material that does not soften/mollify or end up mouldable when heated [8]. Thermosetting resins do not extend the way that elastomers and thermoplastics do when heated above their melting point, such resin plastics attain stiff, hard and rigid after cooling. These resins at initial form before curing (solidification process of plastic) are generally in the form of liquids or low melting point solids. Irreversible transformation process takes place from liquid to solid phase [34]. The thermosets are classified as matrix for fibre-reinforced composites as alkyds, amine, allylics, bakelite, epoxy, polyester, phenolic, polyurethane, silicone and vinyl ester [13]. The classification of thermoset polymers is shown in Fig. 2.

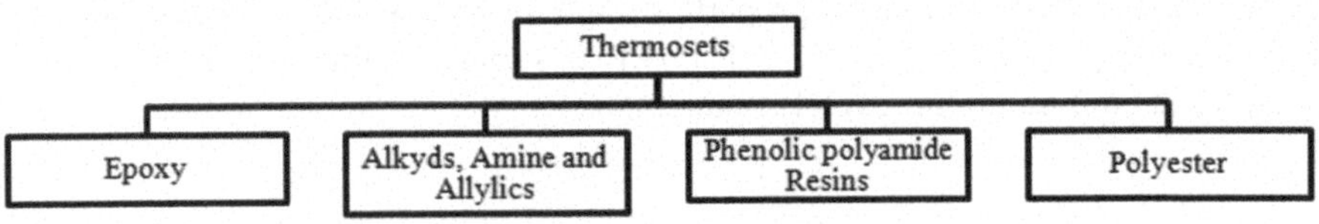

Fig. 2 Classification of thermoset polymers

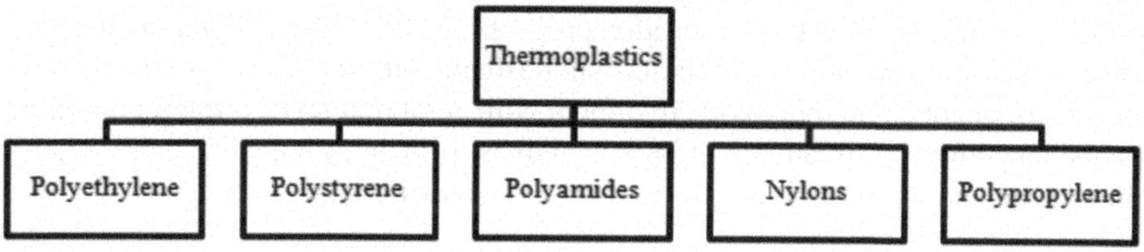

Fig. 3 Classification of thermoplastics polymers

1.2.2 Thermoplastics

The material which is soft and formable when heated above the melting temperature and rigid or hard after cooling is known as thermoplastics [8]. The thermoplastics are classified as matrix for fibre-reinforced composites as acrylics, acetals, cellulosics, fluorocarbons, polyamides, polycarbonates, polyethelyne, polypropelene, polystyrenes and polyvinyl chloride [13, 33]. The classification of thermoplastics polymers is represented in Fig. 3.

Utilization of natural fibre composites in any industries depends upon the processing methods/technologies. Accordingly, some of the processing techniques for thermoplastic polymers are compression moulding, injection moulding, extrusion, LFT-D-method and thermoforming methods. Similarly, for thermosets, processing techniques consists of resin transfer moulding, sheet moulding compound, compression moulding, pultrusion [13, 35, 36].

Bledzki and Gassan [15] claimed that the natural fibres may also be processed by sheet moulding compound (SMC)/bulk moulding compound (BMC) techniques. Polymer matrix (thermoplastics, thermosets and biodegradables) is subjected to physical, chemical treatments for the development of fibre–matrix interaction and mechanical properties. Because of various weight savings, low price of the raw constituent materials, the automobile industries have begun to use NFRCs as exceedingly different kinds of exterior and interior panel applications.

1.3 Pre-treatments of Pineapple Leaf Fibre Composites

Utilization of PALF composites is tremendously increasing in recent years due to its ease, denseness, biodegradability, enhanced properties and many other applications in automotive, marine and aerospace industries. Utilization of PALF composites has tremendously increased in recent years especially in automotive, marine and aerospace industries due to its easy availability, density, biodegradability, enhanced properties in spite of their drawbacks such as lack of good interfacial adhesion, poor resistance to moisture absorption and low melting point [18]. To rectify these drawbacks, pre-treatment processes are necessary, through this treatment, the fibres surface can be cleaned, surface roughness can be improved and chemically modify the surface to reduce the moisture absorption process [37, 38]. Chemical treatments such as mercerization, acrylation, acetylation, silane treatment, iso-cyanate treatment, permanganate treatment and peroxide treatment with combination of different coupling agents and other pre-treatments [34, 38] will help in improving the fibre–matrix adhesion and strength of pineapple-reinforced polymer composites.

2 Fabricating Techniques PALFs Composites

Composites are fabricated using different techniques. Manufacturing of natural fibre composites is one of a complex processes, and it requires the synchronous simultaneous deliberation of several parameters like component geometry, layering sequence, production volume, reinforcement and matrix types, tooling requirements and process economics. To meet specific design, manufacturing challenges depend on the natural fibres and reinforced materials, selection of materials, method for a particular part/part design and its application. Fabrication of these composites involves type of the resin/reinforcement and moulding. Some of the techniques used to fabricate the pineapple leaf fibre-reinforced polymer composites are compression moulding [39, 40], injection moulding [24, 27, 28, 41, 37, 40], extrusion [32, 40], autoclave moulding [30, 40], hand lay-up method [26, 36, 40, 42, 43] and some other techniques.

3 Properties of Pineapple Leaf Fibre Composites

Properties of PALF composites make it suitable for automotive applications. Mishra et al. [22] have explained the chemical composition, structural parameters and some important properties of PALF and sisal fibres with their bio-composites. The chemical composition comparison of PALF and sisal fibres is shown in Table 1.

Wanjun et al. [32] have reported that the mechanical properties of bio-composites are increased when fibre substance increases with the presence of the compatibilizer. Environmental Scanning Electron Microscopy (ESEM) studies confirms that

Table 1 Chemical composition comparison of PALF and sisal fibres [22]

	PALF	SISAL
Cellulose (wt%)	70–82	67–78
Lignin (wt%)	5–12	8–12
Hemicellulose (wt%)	–	10–14.2
Pectin (wt%)	–	10
Wax (wt%)	–	2
Microfibrillar-spiral angle (°)	14	20
Moisture content (wt%)	11.8	11

the increase in fibre loading affects the fibre dispersion in the matrix. With the addition of compatibilizer the fibre dispersion can be controlled. Arib [4] have demonstrated that proper addition fibre-volume-fraction of pineapple leaf fibre-reinforced polypropylene composites will improve the tensile modulus and tensile strength of the composites [4]. Yet, the flexural modulus and stress of the composites will increase with increase of volume fraction.

Kaewpirom and Worrarat [28] reported that mechanical properties of PALF, at various places of the leaf length, were essentially unique, and likewise, tensile and thermal properties of PALF/PLA composites can be appreciably improved by increasing PALF loading.

Ramnath et al. [42] studied the flexural properties of the composites. The composite is manufactured by hand lay-up method, pineapple fibre incorporated with epoxy resin of three layers and then placed in between the two layers of glass fibre-reinforced epoxy resin mixed with hardener. Three samples were hacked as per ASTM: D790 standard at different parts of the composite and tested (Flexular) then observed that there is no appreciable variation within the properties. The typical break load is 1.29 kN, and furthermore, the deflection is 5.533 mm. The flexural strength is calculated as 78.63 MPa. This means that there is a uniform distribution of the reinforcement fibres and that the fibre-matrix adhesion is uniform all over the composite.

Munawar et al. [25] have carried out the investigations on the materials to find the physical and mechanical properties of polylactide (PLA)/polypropelene (PP)/PALF. From the outcomes, it is clearly observed that polylactide (PLA) has lower melting point compared to polypropelene (PP), and therefore, it requires less vitality to be process. When PALF increases, then tensile and flexural strength will decrease within the PLA/PALF and PP/PALF composites. As a comparison among the usage of PLA and PP as matrix elements, usually PP/PALF indicated that tensile and flexure strengths are lower than the PLA/PALF composites. Amid those four verities (Moris Gajah, Maspine, Josapine and N36) of PALF, the peak values of tensile and flexural strength are obtained at Josapine pineapple fibre-reinforced composite. At 10% of Josapine PALF incorporated with PLA, the tensile strength is obtained 4.2 Mpa, and bending strength is 18.15 Mpa.

Based on the literature review on PALF and its composites based on 162 publications related to chemical, physical, and mechanical properties, it can be inferred that Young's modulus and tensile strength are high compared to other natural fibres having the density similar to the PALF. The thermal properties, thermal conductivity, dynamic mechanical analysis, electrical properties of PALF-reinforced polymer composites are not concentrated by many researchers [12].

Santosh Kumar et al. [2] have conducted tensile, flexural and hardness test of PALF composites for volume ratios of 10, 20 and 30%. The tensile strength was obtained 26.91, 35.8 MPa and 65.95 MPa, the flexural strength obtained was 38.55 MPa, 58.37 MPa and 121.83 MPa, and the hardness of the specimen was determined (using Rockwell hardness tester) as 40 B, 59 B and 80 B, respectively, for the different volume ratios studied. The maximum tensile, flexural strength and hardness are seen in 30% volume ratio.

Yusri et al. [24] investigated the mechanical properties PALF and pineapple peduncle fibre reinforced with polypropylene by varying fibre-volume-fraction. From the experimental results, it was observed that at higher fibre-volume-fraction, the tensile properties decreased, and the hardness increased. Also, when sodium hydroxide (NaOH) treated PALF was used as reinforcing agent, the mechanical properties improved further.

Table 2 shows the chemical composition of some important natural fibres. Figure 4 shows the graphical representation of the chemical composition (based on avg. values) listed in Table 2. PALF has more cellulose wt%, rice straw has more hemicelluloses wt%, and coir has more lignin wt% comparing to other natural fibres.

Table 3 demonstrates physical and mechanical properties of natural fibres. The average values of fibres from Table 3 are represented in Fig. 5.

Figure 5 shows that cotton and Kenaf have max density, pineapple has max tensile strength, ramie and pineapple have max Young's modulus, and the banana has max elongation at break.

Table 2 Chemical composition of some important natural fibres [7, 13, 22, 44, 33]

Fibre	Cellulose (wt%)	Hemicellulose (wt%)	Lignin (wt%)	Other composition (wt%)
Abaca	56–63	20–25	7–9	3
Bagasse	55.2	16.8	25.3	–
Bamboo	26–43	30	21–31	–
Banana	63-64	30	21–31	
Coir	32–43	0.15–0.25	40–45	–
Curaua	73.6	9.9	7.5	–
Flax	71	18.6–20.6	2.2	1.5
Hemp	68	15	10	0.8
Jute	61–71	14–20	12–13.0	0.5
Kenaf	72	20.3	9	–
Oil palm	65	–	29	–
Pineapple	77–81	–	12.7	–
Ramie	68.6–76.2	13–16	0.6–0.7	0.3
Rice husk	35–45	19–25	20	–
Rice straw	41–57	33	8–19	8–38
Sisal	65	12	9.9	2
Wheat straw	38–45	15–31	12–20	–

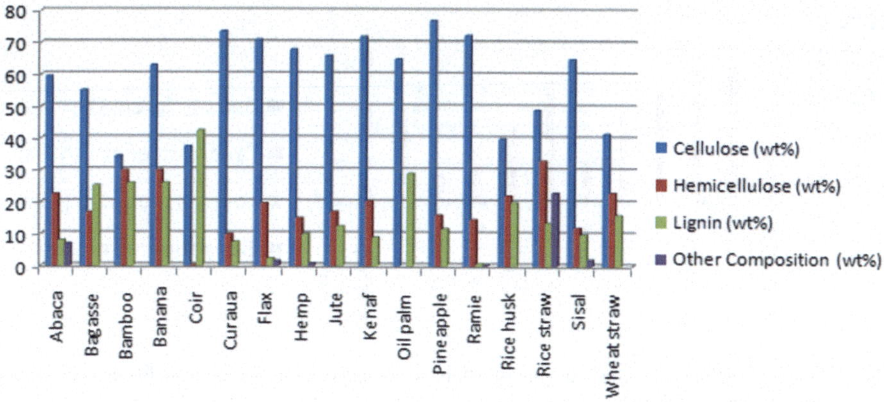

Fig. 4 Average chemical composition, structural parameters of natural fibres are represented in a graphical form

Table 3 Physical and mechanical properties of the some of the natural fibres [1, 3, 5, 6, 8, 9, 13, 14, 19, 20, 21, 34, 35, 36, 44, 33]

Fibre	Density (g/cm^3)	Tensile strength (MPa)	Young's modulus (GPa)	Elongation at break (%)
OPEFB	1.1–1.4	200–300	3200	3
Flax	1.4–1.5	345–1035	70	2.0–3.2
Hemp	1.4–1.5	690–725	27.6–70	1.6–2
Jute	1.3–1.48	393–773	20–26.5	1.5–2
Ramie	1.5	400–938	61.4–128	3.2–3.8
Coir	1.2	175–220	4.0–6.0	20–30
Sisal	1.33–1.5	511–650	9.4–32.0	2.0–2.5
Cotton	1.5–1.6	400	5.5–12.6	7.0–8.0
Kenaf	1.5–1.6	350–900	40–53	1.6–5
Bagasse	1.1–14	120–250	22–26	1.2
Henequen	1.2–1.4	300–750	–	2.0–4.2
pineapple	1.2–1.7	650–1050	82	1.2–3.2
Banana	1.1–1.5	320–500	25–38	53

4 Application of PALF in Automobile Industries

In any automotive industry, one of the vital factors to consider is correct selection of materials for design and manufacturing. The need for environmental protection has motivated researchers and industrialists to replace the usage of synthetic fibres with natural fibres due to their biodegradable nature, easy availability, durability, less abrasiveness, light-in-weight and low cost. NFRC materials are used more

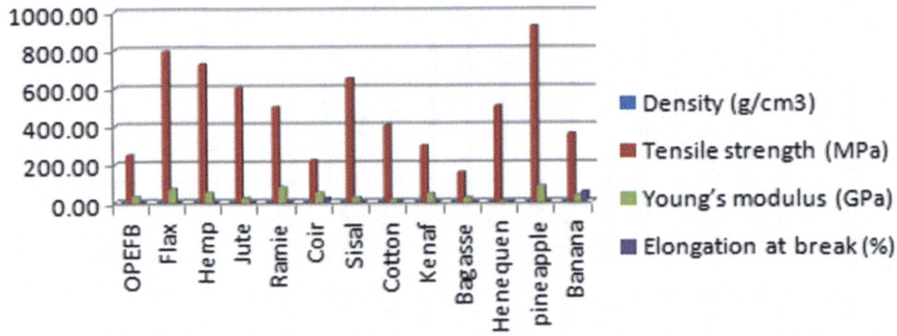

Fig. 5 Average values of physical and mechanical properties of the natural fibres which are represented in a graphical form

in aerospace, automobile industries, marine industries, roofing structures, chemical industries, transportation, logistics industries and interior design [14]. In the automotive industry, the major aspect is fuel efficiency, passenger safety, durability, serviceability, recyclability of their products and other life cycle considerations. For achieving this, in 1930, the first natural fibre-based composites were manufactured for automotive car body parts by Henry Ford, founder of the Ford Motor Company [45]. Later on, other automotive manufacturing companies like Rover group (BMW group), Audi, Mercedes-Benz, Toyota, Mitsubishi, Tata Motors are also concentrating on using NFRCs for manufacturing the body parts with low cost, lightweight, durability, stylish design and safety [19].

Brett and Adas [46] stated that cotton material can be incorporated with the polyester resin for producing the body of East German Trabant car. It is the earliest manufactured vehicle to be built from natural fibres. These cars were still in production up to 1990. Later, in 1990s, natural fibre reinforced with resins was used for producing the car door boards. Ramli et al. [47] expressed that Mercedes-Benz used the epoxy resin consolidated with jute for designing of boor panels to its E-Class model car, and furthermore, Audi fabricated an door trim boards by utilizing polyurethane strengthened flax/sisal fibres. Soya-based foam linings are intended for seats in Volvo C70 and V70 models. Applications of natural fibre composites on automotive industries are listed in Table 4.

Huda et al. [45] stated that automobile manufacturers are utilizing major amounts of composite structures for hybrid-electric vehicles to reduce the production steps, weight, maintenance cost and service life, while increasing design flexibility, durability/toughness and traveller safety broaden driving range. Natural fibre-reinforced polypropylene composites have achieved monetarily fascination in automotive enterprises.

Biagiotti et al. [17] reviewed the natural vegetable fibre composites based on their structure, processing and properties. Mwaikambo [6] proved that the plant fibres have the preferred position over fossil-based fibres due to the higher fibre volume fractions, denseness and high specific stiffness, ease and inexhaustible. [16] have asserted that

Table 4 Applications of natural fibre composites on automotive industries [5, 7, 9, 14, 19, 20, 21, 40, 45, 46, 48, 49]

Automotive manufacturer	Model	Applications
Audi	A2, A3, A4, A6, A8, TT, R-8, 10, 15, 18, Q-2, 3, 5, 7, 8 and S-3, 4, 5	Seat backs, side and back door panels, boot lining, hat rack, spare tyre lining
BMW	3, 5, 7, 8 series and X, i, M—models	Door panels, headliner panel, boot lining, seat backs, bumper, wheel box, noise insulation panels, moulded foot well linings
Citroen	C- 3, 4, 5, DS-3, 4, 5, 6 and 7	Interior door panelling, roof cover
Daimler-Chrysler/Benz	A, B, C, E, G and S-class models, Evo Trucks—MDT—914, 1214 HDT—2523, 3128, 4028 Buses	Door panels, windshield, dashboard, business table, roof cover, sun visor, boot lid finish panel, pillar cover panel
Fiat	Albea, Panda, Punto, Bravo, Marea, Alfa Romeo, FIAT—146, 156, 500	Oil filter housing, electrical junction box, bumper, wheel box, roof cover,
Ford	Mondeo CD 162, Focus, Flex, puma, ecosport, edge, explore Trucks/vans—super duty, ranger, F-150	Door panels, B-pillar, pillar cover panel, boot line
Jeep	Wrangler, grand Cherokee, commander, compass and patriot platform	Door panel inserts, sun visor, interior insulation, insulation, rear storage shelf/panel
Kia	Amanti sedan, Borrego, Forte sedan/hatchback Niro hybrid/electric SUV, Rondo, Sedona minivan	Seat backs, boot lining, hat racks, spare tyre lining, noise insulation panels
Mahindra	Cars—Scorpio, XUV—300, 500, Thar, Bolero, Xylo, Alturas Trucks—bolero, Max, Blazo and buses	Seat padding, natural foams, cargo floor tray, boot lid finish panel
Mercedes-Benz	Trucks and buses	Internal engine cover, engine insulation, sun visor, interior insulation
Mitsubishi	Eclipse cross, Xpander, Pajero/Montero, Lancer, Minicab-MiEV	Hat racks, spare tyre lining, noise insulation panels, moulded foot well linings, door trim

(continued)

Table 4 (continued)

Automotive manufacturer	Model	Applications
Niasan	Micra, Sunny, Terrano, GT-R	Bumper, wheel box, windshield, dashboard
Renault	Clio, Twingo, Duster, kwid	Rear parcel shelf, internal engine cover
Rover	Evoque, Velar, Sport, 2000 and other	Insulation, rear storage shelf/panel
Scania	Busses—F, K, N—series and for trucks—P, G, R, S, L—series	Seat padding, natural foams, cargo floor tray, boot lid finish panel
Skoda	Rapid, Octavia, kodiaq, Superb	Under floor body panels, B-pillar, sliding door inserts, speedometer gears, steering column bush, front fork bush
Toyota	Fortuner, Inova, Brevis, Harrier, Land Cruiser, Celsior, Prius, Raum, Camry	Door panels, seat backs, spare tyre cover, windshield, dashboard, business table, pillar cover panel, door cladding
TATA Motors	Cars—Safari, Indica, Nano, Harrier Buses- Trucks	Door panels (side and back), headliner panels, seat backs, boot lining
Volkswagen	Polo, Vento, Golf, GTI, Passat, Tiguan, Bora	Door panel, seat back, boot lid finish panel, boot liner
Volvo	S—60, 90, XC—60, 40, 90, V—70, 90	Seat padding, natural foams, front fork bush, internal engine cover, engine insulation cargo floor tray

bio-fibre composites are developing as a choice to glass fibre-reinforced composites particularly in automotive industries.

The PALF-reinforced composites are used within the interior and exterior components of automotive like door panels (side and back), headliner panels, seat backs, boot lining, hat racks, spare tyre lining, noise insulation panels, moulded foot well linings, door trim, windshield, dashboard, business table, pillar cowl panel, door protective cover, seatback linings, floor panels, seat bottoms, back cushions, head restraints, below floor body panels, B-pillar, sliding door inserts, speed indicator gears, steering column bush, front fork bush, internal engine cover, engine insulation, oil filter housing, electrical junction box, bumper, wheel box, roof cover, packing trays, door panel inserts, sun visor, interior insulation, insulation, rear storage shelf/panel, seat cushioning, natural foams, lading floor tray, boot lid finish panel and plenty of different components [7, 21]. In many countries, automotive manufactures are planning to produce each and every component of vehicles with recyclable/biodegradable [5, 9].

Polymers and ligno-cellulosic PALF composites can be produced in sheet form for industrial and consumer applications, up to 50% renewable raw material for reducing the consumption of petro-dependent materials in polymers [21]. Load floors layed backside of the vehicle is used as functional weight carrying components which oblige strength and functionality are made of PALF composites [48]. Tata motors aims at manufacturing low price car, to achieve this as an alternate material, and they are now shifted their focus on natural fibres [49]. Glass fibres, carbon fibres and other natural fibres are reinforced with polymers to fabricate the interior and exterior parts with lightweight, stronger, safer and simpler to make the car [20, 49]. Corn scratch, a new material is used to build car body parts of Tata Nano car and other models [49]. Van Eko company manufactures electric bio-scooter has decided to fabricate with natural fibres as reinforcement materials instead of glass fibres [50].

5 Conclusion

The natural fibres are more significant materials for replacement of the non-renewable synthetic/artificial fibres. Such variety of fibres has several features and benefits. Pineapple leaf fibre is a one amongst the natural fibres that contains high cellulose content almost 80% and high crystallinity. Pineapple leaf fibres (PALFs) bolstered composites have several attractive features and benefits, like eco friendly, biodegradability in nature, low cost, tenuity and easy mouldability. This chapter discussed a number of the vital aspects of PALF, extraction, reinforced composite materials and fabricating techniques. The outcomes from this study proves that by the addition of legitimate amount of fibre-volume-fraction of pineapple leaf fibres with polymers can improve the mechanical, physical and thermal properties of the composites. In any case, further investigations need to address vital materials and production hindrances before monetarily available NFRCs are often wide utilized in the automotive sector.

References

1. Mazumdar S (2001) Composites manufacturing: materials, product, and process engineering. CRC Press, Boca Raton
2. Santosh Kumar DS, Praveen BA, Kiran AS, Kempaiah UN (2015) Development of pineapple leaf fibre reinforced epoxy resin composites. Irjet 2(03):2190–2193
3. Arib RMN, Sapuan SM, Hamdan MAMM, Paridah MT, Zaman HMDK (2004) A literature review of pineapple fibre reinforced polymer composites. Polym Polym Compos 12(4):341–348
4. Arib RMN, Sapuan SM, Hamdan MAMM, Paridah MT, Zaman HMDK (2006) Mechanical properties of pineapple leaf fibre reinforced polypropylene composites. Mater Des 27:391–396
5. Holbery J, Houston D (2006) Natural-fibre-reinforced polymer composites in automotive applications. JOM 58(11):80–86
6. Mwaikambo LY (2006) Review of the history, properties and application of plant fibres. AJST 7(2):120–133

7. Friedrich K, Abdulhakim A (2013) Manufacturing aspects of advanced polymer composites for automotive applications. Appl Compos Mater 20:107–128. https://doi.org/10.1007/s10443-012-9258-7
8. Taj S, Munawar MA, Khan S (2007) Review natural fiber-reinforced polymer composites. Proc Pak Acad Sci 44(2):129–144
9. Todor MP, Bulei C, Kiss I (2017) An overview on fiber-reinforced composites used in the automotive industry. Ann Fac Eng Hunedoara-Int J Eng 15(2)
10. Nanthaya K, Taweechai A (2012) Utilisation of pineapple leaf waste for plastic reinforcement: a novel extraction method for short pineapple leaf fibre. Ind Crops Prod 40:55–61. https://doi.org/10.1016/j.indcrop.2012.02.037
11. Nanthaya K, Taweechai A (2014) A new approach to "Greening" plastic composites using pineapple leaf waste for performance and cost effectiveness. Mater Des 55:292–299. https://doi.org/10.1016/j.matdes.2013.10.005
12. Asim M, Khalina A, Jawaid M, Nasir M, Zahra D, Ishak MR, Enamul HM (2015) A review on pineapple leaves fibre and its composites. Int J Polym Sci, 1–16. https://doi.org/10.1155/2015/950567
13. Faruk O, Andrzej K, Bledzki, Hans-Peter F, Sain M (2012) Biocomposites reinforced with natural fibres: 2000–2010. Prog Polym Sci 37:1552–1596. https://doi.org/10.1016/j.progpolymsci.2012.04.003
14. Khan T, Sultan MTBH, Ariffin AH (2018) The challenges of natural fiber in manufacturing, material selection, and technology application: a review. J Reinf Plast Compos 37(11). https://doi.org/10.1177/0731684418756762
15. Bledzki AK, Gassan J (1999) Composites reinforced with cellulose based fibres. Prog Polym Sci 24:221–274
16. Mohanty AK, Misra M, Drzal LT (2002) Sustainable bio-composites from renewable resources: opportunities and challenges in the green materials world. J Polym Environ 10(1–2):19–26
17. Biagiotti J, Puglia D, Kenny JM (2004) A review on natural fibre-based composites-part I: structure, processing and properties of vegetable fibres. J Nat Fibers 1(2):37–68
18. Puglia D, Biagiotti J, Kenny JM (2005) A review on natural fibre-based composites—part II. J Nat Fibers 1(3):23–65
19. Akampumuza O, Wambua PM, Ahmed A, Li W, Qin XH (2017) Review of the applications of biocomposites in the automotive industry. Polym Compos 38(11):2553–2569
20. Baltazar-Y-Jimenez A, Sain M (2012) Natural fibres for automotive applications. Handbook of natural fibres. Wood head Publishing, pp 219–253
21. Dunne R, Desai D, Sadiku R, Jayaramudu J (2016) A review of natural fibres, their sustainability and automotive applications. J Reinf Plast Compos 35(13):1041–1050
22. Mishra S, Mohanty AK, Drzal LT, Misra M, Hinrichsen G (2004) A review on pineapple leaf fibers, sisal fibers and their biocomposites. Macromol Mater Eng 289:955–974
23. Fadzullah SHSMd, Mustafa Z (2016) Fabrication and processing of pineapple leaf fiber reinforced composites. ACME, Book Series, chap. 6, pp 125–147. https://doi.org/10.4018/978-1-5225-0424-5.ch006
24. Yusri Y, Nazuandi BMN, Muhammad SHBA (2016) Pineapple leaf fibre and pineapple peduncle fibre analyzing and characterization for yarn production. ARPN J Eng Appl Sci 11(6):4197–4202
25. Munawar RF, Jamil NH, Shahril MK, Rahim SMSA, Abidin MZZ, Azam MS, Lau Kok-Tee (2015) Development of green composite: pineapple leaf fibers (PALF) reinforced polylactide (PLA). Appl Mech Mater 761:520–525. https://doi.org/10.4028/www.scientific.net/AMM.761.520
26. Devi UL, Bhagawan SS, Thomas S (1997) Mechanical properties of pineapple leaf fiber reinforced polyester composites. Inc J Appl Polym Sci 64:1739–1748
27. George J, Bhacawan SS, Prabhakaran N, Thomas S (1995) Short pineapple-leaf-fiber reinforced low-density polyethylene composites. J Appl Polym Sci 57:843–854
28. Kaewpirom S, Worrarat C (2014) Preparation and properties of pineapple leaf fibre reinforced poly (lactic acid) green composites. Fibres Polym 15(7):1469–1477

29. Luo S, Netravali AN (1999) "Mechanical and thermal properties of environment-friendly "Green" composites made from pineapple leaf fibres and poiy(hydroxybutyrate-co-valerate) resin. Polym Compos 20(3):367–379
30. Cherian BM, Alcides LL, Sivoney FS, Sabu T, Laly AP, Kottaisamy M (2010) Isolation of nanocellulose from pineapple leaf fibres by steam explosion. Carbohydr Polym 81:720–725
31. Ravindra M, Saxena NS, Sreekala MS, Thomas S, Kedar S (2003) Thermal properties of pineapple leaf fibre reinforced composites. Mater Sci Eng A 339(1–2):281–285
32. Liu W, Mishra M, Askeland P, Drzal LT, Mohanty KA (2005) Green composites from soy based plastic and pineapple leaf fibre: fabrication and properties evaluation. Polymer 46:2710–2721
33. John MJ, Anandjiwala RD (2008) Recent developments in chemical modification and characterization of natural fiber-reinforced composites. Polym Compos 29(2):187–207
34. Kalia S, Kaith BS, Kaur I (2009) Pretreatments of natural fibers and their application as reinforcing material in polymer composites—a review. Polym Eng Sci 49(7):1253–1272
35. Ramakrishna M, Vivek K, Yuvraj SN (2009) Recent development in natural fibre reinforced polypropylene composites. J Reinf Plast Compos 28(10):1169–1189
36. Satyanarayana KG, Sukumaran K, Kulkarni AG, Pillai SGK, Rohatgi PK (1986) Fabrication and properties of natural fibre-reinforced polyester composites. Composites 17(4):329–333
37. Threepopnatkul P, Kaerkitcha N, Athipongarporn N, (2009) Effect of surface treatment on performance of pineapple leaf fiber–polycarbonate composites. Compos Part B 40:628–632
38. Alfredo RSN, Marco AMA, Fernanda VDS, Luiz HCM, Jose MM (2013) Characterization and comparative evaluation of thermal, structural, chemical, mechanical and morphological properties of six pineapple leaf fibre varieties for use in composites. Ind Crops Prod 43:529–537
39. Wambua P, Ivens J, Verpoest I (2003) Natural fibres: can they replace glass in fibre reinforced plastics. Compos Sci Technol 63:1259–1264
40. Zhao D, Zhou Z (2014) Applications of light weight composites in automotive industries, chapter -9. In: Yu et al (ed) Lightweight materials from biopolymers and biofibers. ACS symposium series, Washington, DC, pp 143–158
41. George J, Bhagawan SS, Thomas S (1998) Effects of environment on the properties of low density polyethylene composites reinforced with pineapple-leaf fibre. Compos Sci Technol 58:1471–1485
42. Ramnath BV, Krishna CV, Karthik S, Saravanan K, Manickavasagam VM, Elanchezhian C (2014) Evaluation of the flexural properties of pineapple reinforced polymer composite for automotive and electrical applications, Adv Mater Res 893:271–274
43. Manickavasagam VM, Vijaya RB, Elanchezhian C, Jenish J, Jayavel S, Muthukumar V (2014) Investigation on impact and compression properties of pineapple reinforced polymer composite. Appl Mech Mater 591:116–119. https://doi.org/10.4028/www.scientific.net/AMM.591.116
44. Mohammed L, Ansari MNM, Pua G, Jawaid M, Islam MS (2015) A review on natural fiber reinforced polymer composite and its applications. Int J Polym Sci. https://doi.org/10.1155/2015/243947
45. Huda MS, Drzal LT, Ray D, Mohanty AK, Mishra M (2008) Natural-fiber composites in the automotive sector. Properties and performance of natural-fibre composites, Wood Head Publishing, pp 221–268. https://doi.org/10.1533/9781845694593.2.221
46. Brett CS, Adas R (2009) Industrial fibres: recent and current developments. In: Proceedings of the symposium on natural fibres, discover natural fibres, chapter 10 (56), 71–82
47. Ramli N et al (2018) Natural fiber for green technology in automotive industry: a brief review. In: IOP conference series materials science engineering, vol 368, p 012012. https://doi.org/10.1088/1757-899x/368/1/012012
48. Chandramohan D, Bharanichandar J (2013) Natural fiber reinforced polymer composites for automobile accessories. Am J Environ Sci 9(6):494–504
49. Chetan Krishna (2012), Tata motors building cars from corn starch. https://www.drivespark.com/four-wheelers/2012/29-tata-motors-corn-starch-build-cars.html. Accessed 07 Feb 2019

50. Natural fibre-reinforced composite scooter. https://www.plastics.gl/automotive/natural-fibre-reinforced-composite-scooter/. Accessed on 07 Feb 2019
51. Maria CAT, Gabriel OG, Giulio RA, Pedro AN, Frederico MM, Fabio OB, Sergio NM (2015) Evaluation of the diameter influence on the tensile strength of pineapple leaf fibres (PALF) by Weibull method. Mater Res 18(2):185–192

Pineapple Leaf Fibers: Potential Green Resources for Pulp and Paper Production

A. Praveen Kumar

Abstract In recent decades, advances in pulp and paper making involve immense chopping of trees, which consecutively leads to clearing of forests. Rising contest for provisions of wood fibers combined with progressively increasing expenses of wood has caused increased attention in the consumption of agricultural residues for pulp and paper manufacturing in the developed and developing nations. The utilization of natural cellulosic plant residues in pulping and paper production might be necessary since it avoids the necessity for clearance, which presently rises the expenditures of farming and induces ecological deterioration by toxic wastes. The significant goals of this chapter are threefold; (1) to examine the requirements for utilization and improvement of natural cellulosic plant fibers in pulping and paper making; (2) to recognize the various issues related with the utilization of natural plant residues in pulp and paper production, and remedies accessible; and (3) to examine the prospects of various natural cellulosic plant fibers for pulp and paper making and recognize the potential of using pineapple leaf fiber as an alternate source materials in pulp and paper manufacturing mills. Better mechanical characteristics, a renewable resource, and reasonable price are some of the leading aspects that make great prospective of pineapple leaf fibers to be employed as a replacement for conventional wood fibers in pulp and paper production industries.

Keywords Agro waste · Cellulosic fiber · Pulping method · Natural plant fibers · Pineapple leaves · Pulp production

1 Introduction

The continuous supply of wood fibers has been limited all over the world even in the USA [1, 2], India, and China [3]. The European Union also agonizes from lack of wood fibers and initiated its research for unconventional fibers [4]. The leading countries like Germany and Japan are also examining the usage of agricultural residues

A. Praveen Kumar (✉)
Department of Mechanical Engineering, CMR Technical Campus, Hyderabad, Telangana 501401, India
e-mail: praveencmr18@gmail.com

M. Jawaid et al. (eds.), *Pineapple Leaf Fibers*, Green Energy and Technology,
https://doi.org/10.1007/978-981-15-1416-6_15

and natural plant fibers for the manufacturing of pulp and paper [5]. Numerous natural cellulosic plants have been examined for their characteristics and potential of pulp and paper (PP) making over the previous years [6–8]. Various research centers like Indian Grassland and Fodder Research Institute in Dharwad, International Agro-Fiber Research center in Wisconsin, Forest Research Institute of India in Dehradun, and Central Pulp and Paper Research Institute in India have been associated in this research and identified enormous non-wood and natural plant fibers as a prospective source material for the manufacturing of PP [9–11].

Wood fibers are currently the most extensively utilized raw materials in the cellulosic PP making mills. As an alternative of wood fiber, the pulp could be extracted from agro wastes and natural cellulosic plant fibers like straw, flax, kenaf, ramie, grass, bamboo, and bagasse [12, 13]. Numerous research studies in the previous years endorse the potentiality of agricultural residues, biomass, and natural cellulosic plant fibers as source materials for the fabrication of PP [14–16]. The natural plant fibers could show a crucial part in the zone of pulps and paper manufacturing industries. This is because of the reason that they provide a better substitute for considering the shortage of wood fibers, subsequently, they offer fibers with various morphologies, which improve the potentiality of attaining paper with particular characteristics. Owing to the increasing interest about the ecological effect of manufacturing activities particularly in this paper manufacturing sectors, several scientists and researchers have developed additional resolutions to substitute wood with alternate source materials in pulp and paper making industry [17]. Recently, there has been a massive demand for source materials based on cellulosic fibers in pulp and paper making industries.

Straw is one of the main sources of natural cellulosic plant (non-wood) fibers extracted from rice and substantial research work has been conducted to explore its utilization and to examine the various difficulties related with the straw. The existence of silica is the main problem which causes complications in the cleaning of the pulp. The various processes of straw such as bleaching [18], pulping [19], and handling [20] have also been a most important concentration of research. Another potential source material, on which noteworthy investigations have been conducted in various leading countries such as USA, Nigeria, Malaysia, and India, is natural kenaf plant fiber [21, 22]. This is owing to its greater cellulose content and small lignin content. Additional raw materials being studied are hemp, jute, abaca, reed, and bamboo. Consumption of natural cellulosic plant fibers for PP manufacturing in India is not a theme of choice; but a theme of requirement. The plenty of natural plant fibers in some countries is also accountable for its effective utilization in PP manufacturing. Occasionally, the usage in PP production is considered as the best technique to dispose the natural plant fibers.

Several natural plant fibers are in demand for PP making owing to the remarkable characteristics that enhance them to be superior to the conventional wood fibers. Amongst the various cellulosic fibers, abaca fiber is an outstanding source material for making of extraordinary paper quality. Its great strength and lengthy fiber enrich it a better source material for the manufacture of light-weight papers of more permeability and tremendous tear burst and high tensile strength [23]. Moreover, jute fiber possesses several natural benefits over the wood pulp. Reasonably, soft and fibrous jute fibers require a lesser amount of energy to pulp than wood. Due to the lack of lignin content, jute fiber is naturally bright. It does not need either chemical delignification or peroxide bleaching, and jute newsprint does not yellow with age and contact to light as with that prepared from wood [24]. Other fibers like bagasse and reed are greatest at supporting tremendous development to papers and can substitute hardwood chemical pulps for PP [25].

Plenty of cellulose-based natural plant fibers are under rigorous examination owing to their biodegradability and specific mechanical characteristics. The benefits of cellulosic plant residues are their constant supply of the resource, easy usage, and renewable resource [26–28]. Even though natural plant fibers display excellent mechanical and thermal characteristics, it differs with the climate, location, species, plant source, natural features, etc. Pineapple leaf fiber (PALF) is one of the amply existing agricultural residues of India and has not been examined until now as it is essential. The chemical elements of pineapple fiber comprise holocellulose (70–82%), lignin (5–12%), and ash content (1.1%) [29]. A comprehensive study of overall characteristics will highlight reasonable utilization of PALF for pulp and paper making sectors. From the socioeconomic approach, PALF could be a novel source material to the PP manufacturing mills and can be a potential substitute for the conventional exhausting wood fiber.

From the aforementioned studies, it was perceived that the natural cellulosic plant fibers could be employed as an alternative raw material to traditional wood fibers in PP making industries due to its excellent physio-chemical and mechanical characteristics. Therefore, it is essential to ascertain the study of various cellulosic natural plant fibers for its effective usage in pulp and paper production. However, there were very limited studies [30, 31] which examined the influence of pulp extraction in pineapple leaf fibers and its potentiality as a raw material in paper making industries. This chapter is focused on the requirements for utilization and improvement of natural plant (non-wood) fibers in pulping and paper making. Also, this chapter discusses the prospects of various natural cellulosic plant fibers for PP making and recognizes the prospective of utilizing PALF as alternate source materials in PP making industries.

2 Chemical Composition and Structure of Pulp

Both wood and the natural plant materials comprise of comparable chemical components but in different amounts. For instance, natural plant fiber is a biodegradable resource comprising of various chemical constituents and primarily made of cellulose, less amount of hemicellulose, and lignin along with the negligible amount of pectin. Owing to its highest amount of cellulose, natural plant fibers are also called as cellulosic fibers. The total cellulose content in natural plant fibers has been stated to have an average value of about 32.6–88%. Cellulose is the most general organic element on earth, typically about 33% of all plant substance is cellulose and in cotton, it is 90% while wood has an average of 50%. Cellulose occurs in plant cell walls as microfibrils providing a linear and structurally strong framework [32]. These fibrils can be exposed by beating and provide a very large area for bonding. Generally, beating enriches the bonding capability of the fibers during paper making. Lignin is a chemical constituent usually extracted from wood and it is an important portion of the secondary cell walls of natural plant fiber. It covers the places in the cell wall between cellulose, hemicellulose, and pectin constituents and thus, crosslinks various plant polysaccharides transferring strength to the cell wall [33]. Hemicellulose exists in the matrix between cellulose fibrils in the cell wall, and they have been exposed to be closely linked with both cellulose and lignin. The chemical constituents of the various agro-cellulosic plant fibers are presented in Table 1.

The agro-cellulosic plant residues have high inconsistency in characteristics and influenced by chemical composition, cell sizes of fiber, fiber structure, and microfibril angle. The natural cellulosic plant fiber cell contains primary and secondary cell wall, lumen, and middle lamella and the simple structure of natural plant fiber raw materials are displayed in Fig. 1.

Hemicellulose present in the cellulosic fibers serves as a coupling agent between cellulose and lignin. Every single cell has a complicated arrangement comprising of a thin primary cell wall which is the initial layer deposited throughout the growth of a cell surrounding a secondary wall. This wall is composed of three layers and the intermediate dense layer governs the mechanical characteristics of the fiber. The intermediate layer comprises of a chain of spirally looped cellular microfibrils made from lengthy chain cellulose particles.

Table 1 Chemical constituents of selected agro-residue fibers [34]

Fiber type	Cellulose content (%)	Hemicellulose content (%)	Lignin content (%)
Abaca	69–71	21–22	5–6
Banana	60–65	48–59	5.5–10
Coir	36–43	0.15–0.25	41–45
Flax	69–72	18–19	2.5–3
Kenaf	37–63	15–21	18–24
Oil palm	44–50	18–33	17–67
Pineapple leaf	70.5–82	18.7–22	5–12
Rice straw	28–48	23–28	12–16
Sugarcane	55–57	24–25	24–26
Sisal	43–57	21–25	7–9

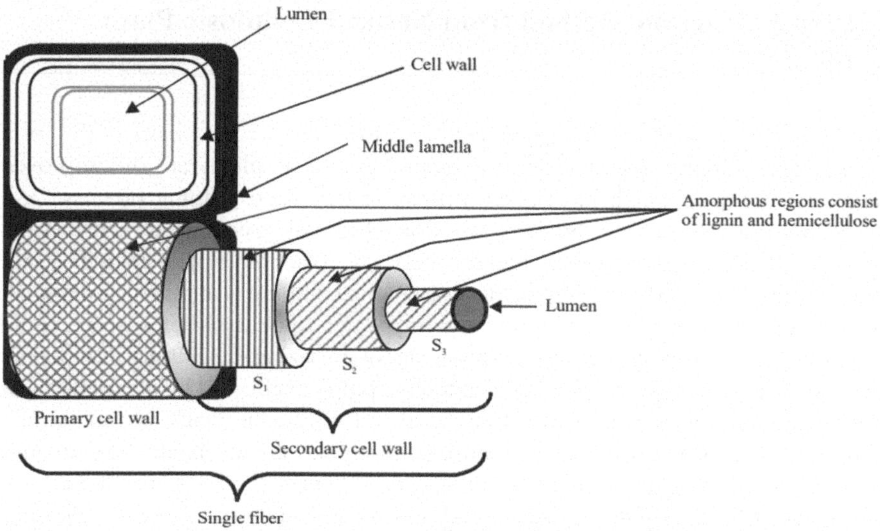

Fig. 1 Configuration of cellulosic plant fiber raw materials [35]

3 The Pulping Characteristics of Natural Cellulosic Plant Fibers

Pulping characteristics denote the physical properties that existing in the pulping, and the level of complication of the pulping procedure. The proficiency of fiber material could be assessed by the ensuing aspects: (1) whether the arrangements and constituents of the fiber are of providently sustainable in pulping process or not; (2) complication on the lignin removal technique and separation of fiber in pulping procedure; (3) the flexibility of pulping techniques, and the accessibility of stock preparation; and (4) the color, drainability, level of bleaching problems, and beating characteristics of pulp. The natural cellulosic plant fiber materials have the ensuing benefits as source material for pulp and paper production: (1) it is the rapid yearly growing resource, and it has lesser lignin content than wood; (2) pulp from natural cellulosic plant fibers could be extracted at low temperatures with lesser quantity of chemicals; (3) a small industrial unit might be viable in production techniques, giving a basic method; (4) the beating of natural cellulosic plant fibers is easy to implement; and (5) from the agricultural perspective, the natural cellulosic plant fiber materials pulping could fetch further economic supports from the farming crops [17, 36, 37].

4 Pulp Extraction Method from Natural Cellulosic Plant Fibers

Pulping is a process of extracting the cellulose from the natural plants in the form of pulp. The pulping procedures of the natural cellulosic plants are the improved techniques of those which have been utilized in the wood pulping process [38]. The pulping procedures can be categorized as chemical, mechanical, and thermal methods. Chemical pulping techniques which have been developed include kraft, organic, sulfite, and soda pulping methods. Chemical pulping is attained by reducing the hemicelluloses and lignin contents into tiny water-soluble particles which could be splashed away from the cellulose without depolymerizing the cellulose content in the fibers. The typical properties of the extracted pulp are effected by the methods of treating particularly the chemical pulping method [39]. So far, many hybrid pulping techniques that utilize a blend of chemical, thermal, and mechanical techniques are engaged in the extraction of the fibers. Few hybrid methods that have been commonly utilized in pulping both wood and natural plant (non-wood) materials comprise thermo-mechanical pulping and thermo-chemical pulping methods. The common method that can be improved for pulping natural cellulosic plant materials is illustrated in Fig. 2.

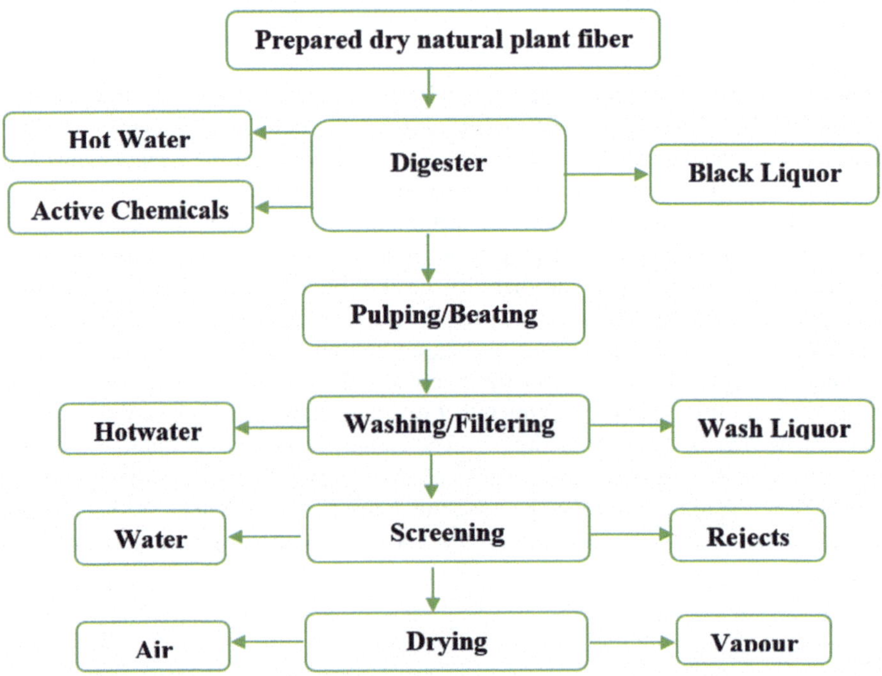

Fig. 2 Pulping flow process for the natural cellulosic plant [40]

5 Cellulosic Fibers as Potential Source for Pulp

Wood fibers are currently the most expansively utilized source material in the pulp and paper manufacturing mills around the world. As an alternative of wood, the pulp might also be prepared from natural cellulosic fibers like kenaf, hemp, grass, ramie, bamboo, and bagasse. Several research studies in previous years endorse the capability of natural cellulosic plant fibers, agricultural, and industrial residues as source materials for PP making [41–43]. The natural fiber yields could show a contributing role in the pulps and paper production sector. This is for the reason that they display a good substitute for compensating the shortage of wood fibers, which improve the prospects of attaining paper with exact characteristics. Natural cellulosic plant fibers offer several benefits which includes fast-growing cycles, moderate irrigation and less consumption of water and fertilization requirements, and less lignin content to reduce chemicals and energy utilized throughout the pulping process. There are plenty of natural cellulosic plant fibers potentially accessible for the PP production mills. In the meantime, all these natural plants comprise cellulose in form of fibers, they stand to be prospective sources for pulp with reduced ecological degradation hazard than the wood fiber which is conventionally the most extensively utilized lignocellulosic material in the making of pulp, furniture, and boards as well as being a source of energy. Therefore, natural cellulosic fibers like kenaf, jute, flax, ramie, banana, reed, and bamboo have been employed as an alternative for wood fiber pulp. As a result, the potentiality of pineapple leaf fibers as source material for PP manufacturing is examined in this chapter.

6 Pineapple Leaf Fiber (PALF)

PALF is one of the agricultural residue resources in farming section, which is extensively cultivated in coastal and humid regions of Asian countries like India, Sri Lanka, and Malaysia. Pineapple fruits are commercially significant and leaves are intended as an unwanted residue of pineapple fruit which is being utilized for extracting cellulosic fibers. The pineapple plant is a dark green-colored short stem and a very shallow root as shown in Fig. 3.

It consists of spiral sword-shaped fibrous leaves with curved ends toward the cross section to sustain the rigidity of the leaves. Moreover, the leaves are lengthy, needle-pointed which endures sharp and spines on margins, and the approximate dimensions are about 0.05–0.08 meters broad and 0.508–1.83 meters long. The natural PALF residue is a multicellular lignocellulosic fiber comprising mostly of cellulose, hemicellulose, and lignin. These fibers are arranged in a ribbon-shaped pattern and contain vascular bundle structure exist in the system of clusters of fibrous cells, which are acquired subsequently mechanical exclusion of all the epidermal tissues. The cells of PALF have a mean diameter of 0.01 mm, length of 4.5 mm, and the width of the cell wall is 0.0083 mm. PALF has the maximum tensile strength and elongation at

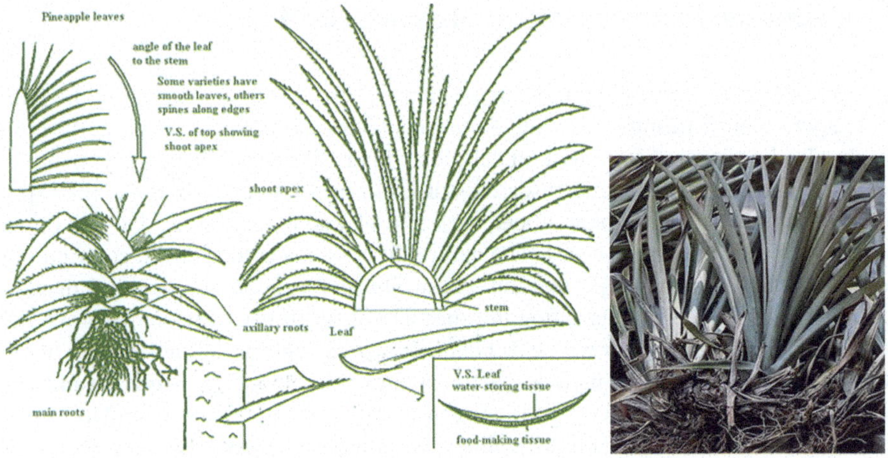

Fig. 3 Pineapple leaf fiber [44]

break is in deviation with other cellulosic fiber like jute, kenaf, and flax fibers. The mechanical characteristics such as ultimate tensile strength, and final deformation of a PALF are in the range of 413–1627 MPa, and 0.8–2.8%, respectively [45]. The chemical constituents of PALF are displayed in Table 2.

From table data, it is noted that the chemical constituents of a PALF are comparable to that of other types of lignocellulosic fibers. The amount of lignin present is slightly greater than that in cotton and sisal, and is lesser than in flax, bagasse, and ramie fibers. Nevertheless, in association to various natural cellulosic plant fibers like coir jute, pineapple leaf fibers have a lesser amount of lignin [29, 46]. As a result, the quantity of required chemical agents for bleaching and pulping methods of pineapple leaf fibers are considerably lesser than that could be desired for further natural plant fibers like bagasse, kenaf and flax is comparable to the characteristics of coir fibers than the properties of banana fibers. Alternatively, the various characteristics of pineapple leaf fiber shown in Table 3 reported that PALF the fiber bundles isolated from new pineapple leaves are better, softer, and weaker than the sisal fiber bundles. Nevertheless, PALF is stronger than the kenaf fiber. The better mechanical characteristics of pineapple leaf fibers are related to its maximum amount of cellulose and relatively low microfibril angle (12–14°) [47].

The strength of the paper is based on the available amount of cellulose in source natural plant fibers. In chemical prospective, natural fiber residues with more than

Table 2 Chemical constituents of PALF [15, 16]

Cellulose content (%)	Hemicellulose (wt%)	Lignin content (%)	Pectin (wt%)	Ash (%)	Fat and wax
69.5	–	4.4	1.2	2.7	4.2
69.5	–	4.4	1.1	0.9	3.3

Table 3 Characteristics of PALF [31]

Density (g/cm^3)	Tensile strength (MPa)	Young's modulus (GPa)	Elongation at break (%)	Dia. (μm)	Microfibril angle
1.526	413	4.2	3.0–4.0	50	14
1.44	170	6.26	1.6	5–30	12

34% of α-cellulose were considered as proficient for PP production. Pineapple plant fibers consist of the maximum fraction of α-cellulose content (70–80%), which is greater than conventional wood fiber (30–60%). One of the techniques to employ this source is to convert it into cellulosic pulp for paper production. PALF is a high-quality natural fiber source but due to lack of proper knowledge on its pulping characteristics, it is left less-utilized. Pineapple plant fibers display better mechanical properties owing to a greater amount of cellulose and lesser microfibril angle when compared to other natural plant fibers. Presently, the majority of the pineapple leaf fiber is either burned or used as fertilizer by agriculturalists, which trashes the prospective source of worthy fibers. Recently, research investigations were conducted to look at the probability of enhancing values to these leaves of PALF [30].

6.1 Potentiality of PALF as Source Material for Paper Production

The aforementioned discussions on chemical and mechanical characteristics of pineapple leaf fibers have made PALF an attractive natural cellulosic plant fiber resource for PP production industries. Further significant benefits in using PALF fibers as source material for pulp and paper are as follows: PALF paper and its related paper products offer an "environmental friendly" since a lesser quantity of chemicals are needed in pulping of PALF than the pulping of wood fiber. Furthermore, hydrogen peroxide is utilized for the bleaching process as an alternative to chlorine which is a foremost ecological concern of paper making industries. Actually, the pulp could be extracted from PALF without consuming chlorine compounds. In the paper production technique, no dioxins or other chlorine compounds are produced and released. Pulping technique of PALF utilizes less amount of energy of about 30% than the conventional wood pulp owing to the less content of lignin in PALF. The quantity of lignin content present in pineapple leaf fiber is about 4–5% which is about half when compared to the conventional wood fiber. Existence of less amount of lignin in PALF makes it simpler to extract the fiber into pulp. Another benefit is that less amount of lignin needs less quantity of chemicals to bleach PALF than the conventional wood fiber. A comprehensive review of the latest investigations and improvements in manufacturing pulp for paper production from pineapple leaf fiber has exposed various realities regarding its application as raw material source.

PALF is a periodic food crop harvested naturally twice a year, which has become the main limitation for uninterrupted supply of raw materials to paper making industries. It was also one of the issues causing the failure of some earlier techniques for paper production using PALF. However, wood fibers from trees can be collected and warehoused throughout the year to meet the demand of paper making industries.

7 Conclusion and Future Perspective

PALF has comparable characteristics and impressive properties with other types of cellulosic plant residues like coir, kenaf, jute, ramie, reed, and bagasse fibers. Better mechanical performance, maximum cellulose content, and less amount of lignin of PALF are the best-desired characteristics for great quality pulp in paper making mills. Natural cellulosic plant residues are very appropriate alternate source materials for PP making mills mostly PALF that are considered the large quantity of agricultural residue and cheap cost materials. Unfortunately, till now, there is no specific data or production scheme for consuming PALF as a substitute for conventional wood fiber in PP making mills. The present study provides valuable facts that will lead to an enhanced consumption of natural cellulosic plant fibers and evidently reported that the natural pineapple leaf fibers can form a potential source material for PP making.

Various significant results have been provided for PALF that explain the potentiality of pine apple leaf fiber as a source material for PP making process. But, still there are many areas that need to be explored, i.e. (1) to increase the quality of the paper, the pineapple leaf fibers can be blended with filler materials. (2) to reduce the usage of energy and chemical in pulping and bleaching process (3) to make the pulping method easy, effective, and economic.

References

1. Atchison JE (1992) US non-wood fiber potential rises as wood costs escalate. Pulp Pap 66(9):139–141
2. Rosenberg J (1996) Alternative fiber sources for newsprint. Ed Publ 130(3):22–26
3. Wyman V (1995) In India, it's either lead, follow closely, or get left behind. Pulp Paper Int 37(6):97–101
4. Chaudhuri PB (1995) Sowing the seeds for a new fiber supply. PPI Pulp Pap Int 37(3):68–69
5. Sameshima K (1994) Japanese local paper mill needs a way to survive; kenaf is one of the hopefuls. TAPPI nonwood plant fiber progress report No. 21: 85–90
6. Alcaide LJ, Baldovin FL, Parra IS (1991) Characterization of cellulose pulp from agricultural residues. TAPPI 74(1):217–221
7. Alcaide LJ, Baldovin FL, Herranz JLF (1993) Evaluation of agricultural residues for paper manufacture. TAPPI 76(3):169–173
8. Robinson F (1988) Kenaf: a new fiber crop for paper production. Calif Agric 42(5):31–32
9. Judt M (1993) Asia leads the way in agricultural fibers. Pulp Pap Int 35(11):72–74
10. Young RA, Akhtar M (eds) (1997) Environmentally friendly technologies for the pulp and paper industry. Wiley

11. Young RA (1997) Processing of agro-based resources into pulp and paper. In: Rowell RM, Young RA, Rowell JK (eds) Paper and composites from agro-based resources. Lewis Publishers, New York, pp 137–245
12. Atchison JE (1976) Agricultural residues and other non-wood plant fibers. Sci 191(4228):768–772
13. Laftah WA, Wan Abdul Rahman WA (2016) Pulping process and the potential of using non-wood pineapple leaves fiber for pulp and paper production: a review. J Nat Fibers 13(1):85–102
14. Mossello AA, Harun J, Tahir PM, Resalati H, Ibrahim R, Shamsi SRF, Mohmamed AZ (2010) A review of literatures related of using kenaf for pulp production (beating, fractionation, and recycled fiber). Mod App Sci 4(9):21–29
15. Ververis C, Georghiou K, Christodoulakis N, Santas P, Santas R (2004) Fiber dimensions, lignin and cellulose content of various plant materials and their suitability for paper production. Ind Crop Prod 19(3):245–254
16. Virk AP, Sharma P, Capalash N (2012) Use of laccase in pulp and paper industry. Biotechnol Progr 28(1):21–32
17. Ashori A (2006a) Nonwood fibers—a potential source of raw material in papermaking. Polym-Plast Technol 45(10): 1133–1136
18. Brink DL, Merriman MM, Radakrishna, K, Berndt H, Reddy M, Yang YS (1988) Rice straw pulping and bleaching. TAPPI nonwood plant fiber pulping progress report No. 18: 1–10
19. Yilmaz Y (1995) Lime-oxygen pulping of wheat straw. Paperi ja puu 77(1–2):51–53
20. Jeyasingam JT (1994) Applying correct raw material preparation methods for straw pulping. TAPPI nonwood plant fiber progress report No. 21: 75–80
21. Pande H, Roy DN (1996) Delignification kinetics of soda pulping of kenaf. J Wood Chem Technol 16(3):311–325
22. Tao W, Moreau JP, Calamari TA (1995) Properties of nonwoven mats from kenaf fibers. TAPPI 78(8):165–169
23. Del Rio JC, Gutierrez A (2006) Chemical composition of abaca (Musa textilis) leaf fibers used for manufacturing of high quality paper pulps. J Agric Food Chem 54(13):4600–4610
24. Jahan MS, Al-Maruf A, Quaiyyum MA (2007) Comparative studies of pulping of jute fiber, jute cutting and jute caddis. Bangladesh J Sci Ind Res 42(4):425–434
25. Hamzeh Y, Ashori A, Khorasani Z, Abdulkhani A, Abyaz A (2013) Pre-extraction of hemicelluloses from bagasse fibers: effects of dry-strength additives on paper properties. Ind Crop Prod 43:365–371
26. Ates S, Ni Y, Akgul M, Tozluoglu A (2008) Characterization and evaluation of Paulownia elongota as a raw material for paper production. Afr J Biotechnol 7(22):4153–4158
27. Belayachi L, Delmas M (1997) Sweet sorghum bagasse: a raw material for the production of chemical paper pulp: effect of depithing. Ind Crop Prod 6(3–4):229–232
28. Shakhes J, Marandi MA, Zeinaly F, Saraian A, Saghafi T (2011) Tobacco residuals as promising lignocellulosic materials for pulp and paper industry. Bio Res 6(4):4481–4493
29. Khalil HSA, Alwani MS, Omar AKM (2006) Chemical composition, anatomy, lignin distribution, and cell wall structure of Malaysian plant waste fibers. Bio Res 1(2):220–232
30. Laftah WA, Rahaman WAWA (2015) Chemical pulping of waste pineapple leaves fiber for kraft paper production. J Mater Res Technol 4(3):254–261
31. Yusof Y, Ahmad MR, Saidin W, Mustapa MS, Tahar MS (2012) Producing paper using pineapple leaf fiber. In Advanced materials research, vol 383. Trans Tech Publications, pp 3382–3386
32. Akin DE (2010) Chemistry of plant fibers. In: Mussig J (eds) Industrial applications of natural fibers: structure, properties and technical applications, Wiley, Ltd., pp 13–22
33. Komuraiah A, Kumar NS, Prasad BD (2014) Chemical composition of natural fibers and its influence on their mechanical properties. Mech Compos Mater 50(3):359–376
34. Asim M, Abdan K, Jawaid M, Nasir M, Dashtizadeh Z, Ishak MR, Hoque ME (2015) A review on pineapple leaves fibre and its composites. Int J Polym Sci 2015:1–16
35. Dungani R, Karina M, Subyakto AS, Hermawan D, Hadiyane A (2016) Agricultural waste fibers towards sustainability and advanced utilization: a review. Asian J Plant Sci 15(1–2):42–55

36. Ashori A (2006b) Pulp and paper from kenaf bast fibers. Fiber Polym 7(1): 26–29
37. Sridach W (2010) The environmentally benign pulping process of non-wood fibers. Suranaree J Sci Technol 17(2):105–123
38. Thompson G, Swain J, Kay M, Forster CF (2001) The treatment of pulp and paper mill effluent: a review. Bioresour Technol 77(3):275–286
39. Gellerstedt G (2009) Chemistry of chemical pulping. Pulping Chem Technol 2:91–120
40. Liu Z, Wang H, Hui L (2018) Pulping and papermaking of non-wood fibers. Pulp Pap Process
41. Jimenez L, Lopez F (1993) Characterization of paper sheets from agricultural residues. Wood Sci Technol 27(6):468–474
42. Monte MC, Fuente E, Blanco A, Negro C (2009) Waste management from pulp and paper production in the European Union. J Waste Manag 29(1):293–308
43. Rowell RM, Rowell J (1996) Paper and composites from agro-based resources. CRC press
44. Sibaly S, Jeetah P (2017) Production of paper from pineapple leaves. J Environ Chem Eng 5(6):5978–5986
45. Satyanarayana KG, Sukumaran K, Mukherjee PS, Pavithran C, Pillai SGK (1990) Natural fibre-polymer composites. Cem Concr Compos 12(2):117–136
46. Reddy N, Yang Y (2005) Biofibers from agricultural byproducts for industrial applications. Trends Biotechnol 23(1):22–27
47. Cherian BM, Leao AL, De Souza SF, Thomas S, Pothan LA, Kottaisamy M (2010) Isolation of nanocellulose from pineapple leaf fibres by steam explosion. Carbohydr Polym 81(3):720–725

Performance of Surface Modified Pineapple Leaf Fiber and Its Applications

G. Rajeshkumar, S. Ramakrishnan, T. Pugalenthi and P. Ravikumar

Abstract Development of pineapple leaf fiber (PALF)-based polymer composites has gain interests due to sustainable and environmental benefits when compared with synthetic-based non-degradable fibers. However, the hydrophilic PALF has poor interfacial bonding with the thermosetting and thermoplastic polymers which are hydrophobic. Moreover, this hydrophilic nature of PLAF leads to more moisture absorption rate, which results in degradation of overall properties. This issue can be addressed by modifying the surface of the fibers. Therefore, a comprehensive understanding of the effect of fiber surface modification on various properties and adhesion with polymers is a key for improving the performance of the PALF and its composites. In this context, the performance of surface modified PALF and its applications are elaborately discussed in this chapter.

Keywords Pineapple leaf fiber · Surface modification · Chemical treatments · Moisture absorption · Interfacial adhesion

1 Introduction

The use of renewable sources for the development of fiber-reinforced polymeric composites (FRPC) has gained significant importance due to their advantages like less weight and cost, high specific strength and modulus and eco-friendliness [12,

G. Rajeshkumar (✉)
Department of Mechanical Engineering, PSG Institute of Technology and Applied Research, Coimbatore, Tamilnadu, India
e-mail: grajeshkumar.me@gmail.com

S. Ramakrishnan
Department of Mechanical Engineering, Kongu Engineering College, Erode, Tamilnadu, India

T. Pugalenthi
Department of Mechanical Engineering, Jeppiaar Maamallan Engineering College, Sriperumpudur, Tamilnadu, India

P. Ravikumar
Department of Mechanical Engineering, Kathir College of Engineering, Coimbatore, Tamilnadu, India

M. Jawaid et al. (eds.), *Pineapple Leaf Fibers*, Green Energy and Technology,
https://doi.org/10.1007/978-981-15-1416-6_16

21, 26]. One such renewable source of reinforcements are plant-based natural fibers, which are extracted from various parts of the plants such as leaves, stems, fruits, seeds, etc., [22]. These natural fibers are chemically composed of lignin, cellulose, hemicellulose, pectin, and a minor amount of wax and ash contents. The physicomechanical and thermal properties of these fibers are comparable to that of the glass fibers [13].

From Table 1, it was noted that the leaves of many plants consist of fibers and also more amount of fibers can be obtained from the leaves when compared to the quantity of fibers obtained from the stem, fruits, seeds, etc. In this way, the usage of pineapple leaf fibers (PALF) as reinforcement in polymer matrix has received the interests of researchers due to their outstanding specific properties. However, these PALF also have some shortcomings such as non-uniformity in terms of shape and size, fibers from individual plants possess different properties (normally, it depends on the condition under which the plant grows), low microbial resistance, low degradation temperature, and susceptibility to rotting. This non-uniformity occurs naturally and is common for all the natural fibers; in addition to this, the fiber extraction process also has significant effect on the properties [4, 29, 36].

Another serious drawback in using PALF with polymers is poor interfacial adhesion between them, due to the hydrophilic and hydrophobic nature of PALF and polymers, respectively. Therefore, enhancing the interfacial adhesion between the PALF and polymers has been the focus of many researchers worldwide. To this context, this chapter will provide a reference to the scientists, researchers, academicians, and fiber/composite-based product manufacturers about the details of PALF surface modification and its performance and applications.

2 Pineapple Leaf Fibers (PALF)

The PALF is obtained from the leaves of Anannus comosus, belonging to the Bromeliaceae family. This plant is largely cultivated in tropical countries, primarily for its fruits. It has a short stem which first produces around 25–30 rosette of leaves which elongated latter and bear abundant spirally arranged fibrous leaves. The pineapple leaf is of 5–8 cm wide, 90 cm long and dark green in color. A fresh leaf yields about 2–3% of fiber. These fibers are extracted from the leaves either through water retting/scrapping or by mechanical (decorticated) process. The properties possessed by the mechanically extracted and water retted PALF have considerable variations (Table 2). Few of the literature mentioned that this fiber is highly suitable for reinforcement in polymers on any scale from nano-, micro- to macro-scale [18, 30–32].

The PALF consists of highest percentage of α-cellulose and low percentage of hemicellulose, lignin, fat and wax, pectin, ash, etc., (Table 3). The higher quantity of α-cellulose in PALF supports the higher weight of the fruit [27]. Moreover, these chemical compositions significantly affect the performance of the fibers, while using

Table 1 List of natural fibers and its origin [14, 24, 23]

Fiber source	Species	Origin
Abaca	Musa textiles	Leaf
Alfa	Stippa tenacissima	Grass
Bagasse	–	Grass
Bamboo	(>1250 species)	Grass
Banana	Musa indica	Leaf
Broom root	Muhlenbergia macroura	Root
Cantala	Agave cantala	Leaf
Caroa	Neoglaziovia variegate	Leaf
China jute	Abutilon theophrasti	Stem
Coir	Cocos nucifera	Fruit
Cotton	Gossypium sp.	Seed
Curaua	Ananas erectifolius	Leaf
Date palm	Phoenix dactylifera	Leaf
Flax	Linum usitatissimum	Stem
Hemp	Cannabis sativa	Stem
Henequen	Agave fourcroydes	Leaf
Isora	Helicteres isora	Stem
Istle	Samuela carnerosana	Leaf
Jute	Corchorus capsularis	Stem
Kapok	Ceiba pentranda	Fruit
Kenaf	Hibiscus cannabinus	Stem
Kudzu	Pueraria thunbergiana	Stem
Mauritius hemp	Furcraea gigantea	Leaf
Nettle	Urtica dioica	Stem
Oil palm	Elaeis guineensis	Fruit
Piassava	Attalea funifera	Leaf
Pineapple	Ananus comosus	Leaf
Phormium	Phormium tenas	Leaf
Phoenix sp.	Phoenix sp.	Leaf
Roselle	Hibiscus sabdariffa	Stem
Ramie	Boehmeria nivea	Stem
Sansevieria (bowstring hemp)	Sansevieria	Leaf
Sisal	Agave sisilana	Leaf
Sponge gourd	Luffa cylinderica	Fruit
Straw (cereal)	–	Stalk
Sun hemp	Crorolaria juncea	Stem
Cadillo/Urena	Urena lobata	Stem
Wood	(>10,000 species)	Stem

Table 2 Physicomechanical properties of decorticated and retted PALF [15]

Parameters	Decorticated PALF	Retted PALF
Fineness (tex)	1.4 (15)	1.3 (18)
Breaking tenacity (cN/tex)	21.1 (20)	22.5 (24)
Tensile strain at break (%)	4.3 (35)	3.9 (35)
Initial modulus (cN/tex)	1038 (43)	1009 (57)
Specific work of rupture (mJ/tex-m)	4.9 (52)	5.0 (64)
Bundle strength (g/tex)	22.6	19.2
Diameter (μm)	12.6 (44)	10.2 (32)
Flexural rigidity (mN-mm^2)	3.8 (15)	3.2 (13)
Coefficient of friction (perpendicular)	0.42	0.44
Coefficient of friction (parallel)	0.64	0.64
Whiteness index	44.3	59.2
Brightness index	15.4	41.2
Absorbency (s)	40	2.0
Moisture regain (%)	5.80	8.0

Table 3 Chemical composition of PALF [1, 7, 33, 37]

Cellulose (%)	Hemi cellulose (%)	Hollocellulose (%)	Lignin (%)	Pectin (%)	Ash (%)	Fat and wax (%)	Extractive (%)
68.5	18.8	–	6.04	1.1	0.9	3.2	–
67.12–69.34	–	82.3–85.5	14.5–15.4	–	1.21	–	3.83–0.97
69.5	–	–	4.4	1.2	2.7	4.2	–
73.4	–	80.5	10.5	–	2	–	5.5

it for various applications. On the other hand, the specific strength of PALF supports in improving the physical and mechanical strength of polymers matrix without using any additional processing techniques.

3 Surface Modification of PALF

The inherently polar and hydrophilic nature of PALF and nonpolar characteristics of polymer matrix results in compounding difficulties which leads to improper adhesive bonding between them. This is one of the major drawbacks of using PALF as reinforcement material in polymer matrix [4]. However, the literature mentioned that the advantages of PALF outweigh the disadvantages and most of these issues have remedial measures in the form of fiber surface modifications. The surface modifications techniques include (i) physical treatments, (ii) physicochemical treatments, (iii) chemical treatments, and (iv) thermal treatments [9, 16, 35]. Among various techniques, the chemical modification technique is most widely used to modify the surface of natural fibers, because it is the most convenient method among the other in terms of better properties and economy. Some of the chemical modification techniques used to modify the surface of the PALF is explained below.

The alkaline group mainly sodium hydroxide (NaOH) treatment is the widely used chemical treatment method to improve the interfacial adhesion between the PALF and polymer matrix and as well as the physical, mechanical, and thermal properties of PALF. Ariffin and Yusof [3] analyzed the mechanism of reaction between the PALF and NaOH solution and found that the OH groups are separated from PALF by the action of the Na+2 ions to produce a new component referred as fiber-NaO instead of fiber-OH. In addition to the alkali treatment, dinitrophenylation, nitration, benzoylation, benzoylation-acetylation, sodium hypochlorite, and hydrogen peroxide treatments are also used to modify the surface of PALF [28].

4 Performance of Surface Modified PALF

4.1 Morphological Changes

The morphological changes that occur on the surface of the PALF after the NaOH treatment (3 and 6%) were reported by Asim et al. [5]. It was concluded that 3% of NaOH is not effective to remove the impurities present of the fiber surface, whereas the PALF treated with 6% of NaOH has clean surface. Moreover, the soaking time also has significant effect on morphological changes of PALF. The cemantic material of PALF and other impurities present on the fiber surface gets removed when the PALF is exposed to higher soaking time results in the formation of grooves like structures on the fiber surface which comes in contact with the polymers during fabrication leads to better interfacial bonding. Furthermore, the reduction in diameter value with respect to increase in soaking time is also an evident for material removal (Fig. 1).

Similarly, Lopattananon et al. [17] mentioned that the cellulose chains of PALF are bounded by hemicellulose, pectin, lignin, etc., resulting in multicellular fiber formation and it gets separated after the NaOH treatment because of the removal of

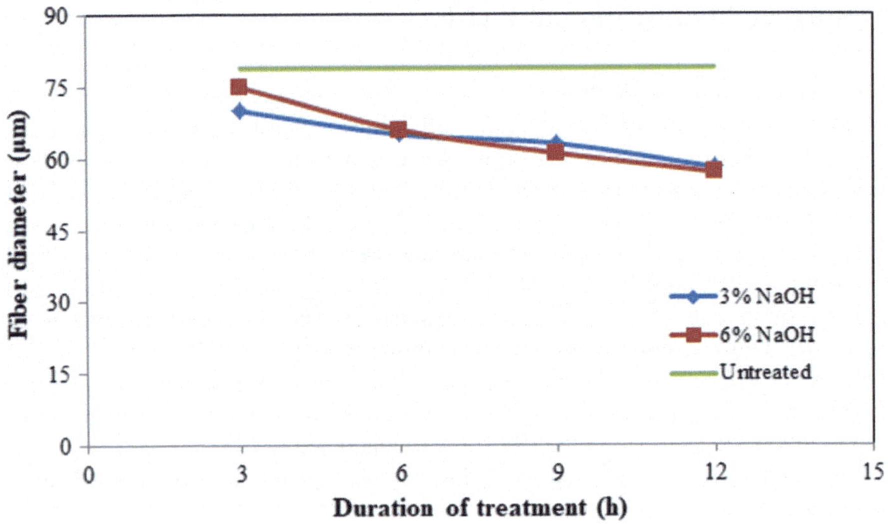

Fig. 1 Diameter of untreated and NaOH treated PALF

such organic materials. This process increases the effective surface area of the fiber which gets bonded with the matrix during fabrication of composites.

4.2 *Tensile Strength*

Figure 2 shows the tensile strength of untreated and NaOH treated PALF. The treated PALF has higher tensile strength when compared to the untreated PALF, because the treated PALF is free from discontinuities and defects which paved the way for failure [2, 5, 22]. The tensile strength of 3 and 6% of NaOH treated PLAF is higher at 6 h of immersion time, while at higher immersion time (9 and 12 h), the tensile strength of PALF gets reduced, which attributed to the fibrillation of fiber. Similar results were reported by Ariffin and Yusof [3].

As per Zin et al. [38] findings, the tensile strength of untreated PALF is 139.90 MPa, while the 6% of NaOH treated PALF records highest tensile strength of 164.55 MPa, which is equivalent to 18% improvement when compared to that of the untreated PALF. The results also mentioned that further increase in the concentration of NaOH (8%) reduced the tensile strength of the fiber due to: (i) excessive removal of lignin and waxy layers and (ii) weakening and damage of fiber because of higher concentration of NaOH.

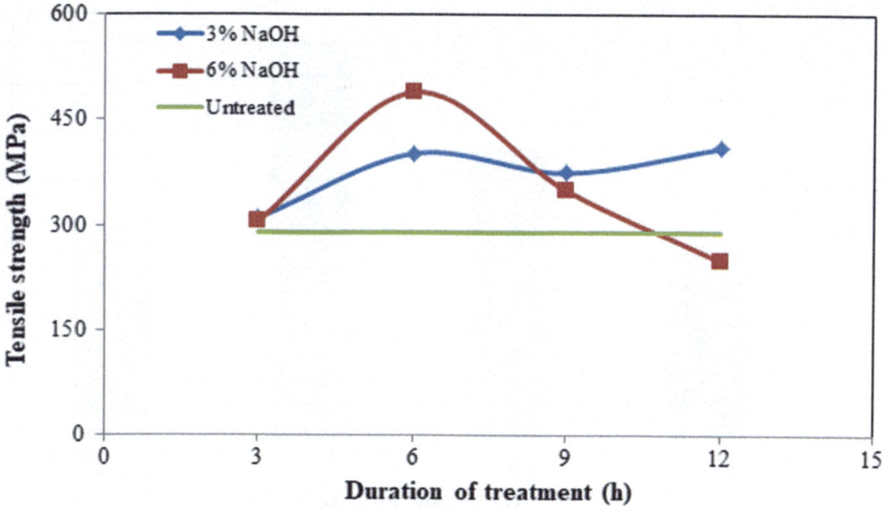

Fig. 2 Tensile strength of untreated and NaOH treated PALF

4.3 Adhesion Analysis

The interfacial adhesion between the fiber and polymer matrix plays a significant role in controlling the macroscopic mechanical properties of FRPC. The Interfacial Shear Strength (IFSS), Single-Fiber Fragmentation Test (SFFT), and Single-Fiber Pullout Test (SFPT) are the types of tests available to analyze the adhesive bonding between the fiber and matrix.

4.3.1 Interfacial Shear Strength (IFSS)

The Interfacial Shear Strength (IFSS) of composite represents the adhesive interlocking between the individual fiber and the matrix. Zin et al. [38] examined the IFSS of untreated and NaOH treated PALF with epoxy matrix (Fig. 3). The untreated fiber has the IFSS of 20.64 MPa, while the fiber treated with 6% of NaOH solution for 1 h has highest IFSS of 42.67 MPa. This is equivalent to 106% improvement when compared to that of the untreated fiber. Moreover, all other treated PALF also shows enhanced IFSS compared to untreated condition. This confirms that the treated PALF has good adhesive bonding with the epoxy matrix. This could be due to: (i) better interlocking adhesion between the treated PALF and epoxy matrix, as the treatment removed the artificial and natural impurities on the fiber surface and made them rough, and (ii) exposure of greater amount of cellulose on the fiber surface.

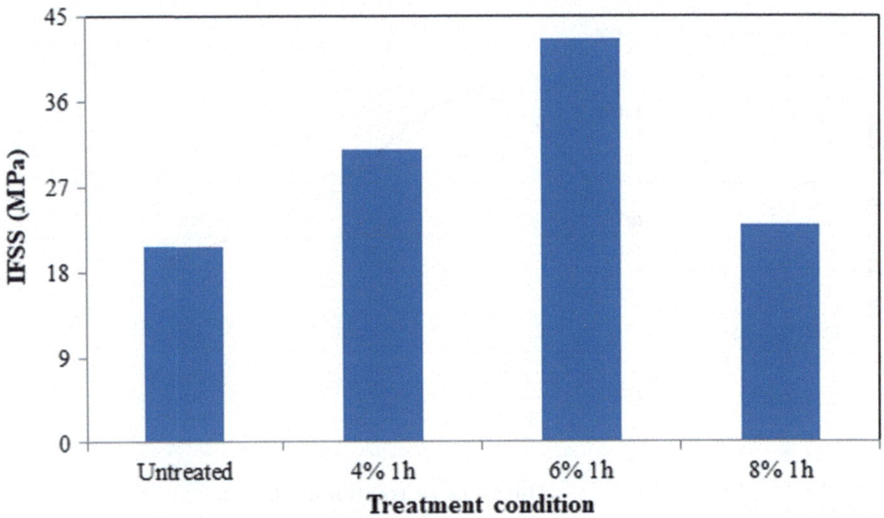

Fig. 3 IFSS of untreated and treated PALF with epoxy polymer

4.3.2 Single-Fiber Pullout Test (SFPT)

Normally, the SFPT is carried out in two ways: (i) clamp one end of the specimen opposite to the fiber loading (tensile loading) end (Fig. 4a) and (ii) specimen is supported at the matrix region nearer to the fiber loading point (Fig. 4b) [34].

Suwanruji et al. [35] analyzed the variations in pullout stress values of untreated and treated PALFs. The experimental results reveal that the pullout stress is higher for treated PLAF samples when compared to untreated one. This confirms that the treated PALF has good interfacial bonding with the polymer matrix. In particular, the poly(methylene(polyphenyl isocyanate)) (PMPPIC) treated fiber records higher pullout stress with polypropylene than the other fibers and it is 77% greater than the untreated PALF and polypropylene (Table 4).

Fig. 4 Clamping methods for SFPT

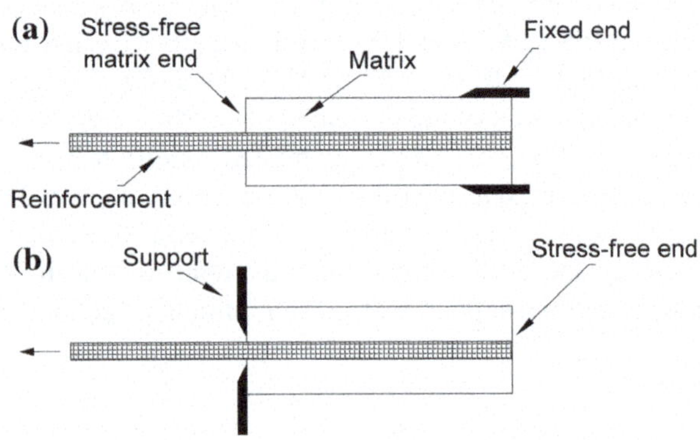

Table 4 Single-fiber pullout stress between PALFs and polypropylene [35]

Condition of PALF	Single-fiber pullout stress (MPa)	% Increase
Untreated	102.59 ± 64.78	–
HDI treated	128.80 ± 74.17	26
APS treated	131.65 ± 39.57	28
MRPS treated	152.15 ± 59.45	48
PMPPIC treated	181.16 ± 56.45	77

where
APS 3-Aminopropyltriethoxysilane; *MRPS* 3-Mercaptopropyltrimethoxysilane; *PMPPIC* poly(methylene(polyphenyl isocyanate)); *HDI* Hexamethylene diisocyanate

4.4 Thermogravimetric Property

Thermogravimetric analysis (TGA) is a technique used to investigate the thermal stability/decomposition of PALF. Table 5 summarizes the thermogravimetric property of untreated and treated PALF. In untreated PALF, the 10% weight loss occurs at 326 °C, while it increased to 340 °C when it was treated with NaOH and silane. 20% and 30% weight loss also follow a similar trend. This indicates the good thermal stability of treated PALF when compared to that of the untreated one. The reason is, in treated PALF, the moisture starts to liberate at higher temperature due to changes that take place in morphology and fine structure. Moreover, the untreated PALF reached the fixed weight loss (i.e., 10, 20, or 30%) levels at minimum temperature when compared to treated fibers. This may be attributed due to the presence of thermally unstable constituents like holocellulose, hemicellulose, ash, etc., whereas the treated fibers are thermally stable because of the absence of such constituents [11].

In general, the decomposition temperature refers to the initial temperature at which the fibers start burning. Zin et al. [38] investigated the thermogravimetric property of untreated and treated PALF and found that the decomposition temperature of untreated PALF is 282 °C, whereas the 6% of NaOH treated PALF gets degraded at 297 °C, which is 5.3% greater than that of the untreated fiber.

Table 5 Thermogravimetric property of untreated and treated PALF [11]

Type of PALF	Temperature (°C)		
	10% weight loss	20% weight loss	30% weight loss
Untreated	326	358	444
NaOH treated	332	357	410
Silane treated	329	357	370
NaOH and Silane treated	340	364	510

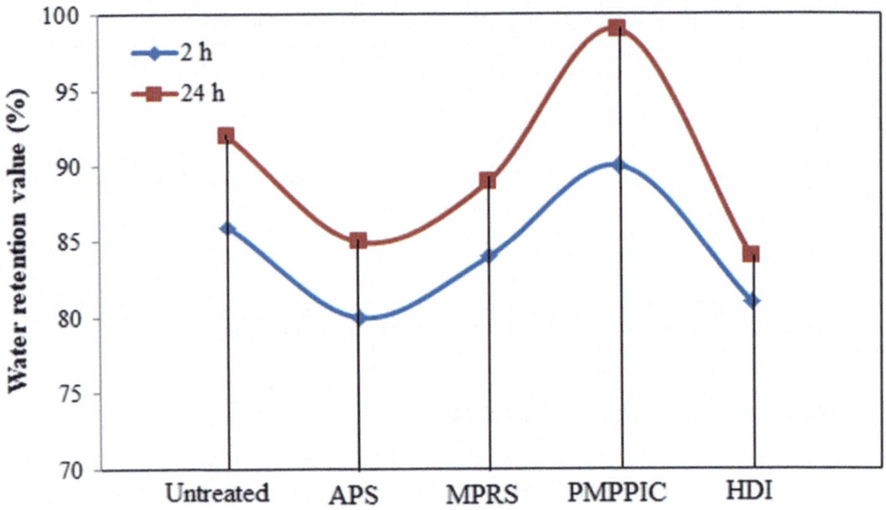

Fig. 5 Percentage water retention of untreated and treated PALF

4.5 Moisture Retention

The moisture absorption is another major drawback of PALF which leads to swelling of composites. This swelling results in degradation of overall properties and reduces dimensional stability [25]. Suwanruji et al. [35] determined the water retention value of untreated and various chemically treated PALF (Fig. 5). The PALF treated with 3-Aminopropyltriethoxysilane (APS), 3-Mercaptopropyltrimethoxysilane (MRPS), 1,6-Diisocyanatohexane or hexamethylene diisocyanate (HDI) shows lower water retention when compared to the untreated PALF at both 2 and 24 h of immersion time. The interaction of –OH groups with reactive groups of coupling agents is responsible for this reduced water absorption rate. Though the MRPS and APS are silane coupling agents (R-Si-X_3), they form silanol being ethoxy and methoxy, respectively, because they have different hydrolysable groups (X). This helps in improving good bonding of PALF with polymer matrices.

5 Applications

One of the South India Textile Research Association (SITRA's) findings under the UNDP/UNIDO assisted project revealed that the PALF could be successfully spun in the cotton spinning system with slight modifications to produce 100% PALF yarn. These yarns are used to make fabrics, fancy carpets, mops, curtains, etc., [18]. Apart from this, the PALF is used for making commercial goods such as table linens, mats, bags, and dresses in the Philippines. Sapuan et al. [29] mentioned that these

PALFs are being used as a textile material in Indonesia and Malaysia. Basu et al. [6] particularly pointed out that the chemically treated PALF is being used in making V-belt cord, transmission cloth, conveyor belt cord, industrial textile, and air-bag tying cords. Moreover, these PALFs are blended with polyester fibers for making needle-punched nonwovens for technical textiles [10]. Nayan et al. [20] indicated that the PALF is used as an inexpensive raw material for pulp and paper applications. Thailand produces pineapple paper fiber, which is later used to develop Pepp chair seats [19]. The other applications of the PALF include particle boards for thermal insulator, reinforcement material for preparing polymer matrix composites, and biomedical applications (Yahya and Yusof [37]. In particular, the nanocellulose isolated PALF promises to be a versatile material having wide range of biotechnology and biomedical applications, such as drug delivery, tissue engineering, medical implants, wound dressing, repair of articular cartilage, vascular grafts, mammary prostheses, urethral catheters, adhesion barriers, artificial skin, and penile prostheses [8].

6 Conclusions and Future Perspective

Over the past two decades, considerable effort has been devoted to enhance the performance of PALF because these fibers are sustainable and mechanically excellent material to reinforce in both the thermoplastic and thermosetting polymers. This chapter has critically addressed the chemical treatment technique followed by the researchers to enhance the properties of the raw PALF. The chemical treatment removes the impurities and other constituents like holocellulose, hemicellulose, lignin, etc., and exposes greater amount of cellulose on the fiber surface. This improves the performance of PALF and its composites. Overall conclusion is that the composites fabricated using surface modified PALF will help in the development of advance composites possessing good mechanical properties, appropriate stiffness, dimensional, and thermal stability. In line with the previous work, an extensive research work is still required to do on surface modified PALF-based hybrid composites to explore the compatibility of PALF with other natural fibers and various polymers.

References

1. Khalil HSA, Alwani MS, Omar AKM (2006) Chemical composition, anatomy, lignin distribution and cell wall structure of Malaysian plant waste fiber. BioResources 1:220–232
2. Alawar A, Hamed AM, Al-Kaabi K (2009) Characterization of treated date palm tree fiber as composite reinforcement. Compos Part B 40(7):601–606
3. Ariffin A, Yusof Y (2017) Effect of extraction process and surface treatment on the mechanical properties in pineapple leaf fibre. MATEC Web Conf 135:00042
4. Asim M, Abdan K, Jawaid M et al (2015) A review on pineapple leaves fibre and its composites. Int J Polym Sci 2015:1–16
5. Asim M, Jawaid M, Abdan K et al (2018) Effect of Alkali treatments on physical and mechanical strength of pineapple leaf fibres. IOP Conf Ser Mater Sci Eng 290(1):012030

6. Basu A, Chellamani KP, Kumar PR (2003) Jute and pineapple leaf fibres for the manufacture of technical textiles. Asian Text J 12:94–96
7. Bhaduri SK, Sen SK, Dasgupta PC (1983) Structural studies of an acidic polysaccharide isolated from the leaf fibre of pineapple (Ananas comosus MERR). Carbohydr Res 121:211–220
8. Cherian BM, Leão AL, de Souza SF et al (2011) Cellulose nanocomposites with nanofibres isolated from pineapple leaf fibers for medical applications. Carbohydr Polym 86(4):1790–1798
9. George J, Bhagawan SS, Thomas S (1997) Improved interactions in chemically modified pineapple leaf fiber reinforced polyethylene composites. Compos Interfaces 5(3):201–223
10. Hayavadana J, Jacob M, Sampath G (2003) Diversified product of pine apple leaf fibres. Man Made Text India 46:301–305
11. Izani MN, Paridah MT, Anwar UMK et al (2013) Effects of fiber treatment on morphology, tensile and thermogravimetric analysis of oil palm empty fruit bunches fibers. Compos Part B Eng 45(1):1251–1257
12. Jain NK, Gupta MK (2018) Hybrid teak/sal wood flour reinforced composites: mechanical, thermal and water absorption properties. Mater Res Exp 5(12):125306
13. Jawaid MHPS, Khalil HA (2011) Cellulosic/synthetic fibre reinforced polymer hybrid composites: a review. Carbohydr Polym 86(1):1–18
14. John MJ, Anandjiwala RD (2008) Recent developments in chemical modification and characterization of natural fiber-reinforced composites. Polym Compos 29(2):187–207
15. Jose S, Das R, Mustafa I, Karmakar S et al (2019) Potentiality of Indian pineapple leaf fiber for apparels. J Nat Fib 16(4):536–544
16. Jose S, Salim R, Ammayappan L (2016) An overview on production, properties, and value addition of pineapple leaf fibers (PALF). J Nat Fib 13(3):362–373
17. Lopattananon N, Panawarangkul K, Sahakaro K et al (2006) Performance of pineapple leaf fiber–natural rubber composites: the effect of fiber surface treatments. J Appl Polym Sci 102(2):1974–1984
18. Mishra S, Mohanty AK, Drzal LT et al (2004) A review on pineapple leaf fibers, sisal fibers and their biocomposites. Macromol Mater Eng 289(11):955–974
19. Nadirah WW, Jawaid M, Al Masri AA et al (2012) Cell wall morphology, chemical and thermal analysis of cultivated pineapple leaf fibres for industrial applications. J Polym Environ 20(2):404–411
20. Nayan NHM, Rahman WAWA, Majid RA (2014) The effect of mercerization process on the structural and morphological properties of pineapple leaf fiber (PALF) pulp. Malaysian J Fundam Appl Sci 10(1)
21. Negawo TA, Polat Y, Buyuknalcaci FN et al (2019) Mechanical, morphological, structural and dynamic mechanical properties of alkali treated Ensete stem fibers reinforced unsaturated polyester composites. Compos Struct 207:589–597
22. Rajeshkumar G, Hariharan V, Scalici T (2016) Effect of NaOH treatment on properties of Phoenix sp. Fiber. J Nat Fib 13(6):702–713
23. Rajeshkumar G, Hariharan V, Sathishkumar TP (2016) Characterization of Phoenix sp. natural fiber as potential reinforcement of polymer composites. J Ind Text 46(3):667–683
24. Rajeshkumar G, Hariharan V, Sathishkumar TP et al (2017) Synergistic effect of fiber content and length on mechanical and water absorption behaviors of Phoenix sp. fiber-reinforced epoxy composites. J Ind Text 47(2):211–232
25. Rajeshkumar G (2018) An experimental study on the interdependence of mercerization, moisture absorption and mechanical properties of sustainable Phoenix sp. fibre-reinforced epoxy composites. J Ind Text https://doi.org/10.1177/1528083718811085
26. Rajeshkumar G, Hariharan V, Saravanakumar SS (2019) Enhancing the free vibration characteristics of epoxy polymers using sustainable Phoenix sp. fiber and nano-clay for machine tool applications. J Nat Fib https://doi.org/10.1080/15440478.2019.1636740
27. Reddy N, Yang Y (2005) Biofibers from agricultural byproducts for industrial applications. Trends Biotechnol 23(1):22–27
28. Samal RK, Bhuyan BL (1994) Chemical modification of lignocellulosic fibers I. Functionality changes and graftcopolymerization of acrylonitrile onto pineapple leaf fibers; their characterization and behavior. J Appl Polym Sci 52(12): 1675–1685

29. Sapuan SM, Mohamed AR, Siregar JP et al (2011) Pineapple leaf fibers and PALF-Reinforced polymer composites. In Cellulose fibers: bio-and nano-polymer composites, Springer, 325–343
30. Senthilkumar K, Rajini N, Saba N et al (2019) Effect of Alkali treatment on mechanical and morphological properties of pineapple leaf fibre/polyester composites. J Polym Environ 27(6):1191–1201
31. Senthilkumar K, Saba N, Chandrasekar M et al (2019) Evaluation of mechanical and free vibration properties of the pineapple leaf fibre reinforced polyester composites. Constr Build Mater 195:423–431
32. Sinha MK (1982) A review of processing technology for the utilisation of agro-waste fibres. Agric Wastes 4(6):461–475
33. Siregar JP, Sapuan SM, Rahman MZA et al (2008) Characterization and chemical composition of short pineapple leaf fibres (PALF). In: Sapuan SM (ed) Proceeding of postgraduate seminar on natural fibre composites. Faculty of Engineering, Universiti Putra Malaysia, Serdang, Selangor, pp 19–24
34. Sørensen BF, Lilholt H (2016) Fiber pull-out test and single fiber fragmentation test-analysis and modelling. IOP Conf Series Mater Sci Eng 139(1):012009
35. Suwanruji P, Tuechart T, Smitthipong W et al (2017) Modification of pineapple leaf fiber surfaces with silane and isocyanate for reinforcing thermoplastic. J Thermoplast Compos Mater 30(10):1344–1360
36. Velusamy K, Navaneethakrishnan P, Rajeshkumar G, Sathishkumar TP (2019) The influence of fiber content and length on mechanical and water absorption properties of Calotropis Gigantea fiber reinforced epoxy composites. J Ind Text 48(8):1274–1290
37. Yahya B, Asia S, Yusof Y (2013) Comprehensive review on the utilization of PALF. Adv Mater Res 701:430–434
38. Zin MH, Abdan K, Mazlan N et al (2018) The effects of alkali treatment on the mechanical and chemical properties of pineapple leaf fibres (PALF) and adhesion to epoxy resin. IOP Conf Series Mater Sci Eng 368(1):012035

The manufacturer's authorised representative in the EU is Springer Nature Customer Service Centre GmbH, Europaplatz 3, 69115 Heidelberg, Germany. If you have any concerns regarding our products, please contact ProductSafety@springernature.com

Printed and bound by CPI Group (UK) Ltd, Croydon, CR0 4YY
25/09/2026
02222901-0010